DYNAMICS OF THE RIGID SOLID WITH GENERAL CONSTRAINTS BY A MULTIBODY APPROACH

DYNAMICS OF THE RIGID SOLID WITH GENERAL CONSTRAINTS BY A MULTIBODY APPROACH

Nicolae Pandrea and

Nicolae-Doru Stănescu

University of Pitești, Pitești, Romania

WILEY

This edition first published 2016

Registered Office
John Wiley & Sons, Ltd, The Atrium, Southern Gate, Chichester, West Sussex, PO19 8SQ, United Kingdom

For details of our global editorial offices, for customer services and for information about how to apply for permission to reuse the copyright material in this book please see our website at www.wiley.com.

Library of Congress Cataloging-in-Publication Data

Pandrea, Nicolae, author.
Dynamics of the rigid solid with general constraints by a multibody approach / Nicolae Pandrea, Nicolae-Doru Stănescu.
pages cm
Includes bibliographical references and index.
ISBN 978-1-118-95438-6 (cloth)
1. Dynamics, Rigid. 2. Multibody systems. 3. Constraints (Physics) I. Stănescu, Nicolae-Doru, author. II. Title.
QA861.P36 2015
531′.3–dc23

2015011298

A catalogue record for this book is available from the British Library.

Set in 11/13pt Times by SPi Global, Pondicherry, India

Printed and bound in Singapore by Markono Print Media Pte Ltd

1 2016

Contents

Preface

This book deals with both holonomic and non-holonomic constraints to study the mechanics of the constrained rigid body. The approach is completely matrix and we study all types of the general constraints that may appear at a rigid solid. The discussion is performed in the most general case, not in particular cases defined by certain types of mechanisms. Our approach is a multibody type one and the obtaining of the matrix of constraints is highlighted in each case discussed in the book. In addition, algorithms for the numerical calculations are given for each type of constraint. The theory is applied to numerical examples which are completely solved, the diagrams resulted being also presented.

The book contains eight chapters as follows. The first chapter is an introduction presenting the elements of mathematical calculation that will be used in the book. The second chapter treats the kinematics of the rigid solid and in this chapter we obtain the distribution of velocities and accelerations for a rigid body. The next chapter is dedicated to the general theorem in the dynamics of the rigid solid, that is, the theorem of momentum, the theorem of the moment of momentum, and the kinetic energy; all these theorems are developed in matrix form. In the fourth chapter are presented the matrix differential equations of motion in the general case of the rigid solid with constraints; the equations of motion are obtained using the general theorems and using the Lagrange equations; a completely new proof is given for the equivalence of these two approaches. In the fifth chapter we discuss the equilibrium of the rigid solid; we introduce the generalized forces and their expressions; as a particular case we study the equilibrium of a rigid solid hanged by springs. The next chapter deals with the motion of the rigid solid having constraints at given proper points; we discuss the rigid body with one fixed point, the rigid body in rotational motion, the rigid body with one or several points situated on given surfaces or curves. In the seventh chapter we discuss the motion of the rigid solid with constraints on given proper curves; the chapter is dedicated to the study of the rigid body at which given curves support on given curves or surfaces. The last chapter is dedicated to the motion of the rigid solid with constraints on the bounded surfaces; in this case the rigid body is supported at fixed points, or it rolls on curves or surfaces.

The authors are grateful to Mrs. Eng. Ariadna–Carmen Stan for her valuable help in the presentation of this book. The excellent cooperation with the team of John Wiley & Sons is gratefully acknowledged.

This book is addressed to a large audience, to all those interested in using models and methods with holonomic and non-holonomic constraints in various fields like: mechanics, physics, civil and mechanical engineering, people involved in teaching, research or design, as well as students.

The book can be also used either as a stand-alone course for the master or PhD students, or as supplemental reading for the courses of computational mechanics, analytical mechanics, multibody mechanics etc. The prerequisites are the courses of elementary algebra and analysis, and mechanics.

Nicolae Pandrea and Nicolae-Doru Stănescu

1

Elements of Mathematical Calculation

This chapter is an introduction presenting the elements of mathematical calculation that will be used in the book.

1.1 Vectors: Vector Operations

A *vector* (denoted by $\mathbf{a}$) is defined by its numerical magnitude or modulus $|\mathbf{a}|$, by the direction Δ, and by sense. The vector is represented (Fig. 1.1) by an orientated segment of straight line.

The sum of two vectors $\mathbf{a}$, $\mathbf{b}$ is the vector $\mathbf{c}$ (Fig. 1.2) represented by the diagonal of the parallelogram constructed on the two vectors; it reads

$$\mathbf{c} = \mathbf{a} + \mathbf{b}. \tag{1.1}$$

The unit vector $\mathbf{u}$ of the vector $\mathbf{a}$ (or of the direction Δ) is defined by the relation

$$\mathbf{u} = \frac{\mathbf{a}}{|\mathbf{a}|}. \tag{1.2}$$

If one denotes by $\mathbf{i}$, $\mathbf{j}$, $\mathbf{k}$ the unit vectors of the axes of dextrorsum orthogonal reference system $Oxyz$, and by a_x, a_y, a_z the projections of vector $\mathbf{a}$ onto the axes, then one may write the analytical expression

$$\mathbf{a} = a_x\mathbf{i} + a_y\mathbf{j} + a_z\mathbf{k}. \tag{1.3}$$

Dynamics of the Rigid Solid with General Constraints by a Multibody Approach, First Edition.
Nicolae Pandrea and Nicolae-Doru Stănescu.

Companion website: www.wiley.com/go/pandrea

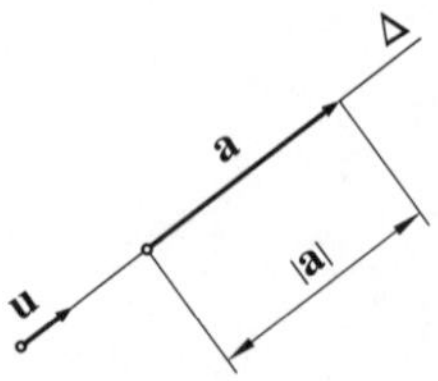

Figure 1.1 Representation of a vector.

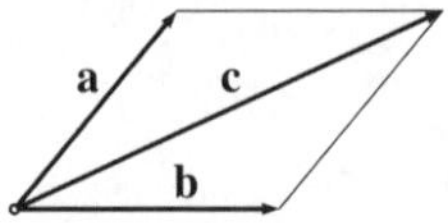

Figure 1.2 The sum of two vectors.

The scalar (dot) product of two vectors is defined by the expression

$$\mathbf{a}\cdot\mathbf{b}=|\mathbf{a}||\mathbf{b}|\cos\alpha, \tag{1.4}$$

where α is the angle between the two vectors.

We obtain the equalities

$$\mathbf{i}\cdot\mathbf{j}=\mathbf{j}\cdot\mathbf{k}=\mathbf{k}\cdot\mathbf{i}=0,\ \mathbf{i}^2=\mathbf{j}^2=\mathbf{k}^2=1 \tag{1.5}$$

and, consequently, one deduces the analytical expressions

$$\mathbf{a}\cdot\mathbf{b}=a_xb_x+a_yb_y+a_zb_z, \tag{1.6}$$

$$|\mathbf{a}|=\sqrt{a_x^2+a_y^2+a_z^2},\ |\mathbf{b}|=\sqrt{b_x^2+b_y^2+b_z^2}, \tag{1.7}$$

$$\cos\alpha=\frac{a_xb_x+a_yb_y+a_zb_z}{\sqrt{a_x^2+a_y^2+a_z^2}\sqrt{b_x^2+b_y^2+b_z^2}}. \tag{1.8}$$

The vector (cross) product of two vectors, denoted by **c**,

$$\mathbf{c}=\mathbf{a}\times\mathbf{b}, \tag{1.9}$$

is the vector perpendicular onto the plan of the vectors **a** and **b**, while the sense is given by the rule of the right screw when the vector **a** rotates over the vector **b** (making the smallest angle); the modulus has the expression

$$|\mathbf{c}|=|\mathbf{a}||\mathbf{b}|\sin\alpha, \tag{1.10}$$

α being the smallest angle between the vectors **a** and **b**.

One obtains the equalities

$$\mathbf{i}\times\mathbf{j}=\mathbf{k},\ \mathbf{j}\times\mathbf{k}=\mathbf{i},\ \mathbf{k}\times\mathbf{i}=\mathbf{j}, \tag{1.11}$$

and the analytical expression

$$\mathbf{a}\times\mathbf{b}=\left(a_y b_z-a_z b_y\right)\mathbf{i}+\left(a_z b_x-a_x b_z\right)\mathbf{j}+\left(a_x b_y-a_y b_x\right)\mathbf{k}. \tag{1.12}$$

The mixed product of three vectors, defined by the relation $\mathbf{a}\cdot(\mathbf{b}\times\mathbf{c})$ and denoted by (**a**, **b**, **c**), leads to the successive equalities

$$(\mathbf{a},\mathbf{b},\mathbf{c})=\mathbf{a}\cdot(\mathbf{b}\times\mathbf{c})=\mathbf{b}\cdot(\mathbf{c}\times\mathbf{a})=\mathbf{c}\cdot(\mathbf{a}\times\mathbf{b})=\begin{vmatrix} a_x & a_y & a_z \\ b_x & b_y & b_z \\ c_x & c_y & c_z \end{vmatrix}. \tag{1.13}$$

The mixed product (**a**, **b**, **c**) is equal to the volume with sign of the parallelepiped constructed having the three vectors as edges (Fig. 1.3). It is equal to zero if and only if the three vectors are coplanar.

The double vector product $\mathbf{a}\times(\mathbf{b}\times\mathbf{c})$ satisfies the equality

$$\mathbf{a}\times(\mathbf{b}\times\mathbf{c})=(\mathbf{a}\cdot\mathbf{c})\mathbf{b}-(\mathbf{a}\cdot\mathbf{b})\mathbf{c}. \tag{1.14}$$

The reciprocal vectors of the (non-coplanar) vectors **a**, **b**, **c** are defined by the expressions

$$\mathbf{a}^*=\frac{\mathbf{b}\times\mathbf{c}}{(\mathbf{a},\mathbf{b},\mathbf{c})},\ \mathbf{b}^*=\frac{\mathbf{c}\times\mathbf{a}}{(\mathbf{a},\mathbf{b},\mathbf{c})},\ \mathbf{c}^*=\frac{\mathbf{a}\times\mathbf{b}}{(\mathbf{a},\mathbf{b},\mathbf{c})}, \tag{1.15}$$

and satisfy the equality

$$(\mathbf{a}^*,\mathbf{b}^*,\mathbf{c}^*)=\frac{1}{(\mathbf{a},\mathbf{b},\mathbf{c})}. \tag{1.16}$$

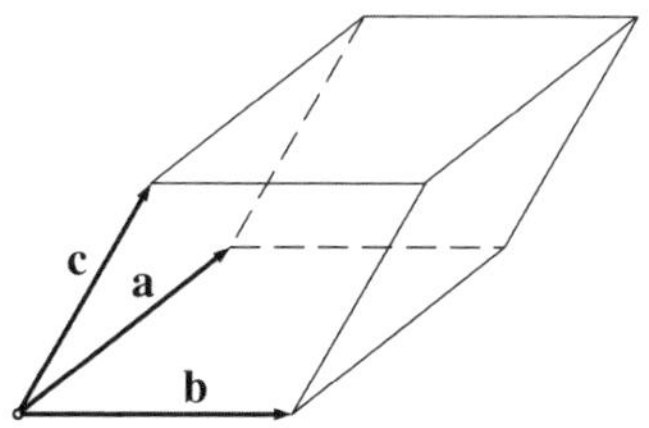

Figure 1.3 The geometric interpretation of the mixed product of three vectors.

An arbitrary vector **v** may be written in the form

$$\mathbf{v} = (\mathbf{v}\cdot\mathbf{a}^*)\mathbf{a} + (\mathbf{v}\cdot\mathbf{b}^*)\mathbf{b} + (\mathbf{v}\cdot\mathbf{c}^*)\mathbf{c}, \tag{1.17}$$

or as

$$\mathbf{v} = (\mathbf{v}\cdot\mathbf{a})\mathbf{a}^* + (\mathbf{v}\cdot\mathbf{b})\mathbf{b}^* + (\mathbf{v}\cdot\mathbf{c})\mathbf{c}^*. \tag{1.18}$$

1.2 Real Rectangular Matrix

By *real rectangular matrix* we understand a table with m rows and n columns ($m \neq n$)

$$[\mathbf{A}] = \begin{bmatrix} a_{11} & a_{12} & \dots & a_{1n} \\ a_{21} & a_{22} & \dots & a_{2n} \\ \dots & \dots & \dots & \dots \\ a_{m1} & a_{m2} & \dots & a_{mn} \end{bmatrix}, \tag{1.19}$$

where the *elements* a_{ij} are real numbers.

Sometimes, we use the abridged notation

$$[\mathbf{A}] = (a_{ij}) \text{ or } [\mathbf{A}] = (a_{ij})_{\substack{1 \le i \le m \\ 1 \le j \le n}}. \tag{1.20}$$

The multiplication between a matrix and a scalar $\lambda \in \mathbb{R}$ is defined by the relation

$$\lambda[\mathbf{A}] = (\lambda a_{ij}), \tag{1.21}$$

while *the sum of two matrices of the same type* (with the same number of rows and the same number of columns) is defined by

$$[\mathbf{A}] + [\mathbf{B}] = (a_{ij} + b_{ij}). \tag{1.22}$$

The zero matrix or the null matrix is the matrix denoted by $[\mathbf{0}]$, which has all its elements equal to zero.

The zero matrix verifies the relations

$$[\mathbf{A}] + [\mathbf{0}] = [\mathbf{0}] + [\mathbf{A}] = [\mathbf{A}]. \tag{1.23}$$

The transpose matrix $[\mathbf{A}]^{\mathrm{T}}$ is the matrix obtained transforming the rows of the matrix $[\mathbf{A}]$ into columns, that is

$$[\mathbf{A}]^{\mathrm{T}} = (a_{ji}). \tag{1.24}$$

The transposing operation has the following properties

$$\left[[\mathbf{A}]^{\mathrm{T}}\right]^{\mathrm{T}} = [\mathbf{A}], [[\mathbf{A}] + [\mathbf{B}]]^{\mathrm{T}} = [\mathbf{A}]^{\mathrm{T}} + [\mathbf{B}]^{\mathrm{T}}, \tag{1.25}$$

where we assumed that the sum can be performed.

The matrix with one column bears the name *column matrix* or *column vector* and it is denoted by $\{\mathbf{A}\}$, that is

$$\{\mathbf{A}\} = [a_{11} \ \ a_{21} \ \ \dots \ \ a_{m1}]^{\mathrm{T}}, \tag{1.26}$$

while the matrix with one row is called *row matrix* or *row vector* and is denoted as

$$[\mathbf{A}] = [a_{11} \ \ a_{12} \ \ \dots \ \ a_{1n}], \tag{1.27}$$

or

$$[\mathbf{A}] = \{\mathbf{A}\}^{\mathrm{T}}, \tag{1.28}$$

where

$$\{\mathbf{A}\} = [a_{11} \ \ a_{12} \ \ \dots \ \ a_{1n}]^{\mathrm{T}}. \tag{1.29}$$

If the matrix $[\mathbf{A}]$ has m rows and n columns, and the matrix $[\mathbf{B}]$ has n rows and p columns, *then* the two matrices *can be multiplied* and the result is a matrix $[\mathbf{C}]$ with m rows and p columns

$$[\mathbf{C}] = [\mathbf{A}][\mathbf{B}], \tag{1.30}$$

where the elements c_{ij}, $1 \le i \le m$, $1 \le j \le p$, of the matrix $[\mathbf{C}]$ satisfy the equality

$$c_{ij} = \sum_{k=1}^{n} a_{ik} b_{kj}, \tag{1.31}$$

that is, the elements of the product matrix are obtained by multiplying the rows of matrix $[\mathbf{A}]$ by the columns of matrix $[\mathbf{B}]$.

The transpose of the product matrix is given by the relation

$$[[\mathbf{A}][\mathbf{B}]]^{\mathrm{T}} = [\mathbf{B}]^{\mathrm{T}}[\mathbf{A}]^{\mathrm{T}}. \tag{1.32}$$

In some cases, there may exist *matrices of matrices* and the multiplication is performed as in the following example

$$\begin{bmatrix} [\mathbf{A}_1] & [\mathbf{A}_2] \\ [\mathbf{A}_3] & [\mathbf{A}_4] \\ [\mathbf{A}_5] & [\mathbf{A}_6] \end{bmatrix} \begin{bmatrix} [\mathbf{B}_1] & [\mathbf{B}_2] \\ [\mathbf{B}_3] & [\mathbf{B}_4] \end{bmatrix} = \begin{bmatrix} [\mathbf{A}_1][\mathbf{B}_1]+[\mathbf{A}_2][\mathbf{B}_3] & [\mathbf{A}_1][\mathbf{B}_2]+[\mathbf{A}_2][\mathbf{B}_4] \\ [\mathbf{A}_3][\mathbf{B}_1]+[\mathbf{A}_4][\mathbf{B}_3] & [\mathbf{A}_3][\mathbf{B}_2]+[\mathbf{A}_4][\mathbf{B}_4] \\ [\mathbf{A}_5][\mathbf{B}_1]+[\mathbf{A}_6][\mathbf{B}_3] & [\mathbf{A}_5][\mathbf{B}_2]+[\mathbf{A}_6][\mathbf{B}_4] \end{bmatrix}, \tag{1.33}$$

where we assumed that the operations of multiplication and addition of matrices can be performed for each separate case.

1.3 Square Matrix

The matrix [**A**] is a *square matrix* if the number of rows is equal to the number of columns; hence

$$[\mathbf{A}] = \begin{bmatrix} a_{11} & a_{12} & \dots & a_{1n} \\ a_{21} & a_{22} & \dots & a_{2n} \\ \dots & \dots & \dots & \dots \\ a_{n1} & a_{n2} & \dots & a_{nn} \end{bmatrix}, \tag{1.34}$$

where the number n is *the dimension* or *the order of the matrix*.

The determinant associated to the matrix [**A**] is denoted by det[**A**].

If $[\mathbf{A}_{ij}]$ is the matrix obtained from the matrix [**A**] by the suppression of the row i and the column j, then *the algebraic complement* a_{ij}^* is given by the expression

$$a_{ij}^* = (-1)^{i+j}\det\left[\mathbf{A}_{ij}\right], 1 \le i, j \le n, \tag{1.35}$$

and the following relation holds true

$$\sum_{k=1}^{n} a_{ik}a_{jk}^* = \sum_{k=1}^{n} a_{kj}a_{ki}^* = \begin{cases} 0 & \text{for } i \neq j \\ \det[\mathbf{A}] & \text{for } i = j \end{cases}. \tag{1.36}$$

The determinants of the matrices satisfy the equalities

$$\det[\mathbf{A}] = \det[\mathbf{A}]^{\mathrm{T}}, \tag{1.37}$$

$$\det[[\mathbf{A}][\mathbf{B}]] = \det[\mathbf{A}] \cdot \det[\mathbf{B}], \tag{1.38}$$

where we assumed that the matrices [**A**] and [**B**] have the same order.

In general, *the multiplication of matrices* is not commutative,

$$[\mathbf{A}][\mathbf{B}] \neq [\mathbf{B}][\mathbf{A}], \tag{1.39}$$

but it is associative and distributive, that is

$$[\mathbf{A}][[\mathbf{B}][\mathbf{C}]] = [[\mathbf{A}][\mathbf{B}]][\mathbf{C}] = [\mathbf{A}][\mathbf{B}][\mathbf{C}], \tag{1.40}$$

$$[\mathbf{A}][[\mathbf{B}] + [\mathbf{C}]] = [\mathbf{A}][\mathbf{B}] + [\mathbf{A}][\mathbf{C}], \tag{1.41}$$

where the matrices [**A**], [**B**] and [**C**] have the same order.

The trace of a matrix, denoted by Tr[**A**] is equal to the sum of the elements situated on the principal diagonal

$$\mathrm{Tr}[\mathbf{A}] = \sum_{i=1}^{n} a_{ii}. \tag{1.42}$$

The diagonal matrix is the matrix with all the elements equal to zero, except *some elements* situated on the principal diagonal.

The unity matrix, generally denoted by [**I**], is *the diagonal matrix* that has all the elements of the principal diagonal equal to unity,

$$[\mathbf{I}] = \begin{bmatrix} 1 & 0 & 0 & \dots & 0 \\ 0 & 1 & 0 & \dots & 0 \\ \dots & \dots & \dots & \dots & \dots \\ 0 & 0 & 0 & \dots & 1 \end{bmatrix}. \tag{1.43}$$

The unity matrix verifies the relations

$$[\mathbf{A}][\mathbf{I}] = [\mathbf{I}][\mathbf{A}] = [\mathbf{A}]. \tag{1.44}$$

The adjunct matrix $\mathbf{A}^*$ is defined by the relation

$$[\mathbf{A}^*] = \left(a_{ij}^*\right). \tag{1.45}$$

The matrix [**A**] is called *singular* if $\det[\mathbf{A}] = 0$; it is called a *non-singular* one if $\det[\mathbf{A}] \neq 0$.

The non-singular matrices [**A**] admit *inverse* matrices $[\mathbf{A}]^{-1}$; the inverse matrices fulfill the conditions

$$[\mathbf{A}]^{-1} = \frac{1}{\det[\mathbf{A}]}[\mathbf{A}^*], \tag{1.46}$$

$$[\mathbf{A}][\mathbf{A}]^{-1} = [\mathbf{A}]^{-1}[\mathbf{A}] = [\mathbf{I}], \tag{1.47}$$

$$\left[[\mathbf{A}]^{\mathrm{T}}\right]^{-1} = \left[[\mathbf{A}]^{-1}\right]^{\mathrm{T}}. \tag{1.48}$$

The matrix [**A**] is called *symmetric* if

$$[\mathbf{A}] = [\mathbf{A}]^{\mathrm{T}}; \tag{1.49}$$

it is called *anti-symmetric* or *skew* if

$$[\mathbf{A}] = -[\mathbf{A}]^{\mathrm{T}}. \tag{1.50}$$

The matrix [**A**] is called *orthogonal* if it fulfills the condition

$$[\mathbf{A}][\mathbf{A}]^{\mathrm{T}} = [\mathbf{I}]. \tag{1.51}$$

The orthogonal matrix [**A**] satisfies the equalities

$$[\mathbf{A}]^{\mathrm{T}} = [\mathbf{A}]^{-1}, \det[\mathbf{A}] = \pm 1. \tag{1.52}$$

The equation of nth degree

$$\det[\lambda[\mathbf{I}] - [\mathbf{A}]] = 0 \tag{1.53}$$

is *the characteristic equation* of the matrix [**A**]; its roots $\lambda_1, \lambda_2, \ldots, \lambda_n$ are called *the eigenvalues of the matrix* [**A**].

The vectors $\left\{\mathbf{v}^{(m_i)}\right\}$ which are obtained from the equality

$$[\mathbf{A}]\left\{\mathbf{v}^{(m_i)}\right\} = \lambda_m\left\{\mathbf{v}^{(m_i)}\right\}, 1 \le m \le k, \tag{1.54}$$

are called *eigenvectors* and, if the matrix [**A**] is a symmetric one, then its eigenvectors are orthogonal

$$\left\{\mathbf{v}^{(r)}\right\}^{\mathrm{T}}\left\{\mathbf{v}^{(s)}\right\} = 0, \text{ if } s \ne r. \tag{1.55}$$

Using the notation

$$b_j = \mathrm{Tr}\left[[\mathbf{A}]^j\right], \tag{1.56}$$

one obtains the characteristic equation

$$\sum_{j=0}^{n} c_{n-j}\lambda^{j} = 0, \tag{1.57}$$

where the coefficients c_j are given by the iterative relations

$$c_0 = 0,\ c_j = -\frac{1}{j}\sum_{k=0}^{j-1} c_k b_{j-k}. \tag{1.58}$$

Observation 1.3.1.

i. The eigenvalues of the matrix [**A**] of order n can be real or complex, distinct or not.
ii. One or more eigenvectors correspond to an eigenvalue λ_m, depending on the order of multiplicity for that eigenvalue.
iii. No matter if the eigenvalue is real or not, keeping into account that the matrix [**A**] has real components, the eigenvectors associated to that eigenvalue are matrices with n rows and one column, with real elements.

Observation 1.3.2. Let us consider that the matrix [**A**] is a square one, of order 3.

i. If the eigenvalues are real and distinct $\lambda_i \in \mathbb{R}$, $\lambda_i \neq \lambda_j$, $i,j \in \{1, 2, 3\}$, $i \neq j$, then the eigenvalues are obtained by solving three matrix equations of the form

$$[\mathbf{A}]\{\mathbf{v}_i\} = \lambda_i\{\mathbf{v}_i\}, i = 1, 2, 3. \tag{1.59}$$

ii. If the eigenvalues are real, but two of them are equal, $\lambda_i \in \mathbb{R}$, $i = 1, 2, 3$, $\lambda_1 = \lambda_2$, $\lambda_3 \neq \lambda_1$, then the eigenvalues result by solving the matrix equations

$$[\mathbf{A}]\{\mathbf{v}_1\} = \lambda_1\{\mathbf{v}_1\}, ([\mathbf{A}] - \lambda_1[\mathbf{I}])\{\mathbf{v}_2\} = \{\mathbf{v}_1\}, [\mathbf{A}]\{\mathbf{v}_3\} = \lambda_3\{\mathbf{v}_3\}. \tag{1.60}$$

iii. If the eigenvalues are real and equal, $\lambda_i = \lambda$, $i = 1, 2, 3$, then the eigenvector are obtained by solving the matrix equations

$$[\mathbf{A}]\{\mathbf{v}_1\} = \lambda\{\mathbf{v}_1\}, ([\mathbf{A}] - \lambda[\mathbf{I}])\{\mathbf{v}_2\} = \{\mathbf{v}_1\}, ([\mathbf{A}] - \lambda[\mathbf{I}])\{\mathbf{v}_3\} = \{\mathbf{v}_2\}. \tag{1.61}$$

iv. If the eigenvalues are one real, $\lambda_1 \in \mathbb{R}$, and two complex conjugate, $\lambda_2 = \alpha + \mathrm{i}\beta$, $\lambda_3 = \alpha - \mathrm{i}\beta$, $\alpha, \beta \in \mathbb{R}$, $\mathrm{i}^2 = -1$, then the eigenvectors result by solving the matrix equations

$$[\mathbf{A}](\{\mathbf{v}_2\} + \mathrm{i}\{\mathbf{v}_3\}) = (\alpha + \mathrm{i}\beta)(\{\mathbf{v}_2\} + \mathrm{i}\{\mathbf{v}_3\}); \tag{1.62}$$

1.4 Skew Matrix of Third Order

Starting from the relation of definition (1.49), it results that a third order skew matrix may be written in the form

$$[\mathbf{B}] = \begin{bmatrix} 0 & -b_3 & b_2 \\ b_3 & 0 & -b_1 \\ -b_2 & b_1 & 0 \end{bmatrix}. \tag{1.63}$$

One associates to the skew matrix [**B**] *the column matrix (vector)*

$$\{\mathbf{b}\} = [b_1 \;\; b_2 \;\; b_3]^{\mathrm{T}} \tag{1.64}$$

and the vector

$$\mathbf{b} = b_1\mathbf{i} + b_2\mathbf{j} + b_3\mathbf{k}. \tag{1.65}$$

It results the equality

$$[\mathbf{B}]\{\mathbf{b}\} = \{\mathbf{0}\}. \tag{1.66}$$

Being given the skew matrices [**A**], [**B**], and the eigenvectors associated to these matrices, then the vector equality

$$\mathbf{a} \times \mathbf{b} = -\mathbf{b} \times \mathbf{a} \tag{1.67}$$

may be put in the matrix expression

$$[\mathbf{A}]\{\mathbf{b}\} = -[\mathbf{B}]\{\mathbf{a}\}. \tag{1.68}$$

For *the skew matrix* [**B**] one may write the following relations (obtained by elementary calculation)

$$\det[\mathbf{B}] = 0, \tag{1.69}$$

$$[\mathbf{B}]^2 = -\left(b_1^2 + b_2^2 + b_3^2\right)[\mathbf{I}] + \{\mathbf{b}\}\{\mathbf{b}\}^{\mathrm{T}}, \tag{1.70}$$

$$[\mathbf{B}]^3 = -\left(b_1^2 + b_2^2 + b_3^2\right)[\mathbf{B}]. \tag{1.71}$$

For the *skew* matrices [**A**], [**B**] and the associated vectors **a**, **b**, denoting the vector product by **c**, $\mathbf{c} = \mathbf{a} \times \mathbf{b}$, and by [**C**] the associated skew matrix, one obtains the relations

$$[\mathbf{A}][\mathbf{B}] = -(a_1b_1 + a_2b_2 + a_3b_3)[\mathbf{I}] + \{\mathbf{b}\}\{\mathbf{a}\}^{\mathrm{T}}, \tag{1.72}$$

$$[\mathbf{B}][\mathbf{A}] = -(a_1b_1 + a_2b_2 + a_3b_3)[\mathbf{I}] + \{\mathbf{a}\}\{\mathbf{b}\}^{\mathrm{T}}, \tag{1.73}$$

$$[\mathbf{C}] = [\mathbf{A}][\mathbf{B}] - [\mathbf{B}][\mathbf{A}] = \{\mathbf{b}\}\{\mathbf{a}\}^{\mathrm{T}} - \{\mathbf{a}\}\{\mathbf{b}\}^{\mathrm{T}}, \tag{1.74}$$

$$\begin{gathered}[\mathbf{C}]^2 = (a_1b_1 + a_2b_2 + a_3b_3)\left[\{\mathbf{a}\}\{\mathbf{b}\}^{\mathrm{T}} - \{\mathbf{b}\}\{\mathbf{a}\}^{\mathrm{T}}\right] - \\ \left(b_1^2 + b_2^2 + b_3^2\right)\{\mathbf{a}\}\{\mathbf{a}\}^{\mathrm{T}} - \left(a_1^2 + a_2^2 + a_3^2\right)\{\mathbf{b}\}\{\mathbf{b}\}^{\mathrm{T}}.\end{gathered} \tag{1.75}$$

If the matrix [**A**] is an *arbitrary* third order one, and the matrices [**B**], [**C**] are *skew* ones, then the matrix

$$[\mathbf{D}] = [\mathbf{A}]^{\mathrm{T}}[\mathbf{B}][\mathbf{A}] \tag{1.76}$$

is a *skew* matrix, and the associated column matrices $\{\mathbf{b}\}$, $\{\mathbf{c}\}$, $\{\mathbf{d}\}$ satisfy the equalities

$$\{\mathbf{d}\} = [\mathbf{A}^*]\{\mathbf{b}\}, \tag{1.77}$$

$$[\mathbf{A}]^{\mathrm{T}}[\mathbf{B}][\mathbf{A}]\{\mathbf{c}\} = -[\mathbf{C}][\mathbf{A}^*]\{\mathbf{b}\}, \tag{1.78}$$

where $[\mathbf{A}^*]$ is the adjunct matrix of the matrix [**A**].

When the matrix [**A**] is orthogonal, one obtains the equalities

$$\{\mathbf{d}\} = [\mathbf{A}]^{\mathrm{T}}\{\mathbf{b}\}, [\mathbf{A}]^{\mathrm{T}}[\mathbf{B}][\mathbf{A}]\{\mathbf{c}\} = -[\mathbf{C}][\mathbf{A}]^{\mathrm{T}}\{\mathbf{b}\}. \tag{1.79}$$

More general, if the matrix [**A**] has k rows and 3 columns, then it results that the kth order square matrix

$$[\mathbf{D}] = [\mathbf{A}]^{\mathrm{T}}[\mathbf{B}][\mathbf{A}] \tag{1.80}$$

is a skew matrix; moreover, it results that if $k = 1$, then the matrix [**D**] is the zero matrix with only one element.

Sometimes, in the analytical calculations, it is useful to use the skew matrices associated to the unit vectors **i**, **j**, **k**,

$$[\mathbf{U}_1] = \begin{bmatrix} 0 & 0 & 0 \\ 0 & 0 & -1 \\ 0 & 1 & 0 \end{bmatrix}, [\mathbf{U}_2] = \begin{bmatrix} 0 & 0 & 1 \\ 0 & 0 & 0 \\ -1 & 0 & 0 \end{bmatrix}, [\mathbf{U}_3] = \begin{bmatrix} 0 & -1 & 0 \\ 1 & 0 & 0 \\ 0 & 0 & 0 \end{bmatrix}, \tag{1.81}$$

and the column matrices

$$\{\mathbf{u}_1\} = \begin{bmatrix} 1 \\ 0 \\ 0 \end{bmatrix}, \{\mathbf{u}_2\} = \begin{bmatrix} 0 \\ 1 \\ 0 \end{bmatrix}, \{\mathbf{u}_3\} = \begin{bmatrix} 0 \\ 0 \\ 1 \end{bmatrix}, \tag{1.82}$$

respectively.

One obtains the expressions

$$[\mathbf{B}] = \sum_{i=1}^{3} b_i[\mathbf{U}_i], \{\mathbf{b}\} = \sum_{i=1}^{3} b_i\{\mathbf{u}_i\}, \tag{1.83}$$

$$[\mathbf{U}_i]\left[\mathbf{U}_j\right][\mathbf{U}_i] = [\mathbf{0}], (\forall)\ i \neq j, \tag{1.84}$$

$$[\mathbf{U}_1] = [\mathbf{U}_2][\mathbf{U}_3] - [\mathbf{U}_3][\mathbf{U}_2] \tag{1.85}$$

and the analogous,

$$[\mathbf{U}_1][\mathbf{U}_2][\mathbf{U}_3] + [\mathbf{U}_3][\mathbf{U}_2][\mathbf{U}_1] = [\mathbf{0}]. \tag{1.86}$$

and the analogous.

Further Reading

Bloch AM, Baillieul J, Crouch P, Marsden J (2007). Nonholonomic Mechanics and Control (Interdisciplinary Applied Mathematics). Berlin: Springer.
Blundell M, Harty D (2004). The Multibody Systems Approach to Vehicle Dynamics. Amsterdam: Elsevier Butterworth-Heinemann.
Dankowicz HJ (2004). Multibody Mechanics and Visualization. London: Springer.
Den Hartog JP (1961). Mechanics. New York: Dover Publications Inc.
Douglas Gregory R (2006). Classical Mechanics. Cambridge: Cambridge University Press.
Eich-Soellner E, Führer C (2013). Numerical Methods in Multibody Dynamics. Wiesbaden: Springer.
Fasano A, Marmi S, Pelloni B (2006). Analytical Mechanics: An Introduction. Oxford: Oxford University Press.
Ginsberg JH (1998). Advanced Engineering Dynamics. 2nd ed. Cambridge: Cambridge University Press.
Ginsberg J (2007). Engineering Dynamics. Cambridge: Cambridge University Press.
Goldstein H, Poole CP Jr, Safko JL (2001). Classical Mechanics. 3rd ed. Edinburg: Pearson.
Greenwood DT (2006). Advanced Dynamics. Cambridge: Cambridge University Press.
Greiner W (2009). Classical Mechanics: Systems of Particles and Hamiltonian Dynamics. 2nd ed. Heidelberg: Springer.
Hand LN, Finch JD (1998). Analytical Mechanics. Cambridge: Cambridge University Press.

Jazar RN (2010). Theory of Applied Robotics: Kinematics, Dynamics, and Control. 2nd ed. New York: Springer.
Kibble TWB, Berkshire FH (2004). Classical Mechanics. 5th ed. London: Imperial College Press.
Kleppner D, Kolenkow R (2013). An Introduction to Mechanics. 2nd ed. Cambridge: Cambridge University Press.
Lurie AI (2002). Analytical Mechanics. Berlin: Springer.
Meriam JL, Kraige LG (2012). Engineering Mechanics: Dynamics. Hoboken: John Wiley & Sons.
Moon FC (1998). Applied Dynamics: With Applications to Multibody and Mechatronic Systems. New York: John Wiley & Sons, Inc.
Morin D (2008). Introduction to Classical Mechanics: With Problems and Solution. Cambridge: Cambridge University Press.
Murray RM, Li Z, Shankar Sastry S (1994). A Mathematical Introduction to Robotic Manipulation. Boca Raton: CRC Press.
Neimark IuI, Fufaev NA (1972). Dynamics of Nonholonomic Systems (Translations of Mathematical Monographs, V. 33). Providence, Rhode Island: American Mathematical Society.
O'Reilly OM (2008). Intermediate Dynamics for Engineers: A unified Treatment of Newton-Euler and Lagrangian Mechanics. Cambridge: Cambridge University Press
Pandrea N, Stănescu ND (2002). Mecanica. Bucureşti: Editura Didactică şi Pedagogică (in Romanian).
Papastavridis JG (2002). Analytical Mechanics: A Comprehensive Treatise on the Dynamics of Constrained Systems; For Engineers, Physicists, and Mathematicians. Oxford: Oxford University Press.
Pfeiffer F (2008). Mechanical System Dynamics (Lecture Notes in Applied and Computational Mechanics). Berlin: Springer.
Spong MW, Hutchinson S, Vidyasagar M (2005). Robot Modeling and Control. New York: John Wiley & Sons.
Taylor JR (2005). Classical Mechanics. Sausalito: University Science Books.
Uicker JJ, Ravani B, Sheth PN (2013). Matrix Methods in the Design Analysis of Mechanisms and Multibody Systems. Cambridge: Cambridge University Press.
Woodhouse N (2010). Introduction to Analytical Dynamics. London: Springer.

2

Kinematics of the Rigid Solid

The chapter treats the kinematics of the rigid solid. Here we obtain the distribution of velocities and accelerations for a rigid body.

2.1 Finite Displacements of the Points of Rigid Solid

The motion of a rigid solid relative to a tri-orthogonal dextrorsum reference system O_0XYZ is completely determined by the motion of a tri-orthogonal dextrorsum reference system $Oxyz$ jointed to the rigid solid, relative to the reference system O_0XYZ (Fig. 2.1).

One considers that at the initial moment the system $Oxyz$ coincides to the system O_0XYZ. In these conditions, the finite displacement of the point O is $\mathbf{O_0O} = \mathbf{s}_O$, while the finite displacement of an arbitrary point P (Fig. 2.1) is $\mathbf{P_0P} = \mathbf{s}$.

We denote by $\mathbf{i}$, $\mathbf{j}$, $\mathbf{k}$ the unit vectors of the mobile axes, by $\mathbf{i}_0$, $\mathbf{j}_0$, $\mathbf{k}_0$ the unit vectors of the fixed axes, by a_{1i}, a_{2i}, a_{3i}, $i = 1, 2, 3$, the director cosines of the axes Ox, Oy, Oz, by X, Y, Z, x, y, z the coordinates of the point P relative to the two reference systems, and by X_O, Y_O, Z_O the coordinates of the point O relative to the fixed reference system. Keeping into account that the point P_0 has the same position relative to the system O_0XYZ as the point P relative to the system $Oxyz$, one may write the vectors

$$\mathbf{r} = x\mathbf{i} + y\mathbf{j} + z\mathbf{k},\ \mathbf{r}_0 = x\mathbf{i}_0 + y\mathbf{j}_0 + z\mathbf{k}_0,\ \mathbf{s}_O = X_O\mathbf{i}_0 + Y_O\mathbf{j}_0 + Z_O\mathbf{k}_0,\ \mathbf{R} = X\mathbf{i}_0 + Y\mathbf{j}_0 + Z\mathbf{k}_0, \tag{2.1}$$

Dynamics of the Rigid Solid with General Constraints by a Multibody Approach, First Edition.
Nicolae Pandrea and Nicolae-Doru Stănescu.

Companion website: www.wiley.com/go/pandrea

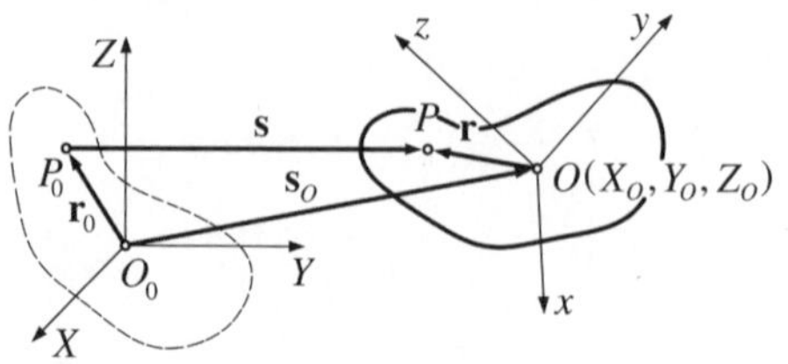

Figure 2.1 Finite displacements of the rigid solid.

the column matrices

$$\{\mathbf{r}\}=[x\ \ y\ \ z]^{\mathrm{T}},\{\mathbf{r}_0\}=[x\ \ y\ \ z]^{\mathrm{T}},\{\mathbf{s}_O\}=[X_O\ \ Y_O\ \ Z_O]^{\mathrm{T}},\{\mathbf{R}\}=[X\ \ Y\ \ Z]^{\mathrm{T}}, \tag{2.2}$$

the rotational matrix

$$[\mathbf{A}]=\begin{bmatrix} a_{11} & a_{12} & a_{13} \\ a_{21} & a_{22} & a_{23} \\ a_{31} & a_{32} & a_{33} \end{bmatrix}, \tag{2.3}$$

and the matrix relation of transformation

$$\{\mathbf{R}\}=\{\mathbf{s}_O\}+[\mathbf{A}]\{\mathbf{r}\}. \tag{2.4}$$

At the initial moment, the point P being situated at the point P_0 (Fig. 2.1), the displacement $\mathbf{s}=\mathbf{P}_0\mathbf{P}$ of this point may be written in the form $\mathbf{s}=\mathbf{R}-\mathbf{r}_0$; keeping into account the notations (2.2), one obtains the matrix expression

$$\{\mathbf{s}\}=\{\mathbf{s}_O\}+[\mathbf{A}]\{\mathbf{r}\}-\{\mathbf{r}\} \tag{2.5}$$

or

$$\{\mathbf{s}\}=\{\mathbf{s}_O\}+[\mathbf{A}]\{\mathbf{r}_0\}-\{\mathbf{r}_0\}, \tag{2.6}$$

where $\{\mathbf{s}\}$ is the column matrix of the projection of vector $\mathbf{s}$ onto the axes of the fixed system O_0XYZ.

2.2 Matrix of Rotation: Properties

2.2.1 General Properties

From the vector relations $\mathbf{i}\cdot\mathbf{j}=\mathbf{j}\cdot\mathbf{k}=\mathbf{k}\cdot\mathbf{i}=0$, $\mathbf{i}^2=\mathbf{j}^2=\mathbf{k}^2=1$ it results the scalar relations

$$\sum_{i=1}^{3} a_{ij}a_{ik}=\begin{cases} 1 & \text{for } j=k, \\ 0 & \text{for } j\neq k, \end{cases} \quad j,k=1,2,3, \tag{2.7}$$

which show that the matrix [**A**] is and orthogonal one; hence

$$[\mathbf{A}]^{-1} = [\mathbf{A}]^{\mathrm{T}}. \tag{2.8}$$

From the vector relation $\mathbf{i} = \mathbf{j} \times \mathbf{k}$ one obtains the scalar relations

$$a_{11} = a_{22}a_{33} - a_{32}a_{23},\ a_{21} = a_{23}a_{31} - a_{21}a_{33},\ a_{31} = a_{21}a_{32} - a_{22}a_{31}; \tag{2.9}$$

developing det[**A**] after the first column, we get

$$\det[\mathbf{A}] = 1. \tag{2.10}$$

Considering the vector defined by the elements of the column matrix $[\mathbf{A}]\{\mathbf{r}\}$ and denoting this vector by $\overline{[\mathbf{A}]\{\mathbf{r}\}}$, from the conditions of distances and angles preservation, it results the expressions

$$\overline{[\mathbf{A}]\{\mathbf{r}\}} = |\mathbf{r}|, \tag{2.11}$$

$$\overline{[\mathbf{A}]\{\mathbf{u}\}} \cdot \overline{[\mathbf{A}]\{\mathbf{v}\}} = \mathbf{u} \cdot \mathbf{v}, \tag{2.12}$$

$$\overline{[\mathbf{A}]\{\mathbf{u}\}} \times \overline{[\mathbf{A}]\{v\}} = \overline{[\mathbf{A}]\{\mathbf{q}\}},\ \text{where } \mathbf{q} = \mathbf{u} \times \mathbf{v}. \tag{2.13}$$

2.2.2 *Successive Displacements*

Let us consider two positions of the rigid solid and two jointed reference systems $O_1x_1y_1z_1$ and $O_2x_2y_2z_2$, respectively. Denoting by $[\mathbf{A}_{10}]$, $[\mathbf{A}_{20}]$ the rotational matrices relative to the fixed reference system O_0XYZ, one obtains the following relations for the column matrices $\{\mathbf{v}^{(1)}\}$, $\{\mathbf{v}^{(2)}\}$, $\{\mathbf{v}^{(0)}\}$ of the projections of an arbitrary vector **v** in the three reference systems

$$\left\{\mathbf{v}^{(0)}\right\} = [\mathbf{A}_{10}]\left\{\mathbf{v}^{(1)}\right\},\ \left\{\mathbf{v}^{(0)}\right\} = [\mathbf{A}_{20}]\left\{\mathbf{v}^{(2)}\right\}. \tag{2.14}$$

It results

$$\left\{\mathbf{v}^{(1)}\right\} = [\mathbf{A}_{10}]^{\mathrm{T}}[\mathbf{A}_{20}]\left\{\mathbf{v}^{(2)}\right\}; \tag{2.15}$$

hence, the matrix of rotation of the reference system $O_2x_2y_2z_2$ relative to the reference system $O_1x_1y_1z_1$ reads

$$[\mathbf{A}_{21}] = [\mathbf{A}_{10}]^{\mathrm{T}}[\mathbf{A}_{20}]. \tag{2.16}$$

From the expression (2.16) it also results

$$[\mathbf{A}_{20}] = [\mathbf{A}_{10}][\mathbf{A}_{21}], \tag{2.17}$$

and therefore, in general, for n positions of the solid rigid, one obtains the matrix relation

$$[\mathbf{A}_{n0}] = [\mathbf{A}_{10}][\mathbf{A}_{21}]\ldots[\mathbf{A}_{n,n-1}]. \tag{2.18}$$

2.2.3 Eigenvalues: Eigenvectors

The eigenvalues λ_1, λ_2, λ_3 of the matrix $[\mathbf{A}]$ are obtained by solving the equation

$$\det[[\mathbf{A}] - \lambda[\mathbf{I}]] = 0, \tag{2.19}$$

where $[\mathbf{I}]$ is the unity matrix.
It results the equation

$$(\lambda - 1)\left[\lambda^2 - (\delta - 1)\cdot\lambda + 1\right] = 0, \tag{2.20}$$

where

$$\delta = \mathrm{Tr}[\mathbf{A}] = a_{11} + a_{22} + a_{33}. \tag{2.21}$$

Keeping into account the relations (2.1), it results $(\delta - 1)^2 \le 4$ and, consequently, one obtains a real solution

$$\lambda_1 = 1 \tag{2.22}$$

and two complex solutions

$$\lambda_{2,3} = \frac{\delta - 1}{2} \pm \mathrm{i}\sqrt{1 - \left(\frac{\delta - 1}{2}\right)^2}. \tag{2.23}$$

To determine the real unitary vector $\{\mathbf{u}\}$ it is necessary to solve the system of equations

$$[\mathbf{A}]\{\mathbf{u}\} = \{\mathbf{u}\}, \tag{2.24}$$

$$\{\mathbf{u}\}^{\mathrm{T}}\{\mathbf{u}\} = 1. \tag{2.25}$$

From the matrix equation (2.24) one obtains the scalar relations

$$u_x(a_{32}+a_{23})=u_y(a_{13}+a_{31})=u_z(a_{21}+a_{12}); \tag{2.26}$$

keeping into account the equalities

$$a_{13}^2-a_{31}^2=a_{21}^2-a_{12}^2=a_{32}^2-a_{23}^2, \tag{2.27}$$

the relations (2.26) become

$$\frac{u_x}{a_{32}-a_{23}}=\frac{u_y}{a_{13}-a_{31}}=\frac{u_z}{a_{21}-a_{12}}. \tag{2.28}$$

The relations (2.28) and (2.27) show that the vectors

$$\mathbf{w}=\frac{1}{2}[(a_{32}-a_{23})\mathbf{i}_0+(a_{13}-a_{31})\mathbf{j}_0+(a_{21}-a_{12})\mathbf{k}_0], \tag{2.29}$$

$$\mathbf{w}^*=\frac{1}{a_{32}+a_{23}}\mathbf{i}_0+\frac{1}{a_{13}+a_{31}}\mathbf{j}_0+\frac{1}{a_{21}+a_{12}}\mathbf{k}_0, \tag{2.30}$$

(which are not unitary ones) verify the matrix equation (2.24), that is, they are eigenvectors of the matrix $[\mathbf{A}]$.

If the vector $\mathbf{w}$ is not equal to zero, then the unitary eigenvector is calculated using the relation

$$\mathbf{u}=\frac{\mathbf{w}}{|\mathbf{w}|}; \tag{2.31}$$

if the vector $\mathbf{w}$ is a null one, then we have to use the relations (2.24) and (2.25).

Example 2.2.1 Determine the real unitary eigenvectors for the rotational matrices

$$[\mathbf{A}_1]=\begin{bmatrix}0 & -1 & 0\\ 0 & 0 & -1\\ 1 & 0 & 0\end{bmatrix}, [\mathbf{A}_2]=\begin{bmatrix}0 & 0 & 1\\ 0 & -1 & 0\\ 1 & 0 & 0\end{bmatrix}, [\mathbf{A}_3]=\begin{bmatrix}1 & 0 & 0\\ 0 & -1 & 0\\ 0 & 0 & -1\end{bmatrix}. \tag{2.32}$$

Solution: Since

$$\det[\mathbf{A}_1]=\det[\mathbf{A}_2]=\det[\mathbf{A}_3]=1, \tag{2.33}$$

it results that the given matrices correspond to dextrorsum tri-orthogonal systems.

For the matrix $[\mathbf{A}_1]$ we get

$$\mathbf{w}_1 = \frac{1}{2}(\mathbf{i}_0 - \mathbf{j}_0 + \mathbf{k}_0),\ \mathbf{u}_1 = \frac{\mathbf{w}_1}{|\mathbf{w}_1|} = \frac{1}{\sqrt{3}}(\mathbf{i}_0 - \mathbf{j}_0 - \mathbf{k}_0). \tag{2.34}$$

For the matrices $[\mathbf{A}_2]$, $[\mathbf{A}_3]$ one obtains the vectors

$$\mathbf{w}_2 = \mathbf{w}_3 = \mathbf{0} \tag{2.35}$$

and then, from the equations (2.24) and (2.25) it results

$$\mathbf{u}_2 = \frac{1}{\sqrt{2}}(\mathbf{i}_0 + \mathbf{k}_0),\ \mathbf{u}_3 = \mathbf{i}_0. \tag{2.36}$$

2.2.4 The Expression of the Matrix of Rotation with the Aid of the Unitary Eigenvector and the Angle of Rotation

One considers the case in which a rigid solid rotates with the angle ξ (Fig. 2.2) about an axis Δ of unit vector $\mathbf{u}$.

By this motion, an arbitrary point P_0 transforms in the point P, the vector $\mathbf{O}_0\mathbf{P}_0$ becomes the vector $\mathbf{O}_0\mathbf{P}$, the vector $\mathbf{CP}_0$ (C being the projection of the point P_0 onto the axis Δ) transforms in the vector $\mathbf{CP}$, while the reference system O_0XYZ becomes the system $Oxyz$, $O \equiv O_0$.

It is obvious that the rotation angle ξ is the angle between the vectors $\mathbf{CP}_0$ and $\mathbf{CP}$.

The Rodrigues relation reads

$$\mathbf{O}_0\mathbf{P} = \mathbf{O}_0\mathbf{P}_0 + \mathbf{u} \times \mathbf{O}_0\mathbf{P}_0 \sin\xi + \mathbf{u} \times (\mathbf{u} \times \mathbf{O}_0\mathbf{P}_0)(1 - \cos\xi), \tag{2.37}$$

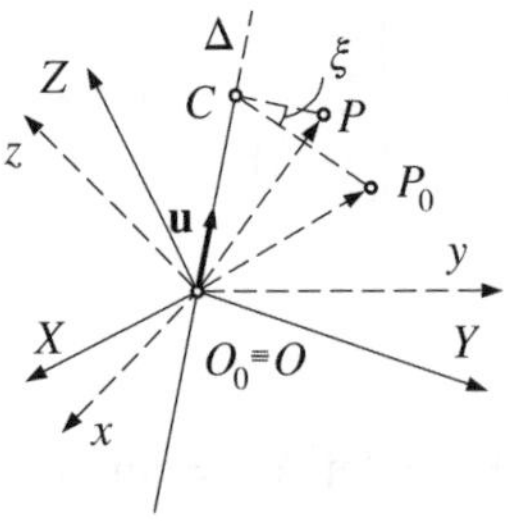

Figure 2.2 The rotation about the axis Δ.

If one considers the notations:

- x, y, z, X, Y, Z – the projections of the vectors $\mathbf{O_0P_0}$ and $\mathbf{O_0P}$ onto the axes of the reference frame O_0XYZ;
- a, b, c – the projections of the unit vector $\mathbf{u}$ onto the axes of the reference system O_0XYZ;
- the matrix [**A**] given by

$$[\mathbf{A}] = \begin{bmatrix} 1-(b^2+c^2)(1-\mathrm{c}\xi) & -c\mathrm{s}\xi+ab(1-\mathrm{c}\xi) & b\mathrm{s}\xi+bc(1-\mathrm{c}\xi) \\ c\mathrm{s}\xi+ab(1-\mathrm{c}\xi) & 1-(a^2+c^2)(1-\mathrm{c}\xi) & -a\mathrm{s}\xi+bc(1-\mathrm{c}\xi) \\ -b\mathrm{s}\xi+ac(1-\mathrm{c}\xi) & a\mathrm{s}\xi+bc(1-\mathrm{c}\xi) & 1-(a^2+b^2)(1-\mathrm{c}\xi) \end{bmatrix}, \tag{2.38}$$

  where the functions $\cos\xi$ and $\sin\xi$ were marked by $\mathrm{c}\xi$ and $\mathrm{s}\xi$, respectively;
- the column matrices

$$\{\mathbf{r}_0\} = [X \;\; Y \;\; Z]^{\mathrm{T}},\; \{\mathbf{r}\} = [x \;\; y \;\; z]^{\mathrm{T}}, \tag{2.39}$$

then, from the expression (2.37) one obtains the matrix relation

$$\{\mathbf{r}_0\} = [\mathbf{A}]\{\mathbf{r}\}, \tag{2.40}$$

which shows that the matrix [**A**] defined by the expression (2.38) is just *the matrix of rotation* given by the expression (2.3).

Calculating the eigenvector **w** given by the equality (2.29), one obtains

$$\mathbf{w} = (a\mathbf{i}_0 + b\mathbf{j}_0 + c\mathbf{k}_0)\sin\xi \tag{2.41}$$

or

$$\mathbf{w} = \mathbf{u}\sin\xi, \tag{2.42}$$

wherefrom it results

$$\sin\xi = |\mathbf{w}|. \tag{2.43}$$

Making the sum of the elements situated on the principal diagonal for the forms (2.3) and (2.38) of the matrix [**A**], and using the notation (2.21), one deduces the expression

$$\cos\xi = \frac{\delta-1}{2}; \tag{2.44}$$

hence, the complex eigenvalues (2.23) of the matrix [**A**] are

$$\lambda_{2,3} = \cos\xi \pm \mathrm{i}\sin\xi. \tag{2.45}$$

With the aid of the skew matrix corresponding to the vector **u**

$$[\mathbf{U}] = \begin{bmatrix} 0 & -c & b \\ c & 0 & -a \\ -b & a & 0 \end{bmatrix}, \tag{2.46}$$

the matrix of rotation (2.38) may be written in the form

$$[\mathbf{A}] = [\mathbf{I}] + [\mathbf{U}]\sin\xi + 2[\mathbf{U}]^2\sin^2\frac{\xi}{2}, \tag{2.47}$$

while the transpose matrix (which is exactly the inverse $[\mathbf{A}]^{-1}$) reads

$$[\mathbf{A}]^{\mathrm{T}} = [\mathbf{I}] - [\mathbf{U}]\sin\xi + 2[\mathbf{U}]^2\sin^2\frac{\xi}{2}. \tag{2.48}$$

Using the identities

$$[\mathbf{U}]^3 = -[\mathbf{U}],\ [\mathbf{U}]^4 = -[\mathbf{U}]^2, \tag{2.49}$$

one deduces the equality

$$[\mathbf{A}]^n = [\mathbf{I}] + [\mathbf{U}]\sin\xi + 2[\mathbf{U}]^2\sin^2\left(\frac{n\xi}{2}\right). \tag{2.50}$$

If the matrix of rotation is given by the relation (2.3), then the principal unitary vector **u** is deduced by the previous exposed methods, while the rotational angle is deduced from the relation (2.44).

Example 2.2.2 Determine the axis and angle of rotation for the matrix

$$[\mathbf{A}] = \begin{bmatrix} 0 & 0 & 1 \\ 1 & 0 & 0 \\ 0 & 1 & 0 \end{bmatrix}. \tag{2.51}$$

Solution: It successively results

$$\cos\xi = -\frac{1}{2}, \xi = \frac{2\pi}{3},\ \mathbf{w} = \frac{1}{2}(\mathbf{i}_O + \mathbf{j}_O + \mathbf{k}_O),\ \mathbf{u} = \frac{1}{\sqrt{3}}(\mathbf{i}_O + \mathbf{j}_O + \mathbf{k}_O). \tag{2.52}$$

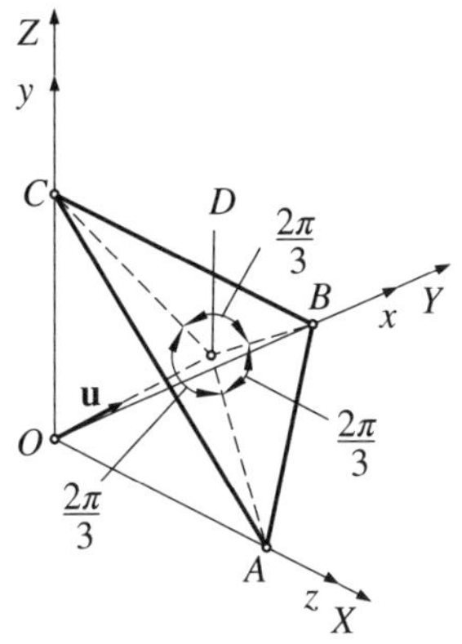

Figure 2.3 Determination of the axis and angle of rotation in Example 2.2.2.

The representation of this rotation is given in Fig. 2.3, where we considered that $O_0A = O_0B = O_0C$.

It results that the axis of rotation intersects the plan of the triangle ABC at the point D, while by this rotation the point A comes to the point B, the point B to the point C, and the point C to the point A.

For our analysis is interesting the case in which the rotational angle ξ is equal to π, that is, the case in which $\mathbf{w} = \mathbf{0}$, $a_{11} + a_{22} + a_{33} = -1$. In this situation, from the relation (2.29) one deuces the equalities $a_{23} = a_{32}$, $a_{13} = a_{31}$, $a_{21} = a_{12}$ and it results that the matrix $[\mathbf{A}]$ becomes a symmetric one.

Considering that $a_{33} \neq -1$, from the first two scalar equations (2.24) and from the equation (2.25) one deduces the solution

$$\{\mathbf{u}\} = \left[\frac{a_{13}}{\sqrt{2(1+a_{33})}} \ \frac{a_{23}}{\sqrt{2(1+a_{33})}} \ \frac{1+a_{33}}{\sqrt{2(1+a_{33})}}\right]^{\mathrm{T}}. \tag{2.53}$$

When $a_{33} = -1$, from the equalities

$$a_{13}^2 + a_{23}^2 + a_{33}^2 = 1, a_{11} + a_{22} + a_{33} = -1, \tag{2.54}$$

it results

$$a_{13} = a_{23} = 0, a_{22} = -a_{11}. \tag{2.55}$$

If $a_{11} \neq 1$, then one obtains the solution

$$\{\mathbf{u}\} = \left[\frac{a_{12}}{\sqrt{2(1-a_{11})}} \ \frac{1-a_{11}}{\sqrt{2(1-a_{11})}} 0\right]^{\mathrm{T}}, \tag{2.56}$$

and if $a_{11}=1$, then one obtains $a_{12}=0$ and the solution

$$\{\mathbf{u}\}=[1\ \ 0\ \ 0]^{\mathrm{T}}. \tag{2.57}$$

Example 2.2.3 Determine the unitary eigenvector for the matrix

$$[\mathbf{A}]=\begin{bmatrix} -\cos^2\varphi-\sin^2\varphi\cos\theta & \sin\varphi\cos\varphi(1-\cos\theta) & \sin\varphi\sin\theta \\ \sin\varphi\cos\varphi(1-\cos\theta) & -\sin^2\varphi-\cos^2\varphi\cos\theta & \cos\varphi\sin\theta \\ \sin\theta\sin\varphi & \sin\theta\cos\varphi & \cos\theta \end{bmatrix}.$$

Solution: One may easily verify the conditions $\mathbf{w}=\mathbf{0}$, $\xi=\pi$, $[\mathbf{A}]=[\mathbf{A}]^{\mathrm{T}}$.

If $\theta\neq\pi$, then one obtains $\{\mathbf{u}\}=\begin{bmatrix}\sin\varphi\sin\frac{\theta}{2} & \cos\varphi\sin\frac{\theta}{2} & \cos\frac{\theta}{2}\end{bmatrix}^{\mathrm{T}}$, while if $\theta=\pi$, then it results $\{\mathbf{u}\}=[\sin\varphi\ \ \cos\varphi\ \ 0]^{\mathrm{T}}$.

2.2.5 *Symmetries: Decomposition of the Rotation into Two Symmetries*

In the case of a symmetry the angle of rotation is $\xi=\pi$, and the matrix of rotation reads

$$[\mathbf{A}]=[\mathbf{I}]+2[\mathbf{U}]^2. \tag{2.58}$$

If one considers the straight lines Δ_1, Δ_2 concurrent at the point O_0 with the straight line Δ and perpendicular to it (Fig. 2.4), the angle between the straight lines Δ_1, Δ_2 being equal to $\frac{\xi}{2}$, and denotes by $\mathbf{u}_1$, $\mathbf{u}_2$ and $\mathbf{u}$ the unit vectors of these straight lines, then one obtains the vector relation

$$\mathbf{u}\sin\frac{\xi}{2}=\mathbf{u}_1\times\mathbf{u}_2; \tag{2.59}$$

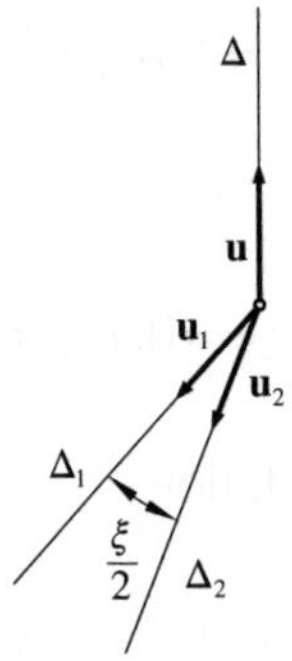

Figure 2.4 Decomposition of the rotation into two symmetries.

the last expression leads to the following relation between the attached skew matrices

$$[\mathbf{U}]\sin\frac{\xi}{2} = [\mathbf{U}_1][\mathbf{U}_2] - [\mathbf{U}_2][\mathbf{U}_1]. \tag{2.60}$$

We will prove that successive symmetries about the axes Δ_1, Δ_2 lead to o rotation of angle ξ about the axis Δ.

To this end one uses the identities

$$\begin{aligned}
[\mathbf{U}_i]^2 &= -[\mathbf{I}] + \{\mathbf{u}_i\}\{\mathbf{u}_i\}^{\mathrm{T}},\\
[\mathbf{U}]^2\sin^2\frac{\xi}{2} &= \left[\{\mathbf{u}_1\}\{\mathbf{u}_2\}^{\mathrm{T}} + \{\mathbf{u}_2\}\{\mathbf{u}_1\}^{\mathrm{T}}\right]\cos\frac{\xi}{2} - \{\mathbf{u}_1\}\{\mathbf{u}_1\}^{\mathrm{T}} - \{\mathbf{u}_2\}\{\mathbf{u}_2\}^{\mathrm{T}},\\
[\mathbf{U}_1]^2 + [\mathbf{U}_2]^2 &= -2[\mathbf{I}] - [\mathbf{U}]^2\sin^2\frac{\xi}{2} + \left[\{\mathbf{u}_1\}\{\mathbf{u}_2\}^{\mathrm{T}} + \{\mathbf{u}_2\}\{\mathbf{u}_1\}^{\mathrm{T}}\right],\\
[\mathbf{U}_1]^2[\mathbf{U}_2]^2 &= [\mathbf{I}] + [\mathbf{U}]^2\sin^2\frac{\xi}{2} - \{\mathbf{u}_2\}\{\mathbf{u}_1\}^{\mathrm{T}}\cos\frac{\xi}{2}
\end{aligned} \tag{2.61}$$

and results the equality

$$\left[[\mathbf{I}] + 2[\mathbf{U}_1]^2\right]\left[[\mathbf{I}] + 2[\mathbf{U}_2]^2\right] = [\mathbf{I}] + [\mathbf{U}]\sin\xi + 2[\mathbf{U}]^2\sin^2\frac{\xi}{2}, \tag{2.62}$$

which proves that the rotation of angle ξ about the axis Δ is equivalent to the successive symmetries about the axes Δ_1, Δ_2.

2.2.6 *Rotations About the Axes of Coordinates*

If one makes a rotation of angle ξ_1 about the axis O_0X ($a=1$, $b=0$, $c=0$), then one obtains from the relation (2.38) the matrix of rotation $[\boldsymbol{\xi}_1]$,

$$[\boldsymbol{\xi}_1] = \begin{bmatrix} 1 & 0 & 0 \\ 0 & \cos\xi_1 & -\sin\xi_1 \\ 0 & \sin\xi_1 & \cos\xi_1 \end{bmatrix}. \tag{2.63}$$

Analogically, one obtains the matrices of rotation about the axes O_0Y and O_0Z,

$$[\boldsymbol{\xi}_2] = \begin{bmatrix} \cos\xi_2 & 0 & \sin\xi_2 \\ 0 & 1 & 0 \\ -\sin\xi_2 & 0 & \cos\xi_2 \end{bmatrix},\ [\boldsymbol{\xi}_3] = \begin{bmatrix} \cos\xi_3 & -\sin\xi_3 & 0 \\ \sin\xi_3 & \cos\xi_3 & 0 \\ 0 & 0 & 1 \end{bmatrix}. \tag{2.64}$$

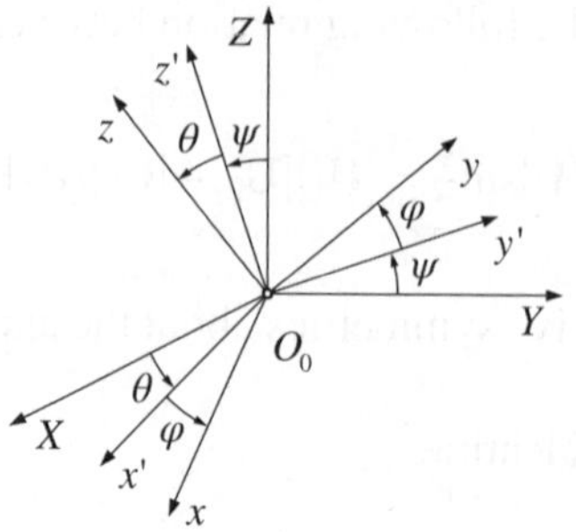

Figure 2.5 Successive rotations about the axes of coordinates.

From the position O_0XYZ (Fig. 2.5) one comes to the position O_0xyz by successive rotations: by angle ψ about the axis O_0X to the position $O_0xy'z'$, by angle θ about the axis O_0y' to the position $O_0x''y'z$, and by angle φ about the axis O_0z to the position O_0xyz.

The matrix of rotation of the reference system O_0xyz relative to the reference system O_0XYZ reads

$$[\mathbf{A}] = [\boldsymbol{\psi}][\boldsymbol{\theta}][\boldsymbol{\varphi}], \tag{2.65}$$

where the matrices $[\boldsymbol{\psi}]$, $[\boldsymbol{\theta}]$ and $[\boldsymbol{\varphi}]$ are given by the relations (2.63) and (2.64), that is

$$[\boldsymbol{\psi}] = \begin{bmatrix} 1 & 0 & 0 \\ 0 & \cos\psi & -\sin\psi \\ 0 & \sin\psi & \cos\psi \end{bmatrix},\ [\boldsymbol{\theta}] = \begin{bmatrix} \cos\theta & 0 & \sin\theta \\ 0 & 1 & 0 \\ -\sin\theta & 0 & \cos\theta \end{bmatrix},\ [\boldsymbol{\varphi}] = \begin{bmatrix} \cos\varphi & -\sin\varphi & 0 \\ \sin\varphi & \cos\varphi & 0 \\ 0 & 0 & 1 \end{bmatrix}. \tag{2.66}$$

In this case the rotation of axes is of (1, 2, 3) type, and the angles ψ, θ and φ are *the Bryan angles*.

In the case of the rotation of (3, 1, 3) type, the angles ψ, θ and φ are *the Euler* angles, and the matrix $[\mathbf{A}]$ is given by the relation (2.65), where

$$[\boldsymbol{\psi}] = \begin{bmatrix} \cos\psi & -\sin\psi & 0 \\ \sin\psi & \cos\psi & 0 \\ 0 & 0 & 1 \end{bmatrix},\ [\boldsymbol{\theta}] = \begin{bmatrix} 1 & 0 & 0 \\ 0 & \cos\theta & -\sin\theta \\ 0 & \sin\theta & \cos\theta \end{bmatrix},\ [\boldsymbol{\varphi}] = \begin{bmatrix} \cos\varphi & -\sin\varphi & 0 \\ \sin\varphi & \cos\varphi & 0 \\ 0 & 0 & 1 \end{bmatrix}; \tag{2.67}$$

it results

$$\mathbf{A} = \begin{bmatrix} c\psi c\varphi - s\psi c\theta s\varphi & -c\psi s\varphi - s\psi c\theta c\varphi & s\psi s\theta \\ s\psi c\varphi + c\psi c\theta s\varphi & -s\psi s\varphi + c\psi c\theta c\varphi & -c\psi s\theta \\ s\theta s\varphi & s\theta c\varphi & c\theta \end{bmatrix}. \tag{2.68}$$

In the general case, the angles ψ, θ and φ are called the angles of rotation about the axes specified by the type of rotation.

There are 12 distinct variants of successive rotations defined by the indices (1, 2, 1), (1, 2, 3), (1, 3, 1), (1, 3, 2), (2, 1, 2), (2, 1, 3), (2, 3, 1), (2, 3, 2), (3, 1, 2), (3, 1, 3), (3, 2, 1), (3 2 3).

Thus, for the (2, 3, 1) type rotation, the matrix of rotation is given by the expression (2.65), where

$$[\boldsymbol{\psi}] = \begin{bmatrix} \cos\psi & 0 & \sin\psi \\ 0 & 1 & 0 \\ -\sin\psi & 0 & \cos\psi \end{bmatrix}, [\boldsymbol{\theta}] = \begin{bmatrix} \cos\theta & -\sin\theta & 0 \\ \sin\theta & \cos\theta & 0 \\ 0 & 0 & 1 \end{bmatrix}, [\boldsymbol{\varphi}] = \begin{bmatrix} 1 & 0 & 0 \\ 0 & \cos\varphi & -\sin\varphi \\ 0 & \sin\varphi & \cos\varphi \end{bmatrix} \tag{2.69}$$

and it results

$$[\mathbf{A}] = \begin{bmatrix} c\psi c\theta & -c\psi s\theta c\varphi + s\psi s\varphi & c\psi s\theta s\varphi + s\psi c\varphi \\ s\theta & c\theta c\varphi & -c\theta s\varphi \\ -s\psi c\theta & s\psi s\theta c\varphi + c\psi s\varphi & -s\psi s\theta s\varphi + c\psi c\varphi \end{bmatrix}. \tag{2.70}$$

If one calls the abbreviated notations

$$\cos\psi = c_1, \sin\psi = s_1, \cos\theta = c_2, \sin\theta = s_2, \cos\varphi = c_3, \sin\varphi = s_3, \tag{2.71}$$

then one obtains the matrices

$$[\boldsymbol{\psi}] = \begin{bmatrix} c_1 & 0 & s_1 \\ 0 & 1 & 0 \\ -s_1 & 0 & c_1 \end{bmatrix}, [\boldsymbol{\theta}] = \begin{bmatrix} c_2 & -s_2 & 0 \\ s_2 & c_2 & 0 \\ 0 & 0 & 1 \end{bmatrix}, [\boldsymbol{\varphi}] = \begin{bmatrix} 1 & 0 & 0 \\ 0 & c_3 & -s_3 \\ 0 & s_3 & c_3 \end{bmatrix}, \tag{2.72}$$

$$[\mathbf{A}] = \begin{bmatrix} c_1c_2 & -c_1s_2c_3 + s_1s_3 & c_1s_2s_3 + s_1c_3 \\ s_2 & c_2c_3 & -c_2s_3 \\ -s_1c_2 & s_1s_2c_3 + c_1s_3 & -s_1s_2s_3 + c_1c_3 \end{bmatrix}. \tag{2.73}$$

2.3 Minimum Displacements: The Chasles Theorem

One considers the real unitary eigenvector **u** of matrix [**A**]; it is known that this vector has the property

$$\overline{[\mathbf{A}]\{\mathbf{u}\}} = \mathbf{u}. \tag{2.74}$$

Writing the relation (2.6) in the vector form

$$\mathbf{s}=\mathbf{s}_O+\overline{[\mathbf{A}]\{\mathbf{r}_0\}}-\mathbf{r}_0, \tag{2.75}$$

scalar multiplying this relation by $\mathbf{u}$, and keeping into account the equalities

$$\mathbf{u}\overline{[\mathbf{A}]\{\mathbf{r}_0\}}=\overline{[\mathbf{A}]\{\mathbf{u}\}}\cdot\overline{[\mathbf{A}]\{\mathbf{r}_0\}}=\mathbf{u}\cdot\mathbf{r}_0, \tag{2.76}$$

one gets the relation

$$\mathbf{s}\cdot\mathbf{u}=\mathbf{s}_O\cdot\mathbf{u}, \tag{2.77}$$

which shows the equality of the projections of displacements of the points of rigid solid onto the direction of the unit vector $\mathbf{u}$.

Moreover, it results the existence of *the minimum displacements* $\mathbf{s}_{\min}$ parallel to the unit vector $\mathbf{u}$; these displacements may by written as

$$\mathbf{s}_{\min}=(\mathbf{s}_O\cdot\mathbf{u})\mathbf{u}. \tag{2.78}$$

The position vectors $\mathbf{r}_0$ for the points which have minimum displacements verify the vector relation (2.75), and the matrix relation

$$\{\mathbf{s}_{\min}\}=\{\mathbf{s}_O\}+[\mathbf{A}]\{\mathbf{r}_0\}-\{\mathbf{r}_0\}, \tag{2.79}$$

respectively.

Multiplying the relation (2.79) at the left by $[\mathbf{A}]^{\mathrm{T}}$, keeping into account the equality $[\mathbf{A}]\{\mathbf{s}_{\min}\}=\{\mathbf{s}_{\min}\}$ and the notation

$$\{\mathbf{s}_O^*\}=[\mathbf{A}]^{\mathrm{T}}\{\mathbf{s}_O\}, \tag{2.80}$$

one obtains the expression

$$\{\mathbf{s}_{\min}\}=\{\mathbf{s}_O^*\}+\{\mathbf{r}_0\}-[\mathbf{A}]^{\mathrm{T}}\{\mathbf{r}_0\}. \tag{2.81}$$

Summing the relations (2.79) and (2.81), it results the equality

$$2\{\mathbf{s}_{\min}\}=\{\mathbf{s}_O\}+\{\mathbf{s}_O^*\}+\left[[\mathbf{A}]-[\mathbf{A}]^{\mathrm{T}}\right]\{\mathbf{r}_0\}, \tag{2.82}$$

which may be written in the vector form

$$\{\mathbf{s}_{\min}\}=\frac{\mathbf{s}_O+\mathbf{s}_O^*}{2}+\mathbf{w}\times\mathbf{r}_0, \tag{2.83}$$

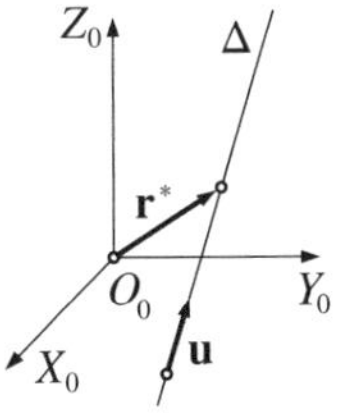

Figure 2.6 The axis of the helical motion.

where

$$\mathbf{w} = \frac{1}{2}\left[(a_{32} - a_{23})\mathbf{i}_0 + (a_{13} - a_{31})\mathbf{j}_0 + (a_{21} - a_{12})\mathbf{k}_0\right]. \tag{2.84}$$

If we vector multiply the relation (2.83) at the left by $\mathbf{w}$, and we denote

$$\mathbf{r}^* = \frac{\mathbf{w} \times \left(\mathbf{s}_O + \mathbf{s}_O^*\right)}{2\mathbf{w}^2}, \tag{2.85}$$

then we obtain the expression

$$\mathbf{r} = \mathbf{r}^* + \lambda\mathbf{u},\ \lambda \in \mathbb{R}, \tag{2.86}$$

which represents the vector equation of a straight line Δ of unit vector $\mathbf{u}$ (Fig. 2.6); this straight line passes through the end of the vector $\mathbf{r}^*$. The straight line Δ is called *the axis of the finite helical motion*.

Consequently, it results (the Chasles theorem) that the rigid may comes from the initial position to the final position (Fig. 2.7) by a finite helical motion of axis Δ, the angle of rotation ξ about this axis being given by the relation $\sin\xi = |\mathbf{w}|$ or by the relation $2\cos\xi = a_{11} + a_{22} + a_{33} - 1$, while the displacement along the axis Δ is equal to the minimum displacement $\mathbf{s}_{\min}$.

Example 2.3.1 Knowing that the displacement $\mathbf{s}_O$ and the matrix $[\mathbf{A}]$ are given by the relations

$$\mathbf{s}_O = l(2\mathbf{i}_0 + 5\mathbf{j}_0 + 11\mathbf{k}_0),\ [\mathbf{A}] = \frac{1}{25}\begin{bmatrix} 9 & -20 & 12 \\ 20 & 0 & -15 \\ 12 & 15 & 16 \end{bmatrix},$$

determine the axis Δ of the finite helical motion, the minimum displacement $\mathbf{s}_{\min}$ and the angle of rotation ξ.

Solution: From the relation (2.84) one obtains

$$\mathbf{w} = \mathbf{u} = \frac{1}{5}(3\mathbf{i}_0 + 4\mathbf{k}_0);$$

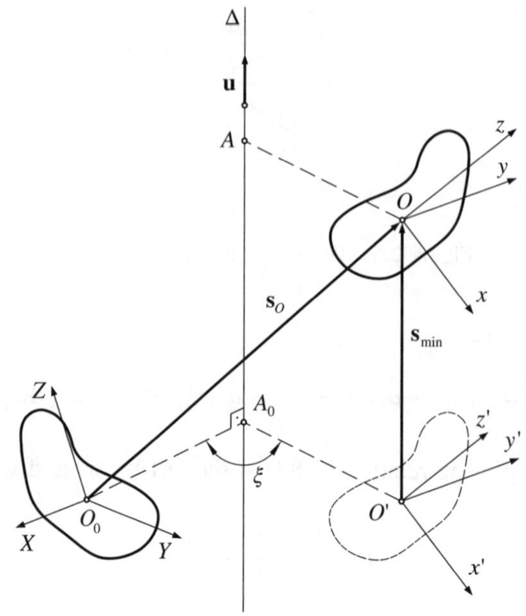

Figure 2.7 The Chasles theorem.

hence

$$\sin\xi = |\mathbf{w}| = 1,\ \xi = \frac{\pi}{2},\ \mathbf{s}_{\min} = \mathbf{s}_O\mathbf{u} = 10l.$$

From the relations (2.80) and (2.85) it results

$$\mathbf{s}_O^* = (10\mathbf{i}_0 + 5\mathbf{j}_0 + 5\mathbf{k}_0)l, \mathbf{r}^* = (-4\mathbf{i}_0 + 3\mathbf{k}_0)l$$

and, consequently, the vector equation of the straight line Δ is

$$\mathbf{r} = (-4\mathbf{i}_0 + 3\mathbf{k}_0)l + \lambda(3\mathbf{i}_0 + 4\mathbf{k}_0).$$

If the vector **w** *is a null one ($\xi = 0$ or $\xi = \pi$)*, then the unit vector **u** is determined from the equation

$$[[\mathbf{A}] - [\mathbf{I}]]\{\mathbf{u}\} = \{\mathbf{0}\}, \tag{2.87}$$

the distance $\mathbf{s}_{\min}$ is obtained from the relation $s_{\min} = \mathbf{s}_O\mathbf{u}$, while the scalar equations of the axis Δ result from the matrix expression (2.79).

Example 2.3.2 Knowing that the displacement $\mathbf{s}_O$ and the matrix [**A**] are given by the relations

$$\mathbf{s}_O = \sqrt{2}(\mathbf{i}_0 + \mathbf{j}_0)l,\ [\mathbf{A}] = \begin{bmatrix} 0 & 0 & 1 \\ 0 & -1 & 0 \\ 1 & 0 & 0 \end{bmatrix},$$

determine the angle the rotation, the real unitary eigenvector **u**, the minimum displacement $s_{\min}$ and the equations of the finite helical axis Δ.

Solution: The angle of rotation obtained from the relation

$$\cos\xi = \frac{a_{11} + a_{22} + a_{33} - 1}{2}$$

is $\xi = \pi$, while the unit vector **u** obtained from the relation (2.87) is

$$\mathbf{u} = \frac{1}{\sqrt{2}}(\mathbf{i}_0 + \mathbf{k}_0);$$

consequently

$$s_{\min} = \mathbf{s}_O \mathbf{u} = l.$$

Further on, from the relation (2.79) it results

$$\frac{1}{\sqrt{2}}\begin{bmatrix}1\\0\\1\end{bmatrix} = l\sqrt{2}\begin{bmatrix}1\\1\\0\end{bmatrix} + \begin{bmatrix}-1 & 0 & 1\\0 & -2 & 0\\1 & 0 & -1\end{bmatrix}\begin{bmatrix}X\\Y\\Z\end{bmatrix}$$

and one obtains the equations of the finite helical axis

$$X - Z = l\frac{\sqrt{2}}{2}, Y = l\frac{\sqrt{2}}{2}.$$

We now consider the case in which the axis Δ of the finite helical motion, the rotation angle ξ and the minimum displacement $s_{\min}$ are known; we want to determine the matrix of rotation [**A**] and the displacement $\mathbf{s}_O$.

In this case, the unit vector **u** of the axis Δ and the rotation angle ξ being known, the matrix [**A**] is determined with the aid of the expression (2.38).

For the determination of the displacement $\mathbf{s}_O$ one considers the vector $\mathbf{A}_0\mathbf{O}_0$ (Fig. 2.7), where A_0 is the point of intersection between the axis Δ and the perpendicular from O_0 on it.

If one rotates the vector $\mathbf{A}_0\mathbf{O}_0$ about the axis Δ by the angle ξ, then one obtains the vector $\mathbf{A}_0\mathbf{O}'$; if one displaces the point O' by the distance $s_{\min}$ into the direction of the axis Δ, then one obtains the point O; we may write the relation

$$\mathbf{s}_O = \mathbf{A}_0\mathbf{O}_0 + \mathbf{A}_0\mathbf{O}' + s_{\min}\mathbf{u}. \tag{2.88}$$

Applying the Rodrigues relation, it results that

$$\mathbf{A_0O'} = \mathbf{A_0O_0} + \mathbf{u} \times \mathbf{A_0O_0} \sin\xi + \mathbf{u} \times (\mathbf{u} \times \mathbf{A_0O_0})(1-\cos\xi); \tag{2.89}$$

since $\mathbf{A_0O_0} \perp \mathbf{u}$ one gets

$$\mathbf{s}_O = s_{\min}\mathbf{u} - \mathbf{A_0O_0}(1-\cos\xi) + \mathbf{u} \times \mathbf{A_0O_0} \sin\xi. \tag{2.90}$$

Defining the axis Δ by the vector equation

$$\mathbf{r} = \boldsymbol{\rho} + \lambda\mathbf{u} \tag{2.91}$$

and writing the equation of the plan perpendicular to Δ and passing through the point O_0,

$$\mathbf{r} \cdot \mathbf{u} = 0, \tag{2.92}$$

from the equations (2.91) and (2.92) one obtains $\lambda = -\boldsymbol{\rho} \cdot \mathbf{u}$ and deduces the expression

$$\mathbf{A_0O_0} = (\boldsymbol{\rho} \cdot \mathbf{u})\mathbf{u} - \boldsymbol{\rho}. \tag{2.93}$$

Example 2.3.3 Determine the matrix of rotation $[\mathbf{A}]$ and the displacement $\mathbf{s}_O$ knowing

$$s_{\min} = 3l, \xi = 90^0, \mathbf{u} = \frac{2\mathbf{i}_0 + 2\mathbf{j}_0 + \mathbf{k}_0}{3}, \boldsymbol{\rho} = 3l(\mathbf{i}_0 + 2\mathbf{j}_0).$$

Solution: It results the successive calculations:

$$[\mathbf{U}] = \frac{1}{3}\begin{bmatrix} 0 & -1 & 2 \\ 1 & 0 & -2 \\ -2 & 2 & 0 \end{bmatrix}, [\mathbf{U}]^2 = \frac{1}{9}\begin{bmatrix} -5 & 4 & 2 \\ 4 & -5 & 2 \\ 2 & 2 & 8 \end{bmatrix}, [\mathbf{A}] = [\mathbf{I}] + [\mathbf{U}]\sin\xi + [\mathbf{U}]^2(1-\cos\xi),$$

$$[\mathbf{A}] = \frac{1}{9}\begin{bmatrix} 4 & 1 & 8 \\ 7 & 4 & -4 \\ -4 & 8 & 1 \end{bmatrix}, \boldsymbol{\rho} \cdot \mathbf{u} = 6l, \mathbf{A_0O_0} = l(1 - 2\mathbf{j}_0 + 2\mathbf{k}_0), \mathbf{s}_O = 3l(\mathbf{i}_0 - \mathbf{j}_0 - \mathbf{k}_0).$$

2.4 Small Displacements

Starting from the matrix expression (2.6) of the displacements of the points of rigid solid (Fig. 2.1)

$$\{\mathbf{s}\} = \{\mathbf{s}_O\} + [\mathbf{A}]\{\mathbf{r}_0\} - \{\mathbf{r}_0\}, \tag{2.94}$$

and using the form (2.47) for the matrix of rotation, one obtains the equality

$$\{\mathbf{s}\} = \{\mathbf{s}_O\} + \sin\xi[\mathbf{U}]\{\mathbf{r}_0\} + 2\sin^2\frac{\xi}{2}[\mathbf{U}]^2\{\mathbf{r}_0\}, \tag{2.95}$$

with the vector correspondent

$$\mathbf{s} = \mathbf{s}_O + (\mathbf{u}\times\mathbf{r}_0)\sin\xi + 2\mathbf{u}\times(\mathbf{u}\times\mathbf{r}_0)\sin^2\frac{\xi}{2}. \tag{2.96}$$

In the case of small displacements one makes the approximation

$$\sin\xi = \xi,\ \sin\frac{\xi}{2} = \frac{\xi}{2} \tag{2.97}$$

and if one makes the notation

$$\boldsymbol{\xi} = \xi\mathbf{u}, \tag{2.98}$$

where $\boldsymbol{\xi}$ is called *the vector of small* rotation, one obtains the vector relation

$$\mathbf{s} = \mathbf{s}_O + \boldsymbol{\xi}\times\mathbf{r}_0 + \frac{1}{2}\boldsymbol{\xi}\times(\boldsymbol{\xi}\times\mathbf{r}_0); \tag{2.99}$$

it has the matrix correspondent

$$\{\mathbf{s}\} = \{\mathbf{s}_O\} + [\boldsymbol{\xi}]\{\mathbf{r}_0\} + \frac{1}{2}[\boldsymbol{\xi}]^2\{\mathbf{r}_0\}, \tag{2.100}$$

where

$$[\boldsymbol{\xi}] = \begin{bmatrix} 0 & -\xi_z & \xi_y \\ \xi_z & 0 & -\xi_x \\ \xi_y & \xi_x & 0 \end{bmatrix}, \tag{2.101}$$

$$\xi_x = \xi a,\ \xi_y = \xi b,\ \xi_z = \xi c. \tag{2.102}$$

If we back down the nonlinear term in ξ, then we obtain the relation

$$\mathbf{s} = \mathbf{s}_O + \boldsymbol{\xi} \times \mathbf{r}_0, \tag{2.103}$$

and the expression

$$\{\mathbf{s}\} = \{\mathbf{s}_O\} + [\boldsymbol{\xi}]\{\mathbf{r}_0\}, \tag{2.104}$$

respectively.

2.5 Velocities of the Points of Rigid Solid

We consider the rigid solid in motion and let $Oxyz$ be the mobile reference system jointed to the rigid solid (Fig. 2.8); O_0XYZ be the fixed reference system; $\mathbf{s}_O$ be the vector position of the point O; $\mathbf{r} = \mathbf{OP}$ be the vector of position for an arbitrary point P of the rigid solid relative to the point O; x, y, z, and X, Y, Z be the coordinates of the point P in the systems $Oxyz$ and O_0XYZ, respectively; X_O, Y_O, Z_O the coordinates of the point O in the reference system O_0XYZ; $[\mathbf{A}]$ be the matrix of rotation.

The vectors of position for the points O, P are

$$\mathbf{O_0P} = X\mathbf{i}_0 + Y\mathbf{j}_0 + Z\mathbf{k}_0, \mathbf{O_0O} = X_O\mathbf{i}_0 + Y_O\mathbf{j}_0 + Z_O\mathbf{k}_0, \tag{2.105}$$

where $\mathbf{i}_0$, $\mathbf{j}_0$, $\mathbf{k}_0$ are the unit vectors of the fixed axes O_0X, O_0Y, O_0Z; by derivation of the previous relations with respect to time, it results that the velocities $\mathbf{v}$, $\mathbf{v}_O$ of the points mentioned above read in the form

$$\mathbf{v} = \dot{X}\mathbf{i}_0 + \dot{Y}\mathbf{j}_0 + \dot{Z}\mathbf{k}_0, \mathbf{v}_O = \dot{X}_O\mathbf{i}_0 + \dot{Y}_O\mathbf{j}_0 + \dot{Z}_O\mathbf{k}_0. \tag{2.106}$$

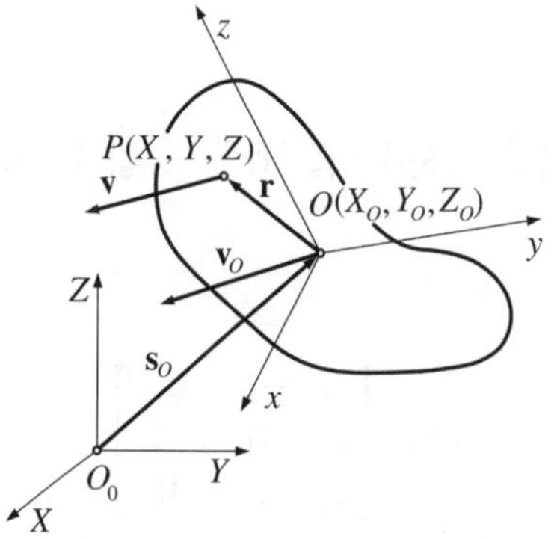

Figure 2.8 The velocities of the points of rigid solid.

Denoting by $v_X = \dot{X}$, $v_Y = \dot{Y}$, $v_Z = \dot{Z}$, $v_{OX} = \dot{X}_O$; $v_{OY} = \dot{Y}_O$, $v_{OZ} = \dot{Z}_O$ the projections of the velocities $\mathbf{v}$ and $\mathbf{v}_O$, respectively, onto the axes of the reference system O_0XYZ, one may write the vector relations

$$\mathbf{v} = v_X\mathbf{i}_0 + v_Y\mathbf{j}_0 + v_Z\mathbf{k}_0,\ \mathbf{v}_O = v_{OX}\mathbf{i}_0 + v_{OY}\mathbf{j}_0 + v_{OZ}\mathbf{k}_0. \tag{2.107}$$

Analogically, if one denotes by v_x, v_y, v_z, and v_{Ox}, v_{Oy}, v_{Oz} the projections of the velocities $\mathbf{v}$ and $\mathbf{v}_O$, respectively, onto the axes of the mobile reference system, one may write the relations

$$\mathbf{v} = v_x\mathbf{i} + v_y\mathbf{j} + v_z\mathbf{k},\ \mathbf{v}_O = v_{Ox}\mathbf{i} + v_{Oy}\mathbf{j} + v_{Oz}\mathbf{k}, \tag{2.108}$$

where $\mathbf{i}$, $\mathbf{j}$ and $\mathbf{k}$ are the unit vectors of the mobile axes Ox, Oy, and Oz, respectively.

Using the matrix notations

$$\{\mathbf{R}\} = [X\ \ Y\ \ Z]^{\mathrm{T}}, \{\mathbf{R}_O\} = [X_O\ \ Y_O\ \ Z_O]^{\mathrm{T}}, \{\mathbf{r}\} = [x\ \ y\ \ z]^{\mathrm{T}}, \{\mathbf{v}^{(0)}\} = [v_X\ \ v_Y\ \ v_Z]^{\mathrm{T}},$$
$$\{\mathbf{v}\} = [v_x\ \ v_y\ \ v_z]^{\mathrm{T}}, \left\{\mathbf{v}_O^{(0)}\right\} = [v_{OX}\ \ v_{OY}\ \ v_{OZ}]^{\mathrm{T}}, \{\mathbf{v}_O\} = [v_{Ox}\ \ v_{Oy}\ \ v_{Oz}]^{\mathrm{T}}, \tag{2.109}$$

recalling the relation (2.4) derived with respect to time, keeping into account that the matrix $\{\mathbf{r}\}$ is constant, one deduces the expression

$$\{\dot{\mathbf{R}}\} = \{\dot{\mathbf{s}}_O\} + [\dot{\mathbf{A}}]\{\mathbf{r}\} = \{\dot{\mathbf{R}}_O\} + [\dot{\mathbf{A}}]\{\mathbf{r}\} \tag{2.110}$$

or

$$\left\{\mathbf{v}^{(0)}\right\} = \left\{\mathbf{v}_O^{(0)}\right\} + [\dot{\mathbf{A}}]\{\mathbf{r}\}. \tag{2.111}$$

With the aid of the matrix relations of transformation

$$\left\{\mathbf{v}^{(0)}\right\} = [\mathbf{A}]\{\mathbf{v}\}, \left\{\mathbf{v}_O^{(0)}\right\} = [\mathbf{A}]\{\mathbf{v}_0\} \tag{2.112}$$

from the equation (2.111) one obtains the equality

$$\{\mathbf{v}\} = \{\mathbf{v}_O\} + [\mathbf{A}]^{\mathrm{T}}[\dot{\mathbf{A}}]\{\mathbf{r}\}. \tag{2.113}$$

The matrix

$$[\boldsymbol{\omega}] = [\mathbf{A}]^{\mathrm{T}}[\dot{\mathbf{A}}] \tag{2.114}$$

is called *the angular velocity matrix relative to the mobile reference system Oxyz* and, as we will prove, it is a skew matrix.

Indeed, the angular velocity matrix may be also written in the form

$$[\boldsymbol{\omega}] = \begin{bmatrix} \mathbf{i}\cdot\dot{\mathbf{i}} & \mathbf{i}\cdot\dot{\mathbf{j}} & \mathbf{i}\cdot\dot{\mathbf{k}} \\ \mathbf{j}\cdot\dot{\mathbf{i}} & \mathbf{j}\cdot\dot{\mathbf{j}} & \mathbf{j}\cdot\dot{\mathbf{k}} \\ \mathbf{k}\cdot\dot{\mathbf{i}} & \mathbf{k}\cdot\dot{\mathbf{j}} & \mathbf{k}\cdot\dot{\mathbf{k}} \end{bmatrix} \tag{2.115}$$

and since

$$\mathbf{i}\cdot\dot{\mathbf{i}} = \mathbf{j}\cdot\dot{\mathbf{j}} = \mathbf{k}\cdot\dot{\mathbf{k}} = 0, \tag{2.116}$$

$$\mathbf{i}\cdot\dot{\mathbf{j}} = -\dot{\mathbf{i}}\cdot\mathbf{j},\ \mathbf{i}\cdot\dot{\mathbf{k}} = -\dot{\mathbf{i}}\cdot\mathbf{k},\ \mathbf{j}\cdot\dot{\mathbf{k}} = -\dot{\mathbf{j}}\cdot\mathbf{k}, \tag{2.117}$$

one obtains the skew matrix

$$[\boldsymbol{\omega}] = \begin{bmatrix} 0 & -\dot{\mathbf{i}}\cdot\mathbf{j} & \dot{\mathbf{k}}\cdot\mathbf{i} \\ \dot{\mathbf{i}}\cdot\mathbf{j} & 0 & -\dot{\mathbf{j}}\cdot\mathbf{k} \\ -\dot{\mathbf{k}}\cdot\mathbf{i} & \dot{\mathbf{j}}\cdot\mathbf{k} & 0 \end{bmatrix}. \tag{2.118}$$

This matrix has as attached vector *the angular velocity vector*

$$\boldsymbol{\omega} = \omega_x\mathbf{i} + \omega_y\mathbf{j} + \omega_z\mathbf{k}, \tag{2.119}$$

where

$$\omega_x = \dot{\mathbf{j}}\cdot\mathbf{k},\ \omega_y = \dot{\mathbf{k}}\cdot\mathbf{i},\ \omega_z = \dot{\mathbf{i}}\cdot\mathbf{j}. \tag{2.120}$$

Keeping into account the previous expressions, the matrix relation (2.113) reads

$$\{\mathbf{v}\} = \{\mathbf{v}_O\} + [\boldsymbol{\omega}]\{\mathbf{r}\} \tag{2.121}$$

and it has the vector correspondent (*the Euler relation*)

$$\mathbf{v} = \mathbf{v}_O + \boldsymbol{\omega}\times\mathbf{r}. \tag{2.122}$$

If we denote by $\{\mathbf{r}^{(0)}\}$ the column matrix of the projections of the vector $\mathbf{r}$ onto the axes of the fixed trihedron O_0XYZ, then, from the relation (2.111) and with the aid of the equality

$$\{\mathbf{r}\} = [\mathbf{A}]^{\mathrm{T}}\left\{\mathbf{r}^{(0)}\right\}, \tag{2.123}$$

we deduce the expression

$$\left\{\mathbf{v}^{(0)}\right\} = \left\{\mathbf{v}_O^{(0)}\right\} + \left[\boldsymbol{\omega}^{(0)}\right]\left\{\mathbf{r}^{(0)}\right\}. \tag{2.124}$$

The angular matrix *relative to the fixed reference system* is the skew matrix

$$\left[\boldsymbol{\omega}^{(0)}\right] = [\dot{\mathbf{A}}][\mathbf{A}]^{\mathrm{T}} = [\mathbf{A}][\boldsymbol{\omega}][\mathbf{A}]^{\mathrm{T}}. \tag{2.125}$$

The matrix equality (2.124) has as vector correspondent the same expression (2.122).

2.6 The Angular Velocity Matrix: Properties

2.6.1 The Matrices of Rotation About the Axes of Coordinates

One considers the rotation of angle α about an axis of coordinates of unit vector $\mathbf{u}_\alpha$, and let $[\boldsymbol{\alpha}]$ be the matrix of rotation, $[\mathbf{U}_\alpha]$ be the skew matrix attached to the unit vector $\mathbf{u}_\alpha$, $[\boldsymbol{\alpha}_p]$ be the derivative of the matrix $[\boldsymbol{\alpha}]$ with respect to α, and $\{\mathbf{u}_\alpha\}$ be the column matrix that corresponds to the unit vector $\mathbf{u}_\alpha$.

One can proved, as it will result from the next example, that the following identities hold true

$$[\boldsymbol{\alpha}_p] = [\boldsymbol{\alpha}][\mathbf{U}_\alpha] = [\mathbf{U}_\alpha][\boldsymbol{\alpha}], \tag{2.126}$$

$$[\boldsymbol{\alpha}_p]^{\mathrm{T}} = -[\mathbf{U}_\alpha][\boldsymbol{\alpha}]^{\mathrm{T}} = -[\boldsymbol{\alpha}]^{\mathrm{T}}[\mathbf{U}_\alpha], \tag{2.127}$$

$$[\boldsymbol{\alpha}]\{\mathbf{u}_\alpha\} = \{\mathbf{u}_\alpha\}. \tag{2.128}$$

For instance, if the axis of rotation is the OY-axis (the second axis), then

$$\mathbf{u}_\alpha = \mathbf{j},\ [\boldsymbol{\alpha}] = \begin{bmatrix} \cos\alpha & 0 & \sin\alpha \\ 0 & 1 & 0 \\ -\sin\alpha & 0 & \cos\alpha \end{bmatrix},\ [\boldsymbol{\alpha}_p] = \begin{bmatrix} -\sin\alpha & 0 & \cos\alpha \\ 0 & 0 & 0 \\ -\cos\alpha & 0 & -\sin\alpha \end{bmatrix},\ [\mathbf{U}_\alpha] = \begin{bmatrix} 0 & 0 & 1 \\ 0 & 0 & 0 \\ -1 & 0 & 0 \end{bmatrix},$$

$$\{\mathbf{u}_\alpha\} = \begin{bmatrix} 0 \\ 1 \\ 0 \end{bmatrix}$$

and the relations (2.126)–(2.128) are verified.

2.6.2 The Angular Velocity Matrix: The Angular Velocity Vector

As it was shown, the matrix of rotation [**A**] writes in the form

$$[\mathbf{A}] = [\boldsymbol{\psi}][\boldsymbol{\theta}][\boldsymbol{\varphi}], \tag{2.129}$$

where ψ, θ, φ are the angles of rotation about the axes defined by chosen variant.

Keeping into account the condition (2.126), it results that one may write the relations

$$[\boldsymbol{\psi}_p] = [\boldsymbol{\psi}][\mathbf{U}_\psi] = [\mathbf{U}_\psi][\boldsymbol{\psi}],\ [\boldsymbol{\theta}_p] = [\boldsymbol{\theta}][\mathbf{U}_\theta] = [\mathbf{U}_\theta][\boldsymbol{\theta}],\ [\boldsymbol{\varphi}_p] = [\boldsymbol{\varphi}][\mathbf{U}_\varphi] = [\mathbf{U}_\varphi][\boldsymbol{\varphi}]. \tag{2.130}$$

Writing the derivative of the matrix [**A**] with respect to time in the form

$$[\dot{\mathbf{A}}] = \dot{\psi}[\boldsymbol{\psi}_p][\boldsymbol{\theta}][\boldsymbol{\varphi}] + \dot{\theta}[\boldsymbol{\psi}][\boldsymbol{\theta}_p][\boldsymbol{\varphi}] + \dot{\varphi}[\boldsymbol{\psi}][\boldsymbol{\theta}][\boldsymbol{\varphi}_p], \tag{2.131}$$

and recalling the relations (2.130), one obtains

$$[\dot{\mathbf{A}}] = \dot{\psi}[\boldsymbol{\psi}][\mathbf{U}_\psi][\boldsymbol{\theta}][\boldsymbol{\varphi}] + \dot{\theta}[\boldsymbol{\psi}][\boldsymbol{\theta}][\mathbf{U}_\theta][\boldsymbol{\varphi}] + \dot{\varphi}[\boldsymbol{\psi}][\boldsymbol{\theta}][\boldsymbol{\varphi}][\mathbf{U}_\varphi]; \tag{2.132}$$

from the equalities (2.132) and (2.129) one deduces the expression of the angular velocity matrix relative to the mobile reference system

$$[\boldsymbol{\omega}] = \dot{\psi}[\boldsymbol{\varphi}]^{\mathrm{T}}[\boldsymbol{\theta}]^{\mathrm{T}}[\mathbf{U}_\psi][\boldsymbol{\theta}][\boldsymbol{\varphi}] + \dot{\theta}[\boldsymbol{\varphi}]^{\mathrm{T}}[\mathbf{U}_\theta][\boldsymbol{\varphi}] + \dot{\varphi}[\mathbf{U}_\varphi]. \tag{2.133}$$

Using the equality (2.133) and recalling the first relation (1.82), one obtains the expression of the column matrix of the angular velocity vector relative to the mobile reference system

$$\{\boldsymbol{\omega}\} = \begin{bmatrix} \omega_x \\ \omega_y \\ \omega_z \end{bmatrix} = \dot{\psi}[\boldsymbol{\varphi}]^{\mathrm{T}}[\boldsymbol{\theta}]^{\mathrm{T}}\{\mathbf{u}_\psi\} + \dot{\theta}[\boldsymbol{\varphi}]^{\mathrm{T}}\{\mathbf{u}_\theta\} + \dot{\varphi}\{\mathbf{u}_\varphi\}. \tag{2.134}$$

Further on, using the notations

$$[\boldsymbol{\theta}_Q] = \left[[\boldsymbol{\theta}]^{\mathrm{T}}\{\mathbf{u}_\psi\}\{\mathbf{u}_\theta\}\{\mathbf{u}_\phi\} \right], \tag{2.135}$$

$$[\mathbf{Q}] = [\boldsymbol{\varphi}]^{\mathrm{T}}[\boldsymbol{\theta}_Q], \tag{2.136}$$

$$\{\dot{\mathbf{q}}_2\} = [\dot{\psi}\ \ \dot{\theta}\ \ \dot{\varphi}]^{\mathrm{T}}, \tag{2.137}$$

one deduces the expression of the column matrix of the angular velocity's components in the mobile reference system *Oxyz*,

$$\{\boldsymbol{\omega}\} = [\mathbf{Q}]\{\dot{\mathbf{q}}_2\}. \tag{2.138}$$

For instance, if the order of rotations is (2, 3, 2), then

$$[\boldsymbol{\varphi}] = \begin{bmatrix} \cos\varphi & 0 & \sin\varphi \\ 0 & 1 & 0 \\ -\sin\varphi & 0 & \cos\varphi \end{bmatrix},\ [\boldsymbol{\theta}] = \begin{bmatrix} \cos\theta & -\sin\theta & 0 \\ \sin\theta & \cos\theta & 0 \\ 0 & 0 & 1 \end{bmatrix},\ \{\mathbf{u}_\psi\} = \{\mathbf{u}_\varphi\} = \begin{bmatrix} 0 \\ 1 \\ 0 \end{bmatrix},\ \{\mathbf{u}_\theta\} = \begin{bmatrix} 0 \\ 0 \\ 1 \end{bmatrix},$$

$$[\boldsymbol{\theta}_Q] = \begin{bmatrix} -\sin\theta & 0 & 0 \\ \cos\theta & 0 & 1 \\ 0 & 1 & 0 \end{bmatrix},\ [\mathbf{Q}] = \begin{bmatrix} -\cos\varphi\sin\theta & -\sin\varphi & 0 \\ \cos\theta & 0 & 1 \\ -\sin\varphi\sin\theta & \cos\varphi & 0 \end{bmatrix};$$

if the order of rotations is (3, 1, 3) (*the Euler angles*), then

$$[\boldsymbol{\varphi}] = \begin{bmatrix} \cos\varphi & -\sin\varphi & 0 \\ \sin\varphi & \cos\varphi & 0 \\ 0 & 0 & 1 \end{bmatrix},\ [\boldsymbol{\theta}] = \begin{bmatrix} 1 & 0 & 0 \\ 0 & \cos\theta & -\sin\theta \\ 0 & \sin\theta & \cos\theta \end{bmatrix},\ \{\mathbf{u}_\psi\} = \{\mathbf{u}_\varphi\} = \begin{bmatrix} 0 \\ 0 \\ 1 \end{bmatrix},\ \{\mathbf{u}_\theta\} = \begin{bmatrix} 1 \\ 0 \\ 0 \end{bmatrix},$$

$$[\boldsymbol{\theta}_Q] = \begin{bmatrix} 0 & 1 & 0 \\ \sin\theta & 0 & 0 \\ \cos\theta & 0 & 1 \end{bmatrix},\ [\mathbf{Q}] = \begin{bmatrix} \sin\varphi\sin\theta & \cos\varphi & 0 \\ \cos\phi\sin\theta & -\sin\varphi & 0 \\ \cos\theta & 0 & 1 \end{bmatrix},$$

and one obtains the known angular velocities

$$\omega_x = \dot{\psi}\sin\varphi\sin\theta + \dot{\theta}\cos\varphi,\ \omega_y = \dot{\psi}\cos\varphi\sin\theta - \dot{\theta}\sin\varphi,\ \omega_z = \dot{\psi}\cos\theta + \dot{\varphi}.$$

2.6.3 *The Matrix of the Partial Derivatives of the Angular Velocity*

Denoting this matrix by $[\widetilde{\boldsymbol{\omega}}]$, it results

$$[\widetilde{\boldsymbol{\omega}}] = \begin{bmatrix} \dfrac{\partial\omega_x}{\partial\psi} & \dfrac{\partial\omega_x}{\partial\theta} & \dfrac{\partial\omega_x}{\partial\varphi} \\ \dfrac{\partial\omega_y}{\partial\psi} & \dfrac{\partial\omega_y}{\partial\theta} & \dfrac{\partial\omega_y}{\partial\varphi} \\ \dfrac{\partial\omega_z}{\partial\psi} & \dfrac{\partial\omega_z}{\partial\theta} & \dfrac{\partial\omega_z}{\partial\varphi} \end{bmatrix} = \begin{bmatrix} \dfrac{\partial\{\boldsymbol{\omega}\}}{\partial\psi} & \dfrac{\partial\{\boldsymbol{\omega}\}}{\partial\theta} & \dfrac{\partial\{\boldsymbol{\omega}\}}{\partial\varphi} \end{bmatrix}; \tag{2.139}$$

considering (*the Lagrange formalism*) that the velocities $\dot{\psi}$, $\dot{\theta}$, $\dot{\varphi}$ are independent with respect to the parameters ψ, θ, φ, from (2.134) one deduces the equalities

$$\frac{\partial\{\boldsymbol{\omega}\}}{\partial\psi} = \{\mathbf{0}\}, \tag{2.140}$$

$$\frac{\partial\{\boldsymbol{\omega}\}}{\partial\theta} = -\dot{\psi}[\boldsymbol{\varphi}]^{\mathrm{T}}[\boldsymbol{\theta}]^{\mathrm{T}}[\mathbf{U}_{\theta}]\{\mathbf{u}_{\psi}\}, \tag{2.141}$$

$$\frac{\partial\{\boldsymbol{\omega}\}}{\partial\varphi} = -\dot{\psi}[\boldsymbol{\varphi}]^{\mathrm{T}}[\mathbf{U}_{\varphi}][\boldsymbol{\theta}]^{\mathrm{T}}\{\mathbf{u}_{\psi}\} - \dot{\theta}[\boldsymbol{\varphi}]^{\mathrm{T}}[\mathbf{U}_{\varphi}][\mathbf{u}_{\theta}]. \tag{2.142}$$

One thus obtains the expression

$$[\widetilde{\boldsymbol{\omega}}] = -\dot{\psi}\left[\{\mathbf{0}\}\ \ [\boldsymbol{\varphi}]^{\mathrm{T}}[\boldsymbol{\theta}]^{\mathrm{T}}[\mathbf{U}_{\theta}]\{\mathbf{u}_{\psi}\}\ \ [\boldsymbol{\varphi}]^{\mathrm{T}}[\mathbf{U}_{\varphi}][\boldsymbol{\theta}]^{\mathrm{T}}\{\mathbf{u}_{\psi}\}\right] - \dot{\theta}\left[\{\mathbf{0}\}\ \ \{\mathbf{0}\}\ \ [\boldsymbol{\varphi}]^{\mathrm{T}}[\mathbf{U}_{\varphi}]\{\mathbf{u}_{\theta}\}\right]. \tag{2.143}$$

With the aid of the matrices $[\boldsymbol{\omega}]$ and $[\mathbf{Q}]$, the matrix $[\widetilde{\boldsymbol{\omega}}]$ may be written in the form

$$[\widetilde{\boldsymbol{\omega}}] = [\dot{\mathbf{Q}}] + [\boldsymbol{\omega}][\mathbf{Q}]. \tag{2.144}$$

Indeed, by direct calculation and keeping into account the relations (1.82), one obtains the equalities

$$\begin{aligned}
[\dot{\mathbf{Q}}] &= -\dot{\varphi}\left[[\mathbf{U}_{\varphi}][\boldsymbol{\varphi}]^{\mathrm{T}}[\boldsymbol{\theta}]^{\mathrm{T}}\{\mathbf{u}_{\psi}\}\ \ [\mathbf{U}_{\varphi}][\boldsymbol{\varphi}]^{\mathrm{T}}\{\mathbf{u}_{\theta}\}\ \ \{\mathbf{0}\}\right] - \dot{\theta}\left[[\boldsymbol{\varphi}]^{\mathrm{T}}[\boldsymbol{\theta}]^{\mathrm{T}}[\mathbf{U}_{\theta}]\{\mathbf{u}_{\psi}\}\ \ \{\mathbf{0}\}\ \ \{\mathbf{0}\}\right],\\
[\boldsymbol{\omega}][\mathbf{Q}] &= -\dot{\psi}\left[\{\mathbf{0}\}\ \ [\boldsymbol{\varphi}]^{\mathrm{T}}[\boldsymbol{\theta}]^{\mathrm{T}}[\mathbf{U}_{\theta}]\{\mathbf{u}_{\psi}\}\ \ [\boldsymbol{\varphi}]^{\mathrm{T}}[\mathbf{U}_{\varphi}][\boldsymbol{\theta}]^{\mathrm{T}}\{\mathbf{u}_{\psi}\}\right]\\
&\quad + \dot{\theta}\left[[\boldsymbol{\varphi}]^{\mathrm{T}}[\boldsymbol{\theta}]^{\mathrm{T}}[\mathbf{U}_{\theta}]\{\mathbf{u}_{\psi}\}\ \ \{\mathbf{0}\}\ \ -[\boldsymbol{\varphi}]^{\mathrm{T}}[\mathbf{U}_{\varphi}]\{\mathbf{u}_{\theta}\}\right]\\
&\quad + \dot{\varphi}\left[[\mathbf{U}_{\varphi}][\boldsymbol{\varphi}]^{\mathrm{T}}[\boldsymbol{\theta}]^{\mathrm{T}}\{\mathbf{u}_{\psi}\}\ \ [\mathbf{U}_{\varphi}][\boldsymbol{\varphi}]^{\mathrm{T}}\{\mathbf{u}_{\theta}\}\ \ \{\mathbf{0}\}\right],
\end{aligned} \tag{2.145}$$

which show that the expression (2.144) is identically verified.

Observing that the matrix $[\mathbf{Q}]$ may be also written in the form

$$[\mathbf{Q}] = \left[\frac{\partial\{\boldsymbol{\omega}\}}{\partial\dot{\psi}}\ \ \frac{\partial\{\boldsymbol{\omega}\}}{\partial\dot{\theta}}\ \ \frac{\partial\{\boldsymbol{\omega}\}}{\partial\dot{\varphi}}\right], \tag{2.146}$$

it results that the equality (2.144) is equivalent to the vector relations

$$\frac{\partial\boldsymbol{\omega}}{\partial\psi} = \frac{\partial}{\partial t}\left(\frac{\partial\boldsymbol{\omega}}{\partial\dot{\psi}}\right) + \boldsymbol{\omega}\times\frac{\partial\boldsymbol{\omega}}{\partial\dot{\psi}},\ \frac{\partial\boldsymbol{\omega}}{\partial\theta} = \frac{\partial}{\partial t}\left(\frac{\partial\boldsymbol{\omega}}{\partial\dot{\theta}}\right) + \boldsymbol{\omega}\times\frac{\partial\boldsymbol{\omega}}{\partial\dot{\theta}},\ \frac{\partial\boldsymbol{\omega}}{\partial\varphi} = \frac{\partial}{\partial t}\left(\frac{\partial\boldsymbol{\omega}}{\partial\dot{\varphi}}\right) + \boldsymbol{\omega}\times\frac{\partial\boldsymbol{\omega}}{\partial\dot{\varphi}}. \tag{2.147}$$

2.7 Composition of the Angular Velocities

We consider the reference systems $O_0x_0y_0z_0$, $O_1x_1y_1z_1$ and the reference system $O_2x_2y_2z_2$ jointed to the rigid solid, and let $[\mathbf{A}_{21}]$ be the matrix of rotation of the reference system $O_2x_2y_2z_2$ relative to the reference system $O_1x_1y_1z_1$ and $[\mathbf{A}_{10}]$ be the matrix of rotation of the reference system $O_1x_1y_1z_1$ relative to the reference system $O_0x_0y_0z_0$.

In these conditions, the matrix of rotation of the reference system $O_2x_2y_2z_2$ relative to the reference system $O_0x_0y_0z_0$ reads

$$[\mathbf{A}_{20}] = [\mathbf{A}_{10}][\mathbf{A}_{21}], \tag{2.148}$$

while the angular velocity matrix $\left[\boldsymbol{\omega}_{20}^{(0)}\right]$ is given by the relation (2.125)

$$\left[\boldsymbol{\omega}_{20}^{(0)}\right] = \left[\dot{\mathbf{A}}_{20}\right][\mathbf{A}_{20}]^{\mathrm{T}}; \tag{2.149}$$

in the conditions of the equality (2.148) the last relation becomes

$$\left[\boldsymbol{\omega}_{20}^{(0)}\right] = \left[\dot{\mathbf{A}}_{10}\right][\mathbf{A}_{21}][\mathbf{A}_{21}]^{\mathrm{T}}[\mathbf{A}_{10}]^{\mathrm{T}} + [\mathbf{A}_{10}]\left[\dot{\mathbf{A}}_{21}\right][\mathbf{A}_{21}]^{\mathrm{T}}[\mathbf{A}_{10}]^{\mathrm{T}} \tag{2.150}$$

or

$$\left[\boldsymbol{\omega}_{20}^{(0)}\right] = \left[\boldsymbol{\omega}_{10}^{(0)}\right] + [\mathbf{A}_{10}]\left[\boldsymbol{\omega}_{21}^{(1)}\right][\mathbf{A}_{10}]^{\mathrm{T}}. \tag{2.151}$$

Further on, using the relation (2.125), it results the equality

$$[\mathbf{A}_{10}]\left[\boldsymbol{\omega}_{21}^{(1)}\right][\mathbf{A}_{10}]^{\mathrm{T}} = \left[\boldsymbol{\omega}_{21}^{(0)}\right]; \tag{2.152}$$

thus, from the relation (2.151) one deduces the matrix expression

$$\left[\boldsymbol{\omega}_{20}^{(0)}\right] = \left[\boldsymbol{\omega}_{10}^{(0)}\right] + \left[\boldsymbol{\omega}_{21}^{(0)}\right], \tag{2.153}$$

from which one obtains the vector relation

$$\boldsymbol{\omega}_{20} = \boldsymbol{\omega}_{10} + \boldsymbol{\omega}_{21}. \tag{2.154}$$

It results that for n reference systems one obtains the vector relation

$$\boldsymbol{\omega}_{n0} = \sum_{i=1}^{n} \boldsymbol{\omega}_{i,i-1}. \tag{2.155}$$

2.8 Accelerations of the Points of Rigid Solid

If one derives the relation (2.122) with respect to time and keeps into account the equality $\dot{\mathbf{r}}=\boldsymbol{\omega}\times\mathbf{r}$ and that the vector $\dot{\mathbf{v}}$ and $\dot{\mathbf{v}}_O$ represent the accelerations $\mathbf{a}$, and $\mathbf{a}_O$ of the points P and O, respectively (Fig. 2.8), while the vector $\dot{\boldsymbol{\omega}}$ represents the angular acceleration $\boldsymbol{\varepsilon}$, then one obtains the vector relation (*the Rivals relation*)

$$\mathbf{a}=\mathbf{a}_0+\boldsymbol{\varepsilon}\times\mathbf{r}+\boldsymbol{\omega}\times(\boldsymbol{\omega}\times\mathbf{r}). \tag{2.156}$$

Denoting by x, y, z, a_x, a_y, a_z, a_{Ox}, a_{Oy}, a_{Oz}, ω_x, ω_y, ω_z, and ε_x, ε_y, ε_z the projections of the vectors $\mathbf{r}$, $\mathbf{a}$, $\mathbf{a}_O$, $\boldsymbol{\omega}$, and $\boldsymbol{\varepsilon}$, respectively, onto the axes of *the mobile reference system Oxyz* and using the notations

$$\{\mathbf{a}\}=[a_x\ \ a_y\ \ a_z]^{\mathrm{T}},\ \{\mathbf{a}_O\}=[a_{Ox}\ \ a_{Oy}\ \ a_{Oz}]^{\mathrm{T}},\ \{\mathbf{r}\}=[x\ \ y\ \ z]^{\mathrm{T}}, \tag{2.157}$$

$$[\boldsymbol{\omega}]=\begin{bmatrix}0 & -\omega_z & \omega_y\\ \omega_z & 0 & -\omega_x\\ -\omega_y & \omega_x & 0\end{bmatrix},\ [\boldsymbol{\varepsilon}]=\begin{bmatrix}0 & -\varepsilon_z & \varepsilon_y\\ \varepsilon_z & 0 & -\varepsilon_x\\ -\varepsilon_y & \varepsilon_x & 0\end{bmatrix}, \tag{2.158}$$

one obtains from the relation (2.156) the matrix expression

$$\{\mathbf{a}\}=\{\mathbf{a}_O\}+\left[[\boldsymbol{\varepsilon}]+[\boldsymbol{\omega}]^2\right]\{\mathbf{r}\}. \tag{2.159}$$

To determine the way in which the accelerations compound, one derives the expression (2.153) with respect to time and it results

$$\left[\boldsymbol{\varepsilon}_{20}^{(0)}\right]=\left[\boldsymbol{\varepsilon}_{10}^{(0)}\right]+\left[\dot{\boldsymbol{\omega}}_{21}^{(0)}\right]. \tag{2.160}$$

Further on, by derivation of the relation

$$\left[\boldsymbol{\omega}_{21}^{(0)}\right]=[\mathbf{A}_{10}]\left[\boldsymbol{\omega}_{21}^{(1)}\right][\mathbf{A}_{10}]^{\mathrm{T}} \tag{2.161}$$

with respect to time, one obtains

$$\left[\dot{\boldsymbol{\omega}}_{21}^{(0)}\right]=[\dot{\mathbf{A}}_{10}]\left[\boldsymbol{\omega}_{21}^{(1)}\right][\mathbf{A}_{10}]^{\mathrm{T}}+[\mathbf{A}_{10}]\left[\dot{\boldsymbol{\omega}}_{21}^{(1)}\right][\mathbf{A}_{10}]^{\mathrm{T}}+[\mathbf{A}_{10}]\left[\boldsymbol{\omega}_{21}^{(1)}\right][\dot{\mathbf{A}}_{10}]^{\mathrm{T}}; \tag{2.162}$$

since

$$[\dot{\mathbf{A}}_{10}]=\left[\boldsymbol{\omega}_{10}^{(0)}\right][\mathbf{A}_{10}],\ [\dot{\mathbf{A}}_{10}]^{\mathrm{T}}=-[\mathbf{A}_{10}]^{\mathrm{T}}[\dot{\mathbf{A}}_{10}][\mathbf{A}_{10}]^{\mathrm{T}},\ [\mathbf{A}_{10}]\left[\boldsymbol{\omega}_{21}^{(1)}\right][\dot{\mathbf{A}}_{10}]^{\mathrm{T}}=\left[\boldsymbol{\varepsilon}_{21}^{0}\right] \tag{2.163}$$

it results

$$\left[\dot{\boldsymbol{\omega}}_{21}^{(0)}\right] = \left[\boldsymbol{\varepsilon}_{21}^{0}\right] + \left[\boldsymbol{\omega}_{10}^{0}\right]\left[\boldsymbol{\omega}_{21}^{0}\right] - \left[\boldsymbol{\omega}_{21}^{0}\right]\left[\boldsymbol{\omega}_{10}^{0}\right] \tag{2.164}$$

and the relation (2.160) becomes

$$\left[\boldsymbol{\varepsilon}_{20}^{0}\right] = \left[\boldsymbol{\varepsilon}_{10}^{0}\right] + \left[\boldsymbol{\varepsilon}_{21}^{0}\right] + \left[\boldsymbol{\omega}_{10}^{0}\right]\left[\boldsymbol{\omega}_{21}^{0}\right] - \left[\boldsymbol{\omega}_{21}^{0}\right]\left[\boldsymbol{\omega}_{10}^{0}\right]. \tag{2.165}$$

Knowing that the last two terms of the relation (2.165) correspond to the vector $\boldsymbol{\omega}_{10} \times \boldsymbol{\omega}_{21}$, one obtains in vector transcription

$$\boldsymbol{\varepsilon}_{20} = \boldsymbol{\varepsilon}_{10} + \boldsymbol{\varepsilon}_{21} + \boldsymbol{\omega}_{10} \times \boldsymbol{\omega}_{21}. \tag{2.166}$$

For $n-1$ mobile reference systems one deduces, by induction, the relation

$$\boldsymbol{\varepsilon}_{n0} = \sum_{i=1}^{n} \boldsymbol{\varepsilon}_{i,i-1} + \sum_{i<j}\sum_{j=2}^{n} \boldsymbol{\omega}_{i,i-1} \times \boldsymbol{\omega}_{j,j-1}. \tag{2.167}$$

Further Reading

Amirouche F (2005). Fundamentals of Multibody Dynamics. Theory and Applications. Boston: Birkhäuser.

Ball RS (1998). A Treatise on the Theory of Screws. Cambridge: Cambridge University Press.

Bloch AM, Baillieul J, Crouch P, Marsden J (2007). Nonholonomic Mechanics and Control (Interdisciplinary Applied Mathematics). Berlin: Springer.

Blundell M, Harty D (2004). The Multibody Systems Approach to Vehicle Dynamics. Amsterdam: Elsevier Butterworth-Heinemann.

Coutinho MG (2001). Dynamic Simulations of Multibody Systems. New York: Springer.

Dankowicz HJ (2004). Multibody Mechanics and Visualization. London: Springer.

de Jalón JG, Bayo E (2011). Kinematic and Dynamic Simulation of Multibody Systems: The Real-Time Challenge. Berlin: Springer Verlag.

Den Hartog JP (1961). Mechanics. New York: Dover Publications Inc.

Douglas Gregory R (2006). Classical Mechanics. Cambridge: Cambridge University Press.

Dresig H, Holzweißig F (2010). Dynamics of Machinery: Theory and Applications. Heidelberg: Springer.

Fasano A, Marmi S, Pelloni B (2006). Analytical Mechanics: An Introduction. Oxford: Oxford University Press.

Gattringer H, Gerstmayr J (2013). Multibody Systems Dynamics, Robotics and Control. Wien: Springer.

Ginsberg JH (1998). Advanced Engineering Dynamics. 2nd ed. Cambridge: Cambridge University Press.

Ginsberg J (2007). Engineering Dynamics. Cambridge: Cambridge University Press.

Goldstein H, Poole CP Jr, Safko JL (2001). Classical Mechanics. 3rd ed. Edinburg: Pearson.

Greenwood DT (2006). Advanced Dynamics. Cambridge: Cambridge University Press.

Greiner W (2009). Classical Mechanics: Systems of Particles and Hamiltonian Dynamics. 2nd ed. Heidelberg: Springer.

Hand LN, Finch JD (1998). Analytical Mechanics. Cambridge: Cambridge University Press.

Jain A (2010). Robot and Multibody Dynamics: Analysis and Algorithms. New York: Springer.

Jazar RN (2011). Advanced Dynamics: Rigid Body, Multibody, and Aerospace Applications. Hoboken: John Wiley & Sons.

Jazar RN (2010). Theory of Applied Robotics: Kinematics, Dynamics, and Control. 2nd ed. New York: Springer.

Karnopp DC, Margolis DL, Rosenberg RC (2012). System Dynamics: Modeling, Simulation, and Control of Mechatronic Systems. Hoboken: John Wiley & Sons.

Kibble TWB, Berkshire FH (2004). Classical Mechanics. 5th ed. London: Imperial College Press.

Kleppner D, Kolenkow R (2013). An Introduction to Mechanics. 2nd ed. Cambridge: Cambridge University Press.

Lagrange JL (1996). Analytical Mechanics. Dordrecht: Kluwer Academic Publishers,

Lurie AI (2002). Analytical Mechanics. Berlin: Springer.

Meriam JL, Kraige LG (2012). Engineering Mechanics: Dynamics. Hoboken: John Wiley & Sons.

Moon FC (1998). Applied Dynamics: With Applications to Multibody and Mechatronic Systems. New York: John Wiley & Sons, Inc.

Morin D (2008). Introduction to Classical Mechanics: With Problems and Solution. Cambridge: Cambridge University Press.

Murray RM, Li Z, Shankar Sastry S (1994). A Mathematical Introduction to Robotic Manipulation. Boca Raton: CRC Press.

Neimark IuI, Fufaev NA (1972). Dynamics of Nonholonomic Systems (Translations of Mathematical Monographs, V. 33). Providence, Rhode Island: American Mathematical Society.

O'Reilly OM (2008). Intermediate Dynamics for Engineers: A unified Treatment of Newton-Euler and Lagrangian Mechanics. Cambridge: Cambridge University Press

Pandrea N (2000). Elemente de mecanica solidelor în coordonate plückeriene. Bucureşti: Editura Academiei Române (in Romanian).

Pandrea N, Stănescu ND (2002). Mecanica. Bucureşti: Editura Didactică şi Pedagogică (in Romanian).

Papastavridis JG (2002). Analytical Mechanics: A Comprehensive Treatise on the Dynamics of Constrained Systems; For Engineers, Physicists, and Mathematicians. Oxford: Oxford University Press.

Pfeiffer F (2008). Mechanical System Dynamics (Lecture Notes in Applied and Computational Mechanics). Berlin: Springer.

Pfeiffer F, Glocker C (1996). Multibody Dynamics with Unilateral Contacts. New York: John Wiley & Sons, Inc.

Roberson RE, Schwertassek R (1988). Dynamics of Multibody Systems. Berlin: Springer Verlag.

Seifried R (2013). Dynamics of Underactuated Multibody Systems: Modeling, Control and Optimal Design. Heidelberg: Springer.

Shabana AA (2013). Dynamics of Multibody Systems. Cambridge: Cambridge University Press.

Spong MW, Hutchinson S, Vidyasagar M (2005). Robot Modeling and Control. New York: John Wiley & Sons.

Stănescu ND (2013). Mecanica sistemelor. Bucureşti: Editura Didactică şi Pedagogică (in Romanian).

Taylor JR (2005). Classical Mechanics. Sausalito: University Science Books.

Udwadia FE, Kalaba RE (2007). Analytical Dynamics: A New Approach. Cambridge: Cambridge University Press.

Uicker JJ, Ravani B, Sheth PN (2013). Matrix Methods in the Design Analysis of Mechanisms and Multibody Systems. Cambridge: Cambridge University Press.

Wittenburg J (2007). Dynamics of Multibody Systems. Berlin: Springer.

Woodhouse N (2010). Introduction to Analytical Dynamics. London: Springer.

3

General Theorems in the Dynamics of the Rigid Solid

The chapter is dedicated to the general theorems in the dynamics of the rigid solid, that is, the theorem of momentum, the theorem of the moment of momentum, and the kinetic energy; all these theorems are developed in matrix form.

3.1 Moments of Inertia

3.1.1 Definitions: Relations Between the Moments of Inertia

One considers a rigid solid, an element of it of mass $\mathrm{d}m$ and volume $\mathrm{d}V$ (Fig. 3.1), and a point A interior to this element; let s be the distance from the point A to a straight line Δ.

The moment of inertia relative to the axis Δ is defined by the integral of volume

$$J = \int s^2 \mathrm{d}m, \tag{3.1}$$

where the integral extends over the whole domain occupied by the rigid solid.

If we mark by s the distance from the point A to another given point, then the relation (3.1) defines the *polar moment of inertia* relative to that point, while if s is the distance from the point A to a plan, then the relation (3.1) defines the *planar moment of inertia* relative to that plan.

Dynamics of the Rigid Solid with General Constraints by a Multibody Approach, First Edition.
Nicolae Pandrea and Nicolae-Doru Stănescu.

Companion website: www.wiley.com/go/pandrea

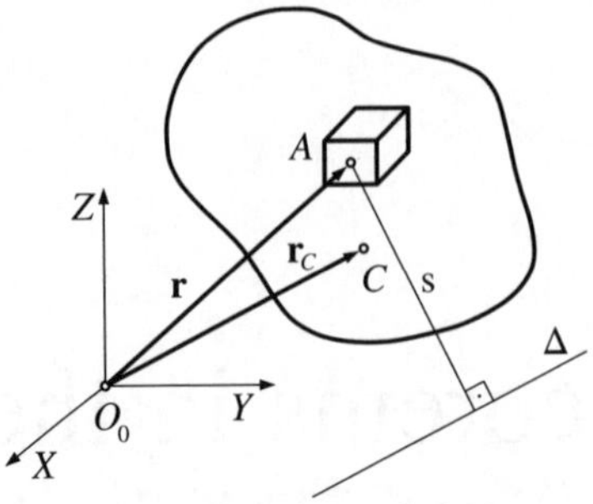

Figure 3.1 Moments of inertia.

Considering a reference system *Oxyz* (Fig. 3.1) and denoting by *x*, *y*, *z* the coordinates of the point *A*, one may write the following moments of inertia:

- polar

$$J_O = \int (x^2 + y^2 + z^2)\,\mathrm{d}m; \tag{3.2}$$

- axial:

$$J_x = \int (y^2 + z^2)\,\mathrm{d}m,\ J_y = \int (x^2 + z^2)\,\mathrm{d}m,\ J_z = \int (x^2 + y^2)\,\mathrm{d}m; \tag{3.3}$$

- planar:

$$J_{xOy} = \int z^2\,\mathrm{d}m,\ J_{xOz} = \int y^2\,\mathrm{d}m,\ J_{yOz} = \int x^2\,\mathrm{d}m; \tag{3.4}$$

- centrifugal:

$$J_{xy} = \int xy\,\mathrm{d}m,\ J_{yz} = \int yz\,\mathrm{d}m,\ J_{xz} = \int xz\,\mathrm{d}m. \tag{3.5}$$

The axial and centrifugal moments of inertia define *the symmetric matrix of moments of inertia*

$$[\mathbf{J}] = \begin{bmatrix} J_x & -J_{xy} & -J_{xz} \\ -J_{yx} & J_y & -J_{yz} \\ -J_{zx} & -J_{zy} & J_z \end{bmatrix}. \tag{3.6}$$

From the previous relations of definition, it results the equalities:

$$J_O = J_{xOy} + J_{xOz} + J_{yOz},\ J_O = \frac{1}{2}\left(J_x + J_y + J_z\right), \tag{3.7}$$

$$J_x = J_{xOy} + J_{xOz},\ J_y = J_{xOy} + J_{yOz},\ J_z = J_{xOz} + J_{yOz}, \tag{3.8}$$

$$J_{xOy} = \frac{J_x + J_y - J_z}{2},\ J_{xOz} = \frac{J_x + J_z - J_y}{2},\ J_{yOz} = \frac{J_y + J_z - J_x}{2}. \tag{3.9}$$

3.1.2 Moments of Inertia for Homogeneous Rigid Solid Bodies

In the case of full homogeneous bodies, the density being a constant $\rho = \frac{m}{V}$, with the mass m, and the volume V of the body, it results the equality $\mathrm{d}m = \rho \mathrm{d}V$ and the general expression of the moments of inertia becomes

$$J = \frac{m}{V} I, \tag{3.10}$$

where

$$I = \int s^2 \mathrm{d}V \tag{3.11}$$

is *the geometric moment of inertia*.

Analogically, for the shell rigid solid one obtains the relations

$$I = \int s^2 \mathrm{d}A,\ J = \frac{m}{A} I, \tag{3.12}$$

where A is the area of shell, and $\mathrm{d}A$ is the element of the area; for the bar rigid solid one gets the relations

$$I = \int s^2 \mathrm{d}l,\ J = \frac{m}{L} I, \tag{3.13}$$

where L is the length of the bar, and $\mathrm{d}l$ is the elementary arc of curve.

3.1.3 Centers of Weight

The center of weight or the center of mass C of the rigid solid (Fig. 3.1) has the vector of position $\mathbf{r}_C$ given by the relation

$$\mathbf{r}_C = \frac{\int \mathbf{r} dm}{m}, \tag{3.14}$$

wherefrom one obtains the coordinates

$$x_C = \frac{\int x dm}{m}, y_C = \frac{\int y dm}{m}, z_C = \frac{\int z dm}{m}. \tag{3.15}$$

The static moments relative to the plans of coordinates are defined by the relations

$$S_{xOy} = \int z dm, S_{xOz} = \int y dm, S_{yOz} = \int x dm \tag{3.16}$$

and satisfy the equalities

$$S_{xOy} = m z_C, S_{xOz} = m y_C, S_{yOz} = m x_C. \tag{3.17}$$

It is obvious that if a plan contains the center of weight, then the static moment relative to that plan is null (theorem of the static moments).

In the case of the homogeneous bodies one deduces the equalities:

- for full bodies:

$$x_C = \frac{\int x dV}{V}, y_C = \frac{\int y dV}{V}, z_C = \frac{\int z dV}{V}; \tag{3.18}$$

- for shells:

$$x_C = \frac{\int x dA}{A}, y_C = \frac{\int y dA}{A}, z_C = \frac{\int z dA}{A}; \tag{3.19}$$

- for bars:

$$x_C = \frac{\int x dl}{l}, y_C = \frac{\int y dl}{l}, z_C = \frac{\int z dl}{l}. \tag{3.20}$$

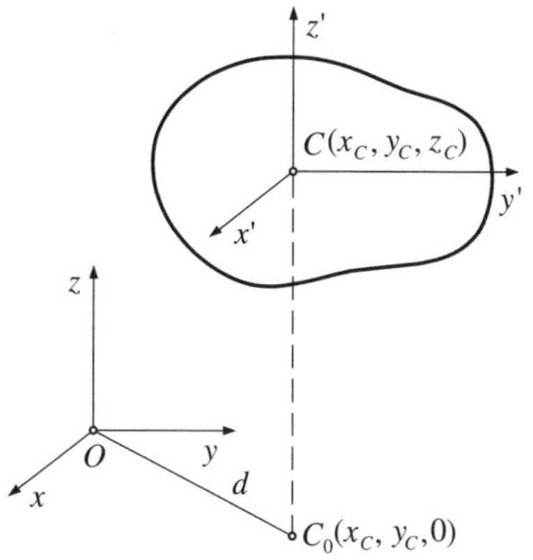

Figure 3.2 The variation of the moments of inertia relative to parallel axes.

3.1.4 Variation of the Moments of Inertia Relative to Parallel Axes

One considers a rigid solid of mass m related to the reference system $Oxyz$ (Fig. 3.2), and let $C(x_C, y_C, z_C)$ be the center of mass, while $Cx'y'z'$ is a reference system with the axes parallel to those of the reference system $Oxyz$.

Between the coordinates x, y, z, and x', y', z' of a point relative to the two reference systems, there exist the relations

$$x = x' + x_C, y = y' + y_C, z = z' + z_C; \tag{3.21}$$

hence

$$J_z = \int \left[(x' + x_C)^2 + (y' + y_C)^2 \right] \mathrm{d}m. \tag{3.22}$$

If one takes into account the equalities

$$\int x' \mathrm{d}m = 0, \int y' \mathrm{d}m = 0 \tag{3.23}$$

(according to the theorem of the static moments), and

$$x_C^2 + y_C^2 = d^2, \tag{3.24}$$

d being the distance between the axes Oz and Cz', then one deduces the expression

$$J_z = J_{z'} + md^2. \tag{3.25}$$

In a more general approach we obtain (Steiner's relation)

$$J_{\Delta} = J_{\Delta_C} + md^2, \tag{3.26}$$

where Δ_C is a straight line that passes through the center of weight, Δ is a straight line parallel to Δ_C, and d is the distance between the two straight lines.

For centrifugal moments of inertia one obtains the expressions

$$J_{xy} = J_{x'y'} + mx_Cy_C,\ J_{xz} = J_{x'z'} + mx_Cz_C,\ J_{yz} = J_{y'z'} + my_Cz_C. \tag{3.27}$$

Using the notation

$$[\mathbf{G}] = \begin{bmatrix} 0 & -z_C & y_C \\ z_C & 0 & -x_C \\ -y_C & x_C & 0 \end{bmatrix}, \tag{3.28}$$

one obtains the relation between the matrices of the moments of inertia

$$[\mathbf{J}] = [\mathbf{J}'] + m[\mathbf{G}]^{\mathrm{T}}[\mathbf{G}]. \tag{3.29}$$

3.1.5 *Variation of the Moments of Inertia Relative to Concurrent Axes*

One considers the rigid solid of mass m in Fig. 3.3, and two reference systems $Oxyz$ and $Ox'y'z'$; let a_{1i}, a_{2i}, a_{3i}, $i = 1, 2, 3$, be the director cosines of the axes Ox', Oy', Oz' relative to the reference system $Oxyz$, and $\{\mathbf{u}_i\}$, $i = 1, 2, 3$, be the column matrices

$$\{\mathbf{u}_i\} = [a_{1i}\ \ a_{2i}\ \ a_{3i}]^{\mathrm{T}}. \tag{3.30}$$

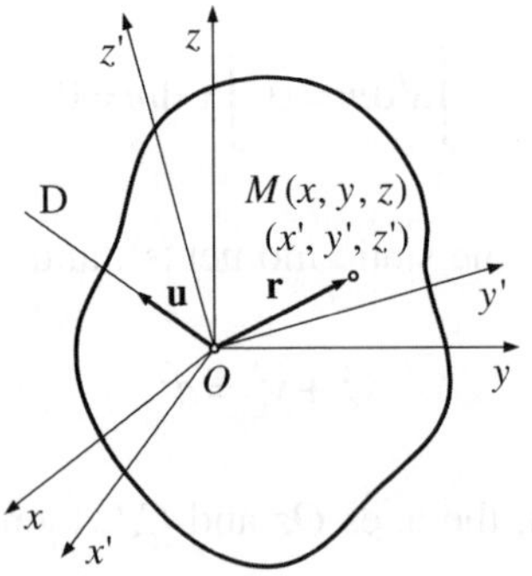

Figure 3.3 The variation of the moments of inertia relative to concurrent axes.

Between the coordinates x, y, z, and x', y', z' of a point M in the reference systems $Oxyz$, and $Ox'y'z'$, respectively, there exists the relations

$$\begin{bmatrix} x' \\ y' \\ z' \end{bmatrix} = [\mathbf{A}] \begin{bmatrix} x \\ y \\ z \end{bmatrix}, \tag{3.31}$$

where $[\mathbf{A}]$ is the matrix of rotation

$$[\mathbf{A}] = [\{\mathbf{u}_1\}\ \{\mathbf{u}_2\}\ \{\mathbf{u}_3\}] = \begin{bmatrix} a_{11} & a_{12} & a_{13} \\ a_{21} & a_{22} & a_{23} \\ a_{31} & a_{32} & a_{33} \end{bmatrix}. \tag{3.32}$$

Keeping into account the relations

$$\{\mathbf{u}_i\}^{\mathrm{T}}\{\mathbf{u}_j\} = \begin{cases} 1 & \text{for } i=j, \\ 0 & \text{for } i \neq j, \end{cases} \tag{3.33}$$

the moment of inertia relative to the axis Ox'

$$J_{x'} = \int \left[(y')^2 + (z')^2\right] \mathrm{d}m, \tag{3.34}$$

writes in the form

$$J_{x'} = \{\mathbf{u}_1\}^{\mathrm{T}}[\mathbf{J}]\{\mathbf{u}_1\}. \tag{3.35}$$

Analogically, one deduces the equalities

$$J_{y'} = \{\mathbf{u}_2\}^{\mathrm{T}}[\mathbf{J}]\{\mathbf{u}_2\}, J_{z'} = \{\mathbf{u}_3\}^{\mathrm{T}}[\mathbf{J}]\{\mathbf{u}_3\}. \tag{3.36}$$

If Δ is a straight line of unit vector $\mathbf{u}(a_1, a_2, a_3)$, then we obtain the expression

$$J_{\Delta} = \{\mathbf{u}\}^{\mathrm{T}}[\mathbf{J}]\{\mathbf{u}\}, \tag{3.37}$$

where

$$\{\mathbf{u}\} = [a_1\ \ a_2\ \ a_3]^{\mathrm{T}}. \tag{3.38}$$

For the centrifugal moments of inertia it results the expressions

$$J_{x'y'} = -\{\mathbf{u}_1\}^{\mathrm{T}}[\mathbf{J}]\{\mathbf{u}_2\},\ J_{x'z'} = -\{\mathbf{u}_1\}^{\mathrm{T}}[\mathbf{J}]\{\mathbf{u}_3\},\ J_{y'z'} = -\{\mathbf{u}_2\}^{\mathrm{T}}[\mathbf{J}]\{\mathbf{u}_3\}; \tag{3.39}$$

hence, there exists the relation

$$[\mathbf{J}'] = [\mathbf{A}]^{\mathrm{T}}[\mathbf{J}][\mathbf{A}]. \tag{3.40}$$

between the matrices of inertia in the two reference systems.

3.1.6 Principal Axes of Inertia: Principal Moments of Inertia

Keeping into account the relation

$$\{\mathbf{u}\}^{\mathrm{T}}\{\mathbf{u}\} - 1 = 0, \tag{3.41}$$

the extreme values of the moment of inertia J_Δ, given by the relation (3.37), are obtained by solving the system of equations deduced by vanishing the partial derivatives of the function

$$F = J_\Delta - \lambda\{\mathbf{u}\}^{\mathrm{T}}\{\mathbf{u}\} + \lambda \tag{3.42}$$

with respect to u_1, u_2 and u_3.

We get the matrix equation

$$[\mathbf{J}]\{\mathbf{u}\} - \lambda\{\mathbf{u}\} = \{\mathbf{0}\}, \tag{3.43}$$

which admits a non-zero solution if

$$\det[[\mathbf{J}] - \lambda[\mathbf{I}]] = 0. \tag{3.44}$$

From the third order equation in λ (3.44) one obtains the eigenvalues λ_1, λ_2, λ_3 for which correspond, from the relation (3.43), the unitary eigenvectors $\{\mathbf{v}_1\}$, $\{\mathbf{v}_2\}$, $\{\mathbf{v}_3\}$; since the matrix $[\mathbf{J}]$ is a symmetric one, one gets that the eigenvectors are orthogonal. It results that *the principal axes of inertia* Δ_1, Δ_2, Δ_3 *are reciprocally perpendicular* and define a dextrorsum reference system.

Corresponding to the eigenvalues, from the relation (3.37) one deduces the *principal moments of inertia*

$$J_1 = \lambda_1,\ J_2 = \lambda_2,\ J_3 = \lambda_3. \tag{3.45}$$

Moreover,

$$J_{\Delta_1\Delta_2} = \{\mathbf{v}_1\}^{\mathrm{T}}[\mathbf{J}]\{\mathbf{v}_2\} = \lambda\{\mathbf{v}_1\}^{\mathrm{T}}\{\mathbf{v}_2\}; \tag{3.46}$$

consequently, the centrifugal moments of inertia relative to the principal axes of inertia are equal to zero.

Denoting by [**B**] the matrix of rotation of the system of principal moments of inertia relative to the reference system *Oxyz*,

$$[\mathbf{B}] = [\{\mathbf{v}_1\}\ \{\mathbf{v}_2\}\ \{\mathbf{v}_3\}], \tag{3.47}$$

one obtains the equality

$$\begin{bmatrix} \lambda_1 & 0 & 0 \\ 0 & \lambda_2 & 0 \\ 0 & 0 & \lambda_3 \end{bmatrix} = [\mathbf{B}]^{\mathrm{T}}[\mathbf{J}][\mathbf{B}]. \tag{3.48}$$

As an application, we consider

$$[\mathbf{J}] = \begin{bmatrix} \frac{6}{4} & \frac{\sqrt{2}}{4} & -\frac{\sqrt{2}}{4} \\ \frac{\sqrt{2}}{4} & \frac{9}{4} & \frac{3}{4} \\ -\frac{12}{4} & \frac{3}{4} & \frac{9}{4} \end{bmatrix},$$

and we obtain the characteristic equation

$$\det[[\mathbf{J}] - \lambda[\mathbf{I}]] = -\lambda^3 + 6\lambda^2 - 11\lambda + 6 = 0,$$

with the solutions $J_1 = \lambda_1 = 1$, $J_2 = \lambda_2 = 2$, $J_3 = \lambda_3 = 3$. It results the matrix

$$[\mathbf{B}] = \begin{bmatrix} \frac{\sqrt{2}}{2} & \frac{\sqrt{2}}{2} & 0 \\ -\frac{1}{2} & \frac{1}{2} & \frac{\sqrt{2}}{2} \\ \frac{1}{2} & -\frac{1}{2} & \frac{\sqrt{2}}{2} \end{bmatrix},$$

which verifies the relation (3.48).

3.2 Momentum: The Theorem of Momentum

We consider a rigid solid of mass m and center of weight C, the reference system $Oxyz$ (Fig. 3.4) jointed to the rigid solid, and the fixed reference system O_0XYZ.

Let P be an arbitrary point of the rigid solid, defined by the vector of position $\mathbf{r}=\mathbf{OP}$, the velocity of the point being $\mathbf{v}$, and let $\mathbf{r}_C=\mathbf{OC}$ be the vector of position of the center of weight C, the velocity of which being $\mathbf{v}_C$. If we take into account the theorem of the static moments ($\int \boldsymbol{\rho}\mathrm{d}m=\mathbf{0}$), where $\boldsymbol{\rho}$ is the vector $\boldsymbol{\rho}=\mathbf{CP}$, and the Euler relation

$$\mathbf{v}=\mathbf{v}_C+\boldsymbol{\omega}\times\boldsymbol{\rho}, \tag{3.49}$$

where $\boldsymbol{\omega}$ is the angular velocity, then *the momentum* $\mathbf{H}=\int \mathbf{v}\mathrm{d}m$ reads

$$\mathbf{H}=m\mathbf{v}_C. \tag{3.50}$$

With the aid of the Euler relation, $\mathbf{v}_C=\mathbf{v}_O+\boldsymbol{\omega}\times\mathbf{r}_C$, where $\mathbf{v}_O$ is the velocity of the point O, one obtains the vector expression of the momentum

$$\mathbf{H}=m(\mathbf{v}_O-\mathbf{r}_C\times\boldsymbol{\omega}). \tag{3.51}$$

Using the notations

$$\{\mathbf{v}_O\}=[v_{Ox}\ \ v_{Oy}\ \ v_{Oz}]^{\mathrm{T}},\{\boldsymbol{\omega}\}=[\omega_x\ \ \omega_y\ \ \omega_z]^{\mathrm{T}},\{\mathbf{H}\}=[H_x\ \ H_y\ \ H_z]^{\mathrm{T}}, \tag{3.52}$$

$$[\mathbf{m}]=\begin{bmatrix} m & 0 & 0\\ 0 & m & 0\\ 0 & 0 & m\end{bmatrix},\ [\mathbf{S}]=\begin{bmatrix} 0 & -mz_C & my_C\\ mz_C & 0 & -mx_C\\ -my_C & mx_C & 0\end{bmatrix}, \tag{3.53}$$

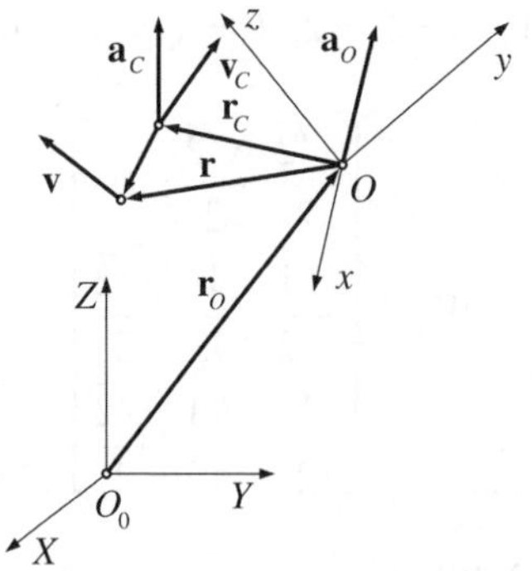

Figure 3.4 The momentum, the moment of momentum, and the kinetic energy of a rigid solid.

where v_{Ox}, v_{Oy}, v_{Oz}, ω_x, ω_y, ω_z, H_x, H_y, H_z, and x_C, y_C, z_C are the projections of the vectors $\mathbf{v}_O$, $\boldsymbol{\omega}$, $\mathbf{H}$, and $\mathbf{r}_C$, respectively, onto the axes of the *mobile* reference system $Oxyz$, we obtain *the matrix form of the momentum*

$$\{\mathbf{H}\} = [\mathbf{m}]\{\mathbf{v}_O\} + [\mathbf{S}]^{\mathrm{T}}\{\boldsymbol{\omega}\}. \tag{3.54}$$

One considers that the forces which act upon the rigid solid reduce at the point O at the torsor $(\mathbf{F},\mathbf{M})$, and denotes by F_x, F_y, F_z, and M_x, M_y, M_z *the projections of the vectors* $\mathbf{F}$, and $\mathbf{M}$, respectively, *onto the axes of the mobile reference system*, and by $\{\mathbf{F}\}$ and $\{\mathbf{M}_O\}$ the associated column matrices.

Knowing that the theorem of momentum is given by the relation

$$\dot{\mathbf{H}} = \mathbf{F} \tag{3.55}$$

and keeping into account the equality (3.50), one obtains the relation

$$m\mathbf{a}_C = \mathbf{F}, \tag{3.56}$$

which defines *the motion of the center of weight.*

Writing the acceleration of the center of weight with the aid of the Rivals relation

$$\mathbf{a}_C = \mathbf{a}_O + \boldsymbol{\varepsilon} \times \mathbf{r}_C + \boldsymbol{\omega} \times (\boldsymbol{\omega} \times \mathbf{r}_C), \tag{3.57}$$

where $\mathbf{a}_O$ is the acceleration of the point O ($\mathbf{a}_O = \dot{\mathbf{v}}_O$), while $\boldsymbol{\varepsilon}$ is the angular acceleration ($\boldsymbol{\varepsilon} = \dot{\boldsymbol{\omega}}$), we deduce *the vector expression of the theorem of momentum*

$$m[\mathbf{a}_O + \boldsymbol{\varepsilon} \times \mathbf{r}_C + \boldsymbol{\omega} \times (\boldsymbol{\omega} \times \mathbf{r}_C)] = \mathbf{F}. \tag{3.58}$$

With the aid of the relations (3.53) and using the notations

$$\{\mathbf{a}_O\} = [a_{Ox} \;\; a_{Oy} \;\; a_{Oz}]^{\mathrm{T}}, \{\boldsymbol{\varepsilon}\} = [\varepsilon_x \;\; \varepsilon_y \;\; \varepsilon_z]^{\mathrm{T}}, \tag{3.59}$$

$$[\boldsymbol{\varepsilon}] = \begin{bmatrix} 0 & -\varepsilon_z & \varepsilon_y \\ \varepsilon_z & 0 & -\varepsilon_x \\ -\varepsilon_y & \varepsilon_x & 0 \end{bmatrix}, [\boldsymbol{\omega}] = \begin{bmatrix} 0 & -\omega_z & \omega_y \\ \omega_z & 0 & -\omega_x \\ -\omega_y & \omega_x & 0 \end{bmatrix}, \tag{3.60}$$

where ε_x, ε_y, ε_z, and a_{Ox}, a_{Oy}, a_{Oz} are the projections of the vectors $\boldsymbol{\varepsilon}$, and $\mathbf{a}_O$, respectively, onto the axes of the mobile reference system, one obtains *the matrix form of the theorem of momentum*

$$[\mathbf{m}]\{\mathbf{a}_O\} + [\mathbf{S}]^{\mathrm{T}}\{\boldsymbol{\varepsilon}\} + [\boldsymbol{\omega}][\mathbf{S}]^{\mathrm{T}}\{\boldsymbol{\omega}\} = \{\mathbf{F}\}. \tag{3.61}$$

3.3 Moment of Momentum: The Theorem of Moment of Momentum

The moment of momentum of the rigid solid relative to the fixed point O_0 (Fig. 3.4) reads successively

$$\mathbf{K}_{O_0} = \int (\mathbf{r}_0 + \mathbf{r}) \times \mathbf{v} \mathrm{d}m = \mathbf{r}_0 \int \mathbf{v} \mathrm{d}m + \int \mathbf{r} \times \mathbf{v} \mathrm{d}m. \tag{3.62}$$

If we keep into account the relations $\int \mathbf{v}\mathrm{d}m = m\mathbf{v}_C$, $\int \mathbf{r}\mathrm{d}m = m\mathbf{r}_C$, and we replace $\mathbf{v}$ by $\mathbf{v}_O + (\boldsymbol{\omega} \times \mathbf{r})$, then we obtain the vector expression

$$\mathbf{K}_{O_0} = \mathbf{r}_0 \times m\mathbf{v}_C + \mathbf{r}_C \times m\mathbf{v}_O + \int \mathbf{r} \times (\boldsymbol{\omega} \times \mathbf{r}) \mathrm{d}m. \tag{3.63}$$

The vector $\int \mathbf{r} \times (\boldsymbol{\omega} \times \mathbf{r})\mathrm{d}m$ has as matrix correspondent the product $[\mathbf{J}_O]\{\boldsymbol{\omega}\}$, where $[\mathbf{J}_O]$ is the matrix of the moments of inertia at the point O; hence the expression (3.63) may be written in the form

$$\mathbf{K}_{O_0} = \mathbf{r}_0 \times m\mathbf{v}_C + \mathbf{r}_C \times m\mathbf{v}_O + \overline{[\mathbf{J}_O]\{\boldsymbol{\omega}\}}. \tag{3.64}$$

Deriving the relation (3.64) with respect to time, and taking into account the equality

$$\mathbf{v}_O \times m\mathbf{v}_C + (\boldsymbol{\omega} \times \mathbf{r}_C) \times m\mathbf{v}_O = \mathbf{0}, \tag{3.65}$$

one obtains

$$\dot{\mathbf{K}}_{O_0} = \mathbf{r}_0 \times m\mathbf{a}_C + \mathbf{r}_C \times m\mathbf{a}_O + \boldsymbol{\omega} \times \overline{[\mathbf{J}_O]\{\boldsymbol{\omega}\}} + \overline{[\mathbf{J}_O]\{\boldsymbol{\varepsilon}\}}. \tag{3.66}$$

The theorem of moment of momentum relative to the fixed point O_0 reads

$$\dot{\mathbf{K}}_{O_0} = \mathbf{M}_O + \mathbf{r}_0 \times \mathbf{F}; \tag{3.67}$$

if one keeps into account the equality (3.56), then one obtains the expression

$$\mathbf{r}_C \times m\mathbf{a}_O + \overline{[\mathbf{J}_O]\{\boldsymbol{\varepsilon}\}} + \boldsymbol{\omega} \times \overline{[\mathbf{J}_O]\{\boldsymbol{\omega}\}} = \mathbf{M}_O, \tag{3.68}$$

which represents *the vector form of the theorem of moment of momentum relative to the point O of the rigid solid.*

In *matrix form* this theorem reads

$$[\mathbf{S}]\{\mathbf{a}_O\} + [\mathbf{J}_O]\{\boldsymbol{\varepsilon}\} + [\boldsymbol{\omega}][\mathbf{J}_O]\{\boldsymbol{\omega}\} = \{\mathbf{M}_O\}. \tag{3.69}$$

Using the notation

$$[\widetilde{\mathbf{M}}] = \begin{bmatrix} [\mathbf{m}] & [\mathbf{S}]^{\mathrm{T}} \\ [\mathbf{S}] & [\mathbf{J}_O] \end{bmatrix}, \tag{3.70}$$

$[\widetilde{\mathbf{M}}]$ being called *the matrix of inertia of the rigid solid*, the expressions (3.61), and (3.69) of the theorem of momentum, and the theorem of moment of momentum, respectively, reunite in the matrix equation

$$[\widetilde{\mathbf{M}}] \begin{bmatrix} \{\mathbf{a}_O\} \\ \{\boldsymbol{\varepsilon}\} \end{bmatrix} + \begin{bmatrix} [\boldsymbol{\omega}][\mathbf{S}]^{\mathrm{T}} \\ [\boldsymbol{\omega}][\mathbf{J}_O] \end{bmatrix} \{\boldsymbol{\omega}\} = \begin{bmatrix} \{\mathbf{F}\} \\ \{\mathbf{M}_O\} \end{bmatrix}. \tag{3.71}$$

3.4 The Kinetic Energy of the Rigid Solid

The kinetic energy of the rigid solid (Fig. 3.4) reads

$$T = \frac{1}{2}\int \mathbf{v}^2 \mathrm{d}m; \tag{3.72}$$

keeping into account the Euler formula $\mathbf{v} = \mathbf{v}_O + \boldsymbol{\omega} \times \mathbf{r}$, one obtains the equality

$$T = \frac{1}{2}\int \left[\mathbf{v}_O^2 + 2\mathbf{v}_O(\boldsymbol{\omega} \times \mathbf{r}) + (\boldsymbol{\omega} \times \mathbf{r})^2\right] \mathrm{d}m \tag{3.73}$$

or

$$T = \frac{1}{2}\left[m\mathbf{v}_O^2 + 2\mathbf{v}_O(\boldsymbol{\omega} \times m\mathbf{r}_C) + \boldsymbol{\omega}\int \mathbf{r} \times (\boldsymbol{\omega} \times \mathbf{r})\mathrm{d}m\right]. \tag{3.74}$$

Using the notations (3.53), one deduces the matrix expression

$$T = \frac{1}{2}\{\mathbf{v}_O\}^{\mathrm{T}}\mathbf{m}\{\mathbf{v}_O\} + \{\mathbf{v}_O\}^{\mathrm{T}}\mathbf{S}^{\mathrm{T}}\{\boldsymbol{\omega}\} + \frac{1}{2}\{\boldsymbol{\omega}\}^{\mathrm{T}}\mathbf{J}_O\{\boldsymbol{\omega}\} \tag{3.75}$$

or

$$T = \frac{1}{2}\left[\{\mathbf{v}_O\}^{\mathrm{T}} \{\boldsymbol{\omega}\}^{\mathrm{T}}\right] [\widetilde{\mathbf{M}}] \begin{bmatrix} \{\mathbf{v}_O\} \\ \{\boldsymbol{\omega}\} \end{bmatrix}. \tag{3.76}$$

Further Reading

Amirouche F (2005). Fundamentals of Multibody Dynamics. Theory and Applications. Boston: Birkhäuser.

Ball RS (1998). A Treatise on the Theory of Screws. Cambridge: Cambridge University Press.

Bloch AM, Baillieul J, Crouch P, Mardsen J (2007). Nonholonomic Mechanics and Control (Interdisciplinary Applied Mathematics). Berlin: Springer.

Blundell M, Harty D (2004). The Multibody Systems Approach to Vehicle Dynamics. Amsterdam: Elsevier Butterworth-Heinemann.

Chaudhary H, Saha SH (2008). Dynamics and Balancing of Multibody Systems (Lecture Notes in Applied and Computational Mechanics). Berlin: Springer.

Cortes J (2002). Geometric, Control and Numerical Aspects of Nonholonomic Systems. Berlin: Springer.

Coutinho MG (2001). Dynamic Simulations of Multibody Systems. New York: Springer.

Dankowicz HJ (2004). Multibody Mechanics and Visualization. London: Springer.

de Jalón JG, Bayo E (2011). Kinematic and Dynamic Simulation of Multibody Systems: The Real-Time Challenge. Berlin: Springer Verlag.

Den Hartog JP (1961). Mechanics. New York: Dover Publications Inc.

Douglas Gregory R (2006). Classical Mechanics. Cambridge: Cambridge University Press.

Dresig H, Holzweißig F (2010). Dynamics of Machinery: Theory and Applications. Heidelberg: Springer.

Fasano A, Marmi S, Pelloni B (2006). Analytical Mechanics: An Introduction. Oxford: Oxford University Press.

Gattringer H, Gerstmayr J (2013). Multibody Systems Dynamics, Robotics and Control. Wien: Springer.

Ginsberg JH (1998). Advanced Engineering Dynamics. 2nd ed. Cambridge: Cambridge University Press.

Ginsberg J (2007). Engineering Dynamics. Cambridge: Cambridge University Press.

Goldstein H, Poole CP Jr, Safko JL (2001). Classical Mechanics. 3rd ed. Edinburg: Pearson.

Greenwood DT (2006). Advanced Dynamics. Cambridge: Cambridge University Press.

Greiner W (2009). Classical Mechanics: Systems of Particles and Hamiltonian Dynamics. 2nd ed. Heidelberg: Springer.

Hand LN, Finch JD (1998). Analytical Mechanics. Cambridge: Cambridge University Press.

Jain A (2010). Robot and Multibody Dynamics: Analysis and Algorithms. New York: Springer.

Jazar RN (2011). Advanced Dynamics: Rigid Body, Multibody, and Aerospace Applications. Hoboken: John Wiley & Sons.

Jazar RN (2010). Theory of Applied Robotics: Kinematics, Dynamics, and Control. 2nd ed. New York: Springer.

Karnopp DC, Margolis DL, Rosenberg RC (2012). System Dynamics: Modeling, Simulation, and Control of Mechatronic Systems. Hoboken: John Wiley & Sons.

Kibble TWB, Berkshire FH (2004). Classical Mechanics. 5th ed. London: Imperial College Press.

Kleppner D, Kolenkow R (2013). An Introduction to Mechanics. 2nd ed. Cambridge: Cambridge University Press.

Lagrange JL (1996). Analytical Mechanics. Dordrecht: Kluwer Academic Publishers,

Lurie AI (2002). Analytical Mechanics. Berlin: Springer.

Meriam JL, Kraige LG (2012). Engineering Mechanics: Dynamics. Hoboken: John Wiley & Sons.

Moon FC (1998). Applied Dynamics: With Applications to Multibody and Mechatronic Systems. New York: John Wiley & Sons, Inc.

Morin D (2008). Introduction to Classical Mechanics: With Problems and Solution. Cambridge: Cambridge University Press.

Murray RM, Li Z, Shankar Sastry S (1994). A Mathematical Introduction to Robotic Manipulation. Boca Raton: CRC Press.
Neimark IuI, Fufaev NA (1972). Dynamics of Nonholonomic Systems (Translations of Mathematical Monographs, V. 33). Providence, Rhode Island: American Mathematical Society.
O'Reilly OM (2008). Intermediate Dynamics for Engineers: A unified Treatment of Newton-Euler and Lagrangian Mechanics. Cambridge: Cambridge University Press
Pandrea N (2000). Elemente de mecanica solidelor în coordonate plückeriene. Bucureşti: Editura Academiei Române (in Romanian).
Pandrea N, Stănescu ND (2002). Mecanica. Bucureşti: Editura Didactică şi Pedagogică (in Romanian).
Papastavridis JG (2002). Analytical Mechanics: A Comprehensive Treatise on the Dynamics of Constrained Systems; For Engineers, Physicists, and Mathematicians. Oxford: Oxford University Press.
Pfeiffer F (2008).Mechanical System Dynamics (Lecture Notes in Applied and Computational Mechanics). Berlin: Springer.
Pfeiffer F, Glocker C (1996). Multibody Dynamics with Unilateral Contacts. New York: John Wiley & Sons, Inc.
Roberson RE, Schwertassek R (1988). Dynamics of Multibody Systems. Berlin: Springer Verlag.
Seifried R (2013). Dynamics of Underactuated Multibody Systems: Modeling, Control and Optimal Design. Heidelberg: Springer.
Shabana AA (2013). Dynamics of Multibody Systems. Cambridge: Cambridge University Press.
Spong MW, Hutchinson S, Vidyasagar M (2005). Robot Modeling and Control. New York: John Wiley & Sons.
Stănescu ND (2013). Mecanica sistemelor. Bucureşti: Editura Didactică şi Pedagogică (in Romanian).
Taylor JR (2005). Classical Mechanics. Sausalito: University Science Books.
Teodorescu PP (2009). Mechanical Systems, Classical Models: Volume 3: Analytical Mechanics (Mathematical and Analytical Techniques with Applications to Engineering). Dordrecht: Springer.
Udwadia FE, Kalaba RE (2007). Analytical Dynamics: A New Approach. Cambridge: Cambridge University Press.
Uicker JJ, Ravani B, Sheth PN (2013). Matrix Methods in the Design Analysis of Mechanisms and Multibody Systems. Cambridge: Cambridge University Press.
Wittenburg J (2007). Dynamics of Multibody Systems. Berlin: Springer.
Woodhouse N (2010). Introduction to Analytical Dynamics. London: Springer.

4

Matrix Differential Equations of the Motion of Rigid Solid

In this chapter are presented the matrix differential equations of motion in the general case of the rigid solid with constraints; the equations of motion are obtained using the general theorems and using the Lagrange equations; a completely new proof is given for the equivalence of these two approaches.

4.1 The Differential Equations Obtained from the General Theorems

4.1.1 General Aspects

The kinematic parameters which define the position of the rigid solid are the coordinates X_O, Y_O, Z_O of the point O (Fig. 3.4), and the angles of successive rotations ψ, θ, φ about the axes of rotation, angles determined by choosing one of the 12 variants described in chapter 2; consequently, the two matrix expressions (3.61) and (3.69), resulted from the theorems of momentum and moment of momentum, must be worked out by using the notations

$$\{\mathbf{s}\} = [X_O \ \ Y_O \ \ Z_O]^{\mathrm{T}}, \{\boldsymbol{\beta}\} = [\psi \ \ \theta \ \ \varphi]^{\mathrm{T}}, \tag{4.1}$$

$$\{\mathbf{q}\} = \begin{bmatrix} \{\mathbf{s}\} \\ \{\boldsymbol{\beta}\} \end{bmatrix} = [X_O \ \ Y_O \ \ Z_O \ \ \psi \ \ \theta \ \ \varphi]^{\mathrm{T}}. \tag{4.2}$$

Dynamics of the Rigid Solid with General Constraints by a Multibody Approach, First Edition.
Nicolae Pandrea and Nicolae-Doru Stănescu.

Companion website: www.wiley.com/go/pandrea

On the other hand, knowing that the generalized forces F_{X_O}, F_{Y_O}, and F_{Z_O}, corresponding to the parameters X_O, Y_O, and Z_O, respectively, are the projections of the resultant of the forces that act upon the rigid solid, onto the fixed axes, one obtains the relations

$$\{\mathbf{F_s}\} = [F_{X_O} \;\; F_{Y_O} \;\; F_{Z_O}]^{\mathrm{T}},\ \{\mathbf{F}\} = [\mathbf{A}]^{\mathrm{T}}\{\mathbf{F_s}\}. \tag{4.3}$$

Analogically, denoting by F_ψ, F_θ, and F_φ, the generalized forces corresponding to the parameters ψ, θ, and φ, respectively, we get

$$\{\mathbf{F}_{\boldsymbol{\beta}}\} = [F_\psi \;\; F_\theta \;\; F_\varphi]^{\mathrm{T}},\ \{\mathbf{F}_{\boldsymbol{\beta}}\} = [\mathbf{Q}]^{\mathrm{T}}\{\mathbf{M}_O\}. \tag{4.4}$$

We have to arrange the expressions (3.61) and (3.69) of the general theorems, such that they contain the matrices $\{\mathbf{s}\}$, $\{\boldsymbol{\beta}\}$, $\{\mathbf{F_s}\}$, and $\{\mathbf{F}_{\boldsymbol{\beta}}\}$.

4.1.2 *The Differential Equations*

Making the replacements

$$\{\mathbf{a}_O\} = [\mathbf{A}]^{\mathrm{T}}\{\ddot{\mathbf{s}}\}, \{\mathbf{F}\} = [\mathbf{A}]^{\mathrm{T}}\{\mathbf{F_s}\}, \tag{4.5}$$

in the equation (3.61) of *the theorem of momentum*, it results

$$[\mathbf{m}][\mathbf{A}]^{\mathrm{T}}\{\ddot{\mathbf{s}}\} + [\mathbf{S}]^{\mathrm{T}}\{\boldsymbol{\varepsilon}\} + [\boldsymbol{\omega}]\mathbf{S}^{\mathrm{T}}\{\boldsymbol{\omega}\} = [\mathbf{A}]^{\mathrm{T}}\{\mathbf{F_s}\}; \tag{4.6}$$

multiplying by [**A**] at the left, we get

$$[\mathbf{m}]\{\ddot{\mathbf{s}}\} + [\mathbf{A}][\mathbf{S}]^{\mathrm{T}}\{\boldsymbol{\varepsilon}\} + [\mathbf{A}][\boldsymbol{\omega}][\mathbf{S}]^{\mathrm{T}}\{\boldsymbol{\omega}\} = \{\mathbf{F_s}\}. \tag{4.7}$$

Further on, making the replacements

$$\{\boldsymbol{\omega}\} = [\mathbf{Q}]\{\dot{\boldsymbol{\beta}}\}, \{\boldsymbol{\varepsilon}\} = [\mathbf{Q}]\{\ddot{\boldsymbol{\beta}}\} + [\dot{\mathbf{Q}}]\{\dot{\boldsymbol{\beta}}\}, \tag{4.8}$$

$$[\mathbf{A}][\boldsymbol{\omega}] = [\dot{\mathbf{A}}], \tag{4.9}$$

one obtains the expression

$$[\mathbf{m}]\{\ddot{\mathbf{s}}\} + [\mathbf{A}][\mathbf{S}]^{\mathrm{T}}[\mathbf{Q}]\{\ddot{\boldsymbol{\beta}}\} + \left[[\mathbf{A}][\mathbf{S}]^{\mathrm{T}}[\dot{\mathbf{Q}}] + [\dot{\mathbf{A}}][\mathbf{S}]^{\mathrm{T}}[\mathbf{Q}]\right]\{\dot{\boldsymbol{\beta}}\} = \{\mathbf{F}_s\}. \tag{4.10}$$

Multiplying the *equation of the moment of momentum* (3.69) by $[\mathbf{Q}]^{\mathrm{T}}$ and taking into account the relation (4.4), one deduces the equality

$$[\mathbf{Q}]^{\mathrm{T}}[\mathbf{S}]\{\mathbf{a}_O\}+[\mathbf{Q}]^{\mathrm{T}}[\mathbf{J}_O]\{\boldsymbol{\varepsilon}\}+[\mathbf{Q}]^{\mathrm{T}}[\boldsymbol{\omega}][\mathbf{J}_O]\{\boldsymbol{\omega}\}=\{\mathbf{F}_{\boldsymbol{\beta}}\} \tag{4.11}$$

or

$$[\mathbf{Q}]^{\mathrm{T}}[\mathbf{S}][\mathbf{A}]^{\mathrm{T}}\{\ddot{\mathbf{s}}\}+[\mathbf{Q}]^{\mathrm{T}}[\mathbf{J}_O]\{\ddot{\boldsymbol{\beta}}\}+\left[[\mathbf{Q}]^{\mathrm{T}}[\mathbf{J}_O]\left[\dot{\mathbf{Q}}\right]+[\mathbf{Q}]^{\mathrm{T}}[\boldsymbol{\omega}][\mathbf{J}_O]\left[\dot{\mathbf{Q}}\right]\right]\{\dot{\boldsymbol{\beta}}\}=\{\mathbf{F}_{\boldsymbol{\beta}}\}. \tag{4.12}$$

Further on, using the notations

$$\left[\mathbf{M}_{\mathbf{q}}\right]=\begin{bmatrix}[\mathbf{m}] & [\mathbf{A}][\mathbf{S}]^{\mathrm{T}}[\mathbf{Q}]\\ [\mathbf{Q}]^{\mathrm{T}}[\mathbf{S}][\mathbf{A}]^{\mathrm{T}} & [\mathbf{Q}]^{\mathrm{T}}[\mathbf{J}_O][\mathbf{Q}]\end{bmatrix}, \tag{4.13}$$

$$\left\{\widetilde{\mathbf{F}}_{\mathbf{s}}\right\}=-\left[[\mathbf{A}][\mathbf{S}]^{\mathrm{T}}\left[\dot{\mathbf{Q}}\right]+\left[\dot{\mathbf{A}}\right][\mathbf{S}]^{\mathrm{T}}[\mathbf{Q}]\right]\{\dot{\boldsymbol{\beta}}\},\ \left\{\widetilde{\mathbf{F}}_{\boldsymbol{\beta}}\right\}=-\left[[\mathbf{Q}]^{\mathrm{T}}[\mathbf{J}_O]\left[\dot{\mathbf{Q}}\right]+[\mathbf{Q}]^{\mathrm{T}}[\boldsymbol{\omega}][\mathbf{J}_O][\mathbf{Q}]\right]\{\dot{\boldsymbol{\beta}}\}, \tag{4.14}$$

$$\left\{\mathbf{F}_{\mathbf{q}}\right\}=\begin{bmatrix}\{\mathbf{F}_{\mathbf{s}}\}\\ \{\mathbf{F}_{\boldsymbol{\beta}}\}\end{bmatrix},\ \left\{\widetilde{\mathbf{F}}_{\mathbf{q}}\right\}=\begin{bmatrix}\left\{\widetilde{\mathbf{F}}_{\mathbf{s}}\right\}\\ \left\{\widetilde{\mathbf{F}}_{\boldsymbol{\beta}}\right\}\end{bmatrix}, \tag{4.15}$$

the equations (4.9) and (4.11) reunite in the matrix differential equation

$$\left[\mathbf{M}_{\mathbf{q}}\right]\{\ddot{\mathbf{q}}\}=\left\{\mathbf{F}_{\mathbf{q}}\right\}+\left\{\widetilde{\mathbf{F}}_{\mathbf{q}}\right\}. \tag{4.16}$$

To be out to elaborate an algorithm of calculation it is necessary to recall the equalities

$$[\mathbf{A}]=[\boldsymbol{\psi}][\boldsymbol{\theta}][\boldsymbol{\varphi}],\ \left[\mathbf{A}_{\psi}\right]=\left[\mathbf{U}_{\psi}\right][\mathbf{A}],\ [\mathbf{A}_{\theta}]=[\boldsymbol{\psi}][\mathbf{U}_{\theta}][\boldsymbol{\psi}]^{\mathrm{T}}[\mathbf{A}],\ \left[\mathbf{A}_{\varphi}\right]=[\mathbf{A}]\left[\mathbf{U}_{\varphi}\right], \tag{4.17}$$

$$\left[\dot{\mathbf{A}}\right]=\dot{\psi}\left[\mathbf{A}_{\psi}\right]+\dot{\theta}[\mathbf{A}_{\theta}]+\dot{\varphi}\left[\mathbf{A}_{\varphi}\right],\ [\boldsymbol{\omega}]=[\mathbf{A}]^{\mathrm{T}}\left[\dot{\mathbf{A}}\right]. \tag{4.18}$$

4.2 The Lagrange Equations in the Case of the Holonomic Constraints

If one denotes by q_k, and F_{q_k}, $k=1, 2, \ldots, n$, the generalized coordinates, and forces, respectively, then the Lagrange equations for a mechanical system in motion writes in the form

$$\frac{\mathrm{d}}{\mathrm{d}t}\left(\frac{\partial T}{\partial \dot{q}}\right)-\frac{\partial T}{\partial q}=F_{q_k}, k=1, 2, \ldots, n, \tag{4.19}$$

where T is the kinetic energy, while F_{q_k} is the generalized force.

By using the matrix notations

$$\left\{\frac{\partial T}{\partial \dot{\mathbf{q}}}\right\} = \left[\frac{\partial T}{\partial \dot{q}_1}\ \frac{\partial T}{\partial \dot{q}_2}\ \cdots\ \frac{\partial T}{\partial \dot{q}_n}\right]^{\mathrm{T}},\ \left\{\frac{\partial T}{\partial \mathbf{q}}\right\} = \left[\frac{\partial T}{\partial q_1}\ \frac{\partial T}{\partial q_2}\ \cdots\ \frac{\partial T}{\partial q_n}\right]^{\mathrm{T}},$$

$$\{\mathbf{F_q}\} = [F_{q_1}\ F_{q_2}\ \ldots\ F_{q_n}]^{\mathrm{T}}, \tag{4.20}$$

the Lagrange equations take the form

$$\frac{\mathrm{d}}{\mathrm{dt}}\left\{\frac{\partial T}{\partial \dot{\mathbf{q}}}\right\} - \left\{\frac{\partial T}{\partial \mathbf{q}}\right\} = \{\mathbf{F_q}\}. \tag{4.21}$$

We have to mention that the generalized force $\{\mathbf{F_q}\}$ consists in both *given forces* and *constraint forces*.

Using the substitutions

$$\{\mathbf{v}_O\} = [\mathbf{A}]^{\mathrm{T}}\{\dot{\mathbf{s}}\},\ \{\boldsymbol{\omega}\} = [\mathbf{Q}]\{\dot{\boldsymbol{\beta}}\}, \tag{4.22}$$

where

$$\{\mathbf{s}\} = [X_O\ Y_O\ Z_O]^{\mathrm{T}},\ \{\boldsymbol{\beta}\} = [\psi\ \theta\ \varphi]^{\mathrm{T}},\ \{\mathbf{q}\} = [X_O\ Y_O\ Z_O\ \psi\ \theta\ \varphi]^{\mathrm{T}}, \tag{4.23}$$

and the notation (4.12), the kinetic energy of the rigid solid, defined by the relation (3.76), reads

$$T = \frac{1}{2}\{\dot{\mathbf{q}}\}^{\mathrm{T}}[\mathbf{M_q}]\{\dot{\mathbf{q}}\}. \tag{4.24}$$

It results

$$\left\{\frac{\partial T}{\partial \dot{\mathbf{q}}}\right\} = [\mathbf{M_q}]\{\dot{\mathbf{q}}\}, \tag{4.25}$$

$$\left\{\frac{\partial T}{\partial \mathbf{q}}\right\} = \frac{1}{2}\left[\{\dot{\mathbf{q}}\}^{\mathrm{T}}\frac{\partial[\mathbf{M_q}]}{\partial X_O}\{\dot{\mathbf{q}}\}\ \ \{\dot{\mathbf{q}}\}^{\mathrm{T}}\frac{\partial[\mathbf{M_q}]}{\partial Y_O}\{\dot{\mathbf{q}}\}\ \ldots\ \{\dot{\mathbf{q}}\}^{\mathrm{T}}\frac{\partial[\mathbf{M_q}]}{\partial \varphi}\{\dot{\mathbf{q}}\}\right], \tag{4.26}$$

and the equation (4.21) becomes

$$[\mathbf{M_q}]\{\ddot{\mathbf{q}}\} + [\dot{\mathbf{M}}_\mathbf{q}]\{\dot{\mathbf{q}}\} - \left\{\frac{\partial T}{\partial \mathbf{q}}\right\} = \{\mathbf{F_q}\}. \tag{4.27}$$

If one uses the notations

$$\left\{\tilde{\tilde{\mathbf{F}}}_{\mathbf{q}}\right\}=-\left[\dot{\mathbf{M}}_{\mathbf{q}}\right]\{\dot{\mathbf{q}}\}+\left\{\frac{\partial T}{\partial \mathbf{q}}\right\}, \tag{4.28}$$

then one obtains the equation

$$\left[\mathbf{M}_{\mathbf{q}}\right]\{\ddot{\mathbf{q}}\}=\left\{\mathbf{F}_{\mathbf{q}}\right\}+\left\{\tilde{\tilde{\mathbf{F}}}_{\mathbf{q}}\right\}. \tag{4.29}$$

The last relation is *identical to the equation* (4.16) if *we prove that*

$$\left\{\tilde{\mathbf{F}}_{\mathbf{q}}\right\}=\left\{\tilde{\tilde{\mathbf{F}}}_{\mathbf{q}}\right\}. \tag{4.30}$$

4.3 The Equivalence between the Differential Equations Obtained from the General Theorems and the Lagrange Equations

4.3.1 The Equivalence for the First Component

We consider the decomposition of the matrix $\left\{\tilde{\tilde{\mathbf{F}}}_{\mathbf{q}}\right\}$ in the components $\left\{\tilde{\tilde{\mathbf{F}}}_{\mathbf{s}}\right\}$, $\left\{\tilde{\tilde{\mathbf{F}}}_{\beta}\right\}$, $\left\{\tilde{\tilde{\mathbf{F}}}_{\mathbf{q}}\right\}=\left[\left\{\tilde{\tilde{\mathbf{F}}}_{\mathbf{s}}\right\}\ \left\{\tilde{\tilde{\mathbf{F}}}_{\beta}\right\}\right]^{\mathrm{T}}$; the proof of the equality (4.30) reduces to the proof of the equalities

$$\left\{\tilde{\mathbf{F}}_{\mathbf{s}}\right\}=\left\{\tilde{\tilde{\mathbf{F}}}_{\mathbf{s}}\right\},\ \left\{\tilde{\mathbf{F}}_{\beta}\right\}=\left\{\tilde{\tilde{\mathbf{F}}}_{\beta}\right\}. \tag{4.31}$$

For this goal, one considers the expression (3.75) for the kinetic energy; this expression may be written in the form

$$T=\frac{1}{2}m\left(\dot{X}_O^2+\dot{Y}_O^2+\dot{Z}_O^2\right)+\left[\dot{X}_O\ \ \dot{Y}_O\ \ \dot{Z}_O\right][\mathbf{A}][\mathbf{S}]^{\mathrm{T}}\{\boldsymbol{\omega}\}+\frac{1}{2}\{\boldsymbol{\omega}\}^{\mathrm{T}}[\mathbf{J}_O]\{\boldsymbol{\omega}\}; \tag{4.32}$$

it successively results

$$\frac{\partial T}{\partial \dot{X}_0}=m\dot{X}_O+[1\ \ 0\ \ 0][\mathbf{A}][\mathbf{S}]^{\mathrm{T}}[\mathbf{Q}]\{\dot{\boldsymbol{\beta}}\}, \tag{4.33}$$

$$\frac{\mathrm{d}}{\mathrm{dt}}\begin{bmatrix}\frac{\partial T}{\partial \dot{X}_O}\\ \frac{\partial T}{\partial \dot{Y}_O}\\ \frac{\partial T}{\partial \dot{Z}_O}\end{bmatrix} = [\mathbf{m}]\{\ddot{\mathbf{s}}\} + [\mathbf{A}][\mathbf{S}]^{\mathrm{T}}[\mathbf{Q}]\{\ddot{\boldsymbol{\beta}}\} + \left[[\dot{\mathbf{A}}][\mathbf{S}]^{\mathrm{T}}[\mathbf{Q}] + [\mathbf{A}][\mathbf{S}]^{\mathrm{T}}[\dot{\mathbf{Q}}]\right]\{\dot{\boldsymbol{\beta}}\}. \tag{4.34}$$

Since

$$\begin{bmatrix}\frac{\partial T}{\partial X_O} & \frac{\partial T}{\partial Y_O} & \frac{\partial T}{\partial Z_O}\end{bmatrix}^{\mathrm{T}} = \{\mathbf{0}\}, \tag{4.35}$$

one obtains the matrix equation

$$[\mathbf{m}]\{\ddot{\mathbf{s}}\} + [\mathbf{A}][\mathbf{S}]^{\mathrm{T}}[\mathbf{Q}]\{\ddot{\boldsymbol{\beta}}\} = \{\mathbf{F}_{\mathrm{s}}\} + \left\{\tilde{\tilde{\mathbf{F}}}_{\mathrm{s}}\right\}, \tag{4.36}$$

where

$$\left\{\tilde{\tilde{\mathbf{F}}}_{\mathrm{s}}\right\} = \left[[\dot{\mathbf{A}}][\mathbf{S}]^{\mathrm{T}}[\mathbf{Q}] + [\mathbf{A}][\mathbf{S}]^{\mathrm{T}}[\dot{\mathbf{Q}}]\right]\{\dot{\boldsymbol{\beta}}\}. \tag{4.37}$$

Comparing this expression to the first relation (4.14), it results that the first relation (4.31) is proved.

4.3.2 *The Equivalence for the Second Component*

The kinetic energy, given by the expression (4.32) may be written in the form

$$T = T_1 + T_2 + T_3, \tag{4.38}$$

where

$$T_1 = \frac{1}{2}m\{\dot{\mathbf{s}}\}^{\mathrm{T}}\{\dot{\mathbf{s}}\},\ T_2 = \{\boldsymbol{\omega}\}^{\mathrm{T}}[\mathbf{S}][\mathbf{A}]^{\mathrm{T}}\{\dot{\mathbf{s}}\},\ T_3 = \frac{1}{2}\{\boldsymbol{\omega}\}^{\mathrm{T}}[\mathbf{J}_O]\{\boldsymbol{\omega}\}. \tag{4.39}$$

Since the term T_1 does not depend on $\{\boldsymbol{\beta}\}$, it results

$$\left\{\tilde{\tilde{\mathbf{F}}}_{\boldsymbol{\beta}}\right\} = \left\{\tilde{\tilde{\mathbf{F}}}_{\boldsymbol{\beta}}^{(2)}\right\} + \left\{\tilde{\tilde{\mathbf{F}}}_{\boldsymbol{\beta}}^{(3)}\right\}, \tag{4.40}$$

where the component $\left\{\overset{\approx(2)}{\mathbf{F}_{\boldsymbol{\beta}}}\right\}$ is obtained in the case when the kinetic energy is equal to T_2, while the component $\left\{\overset{\approx(3)}{\mathbf{F}_{\boldsymbol{\beta}}}\right\}$ is obtained in the case when the kinetic energy is equal to T_3.

A. Calculation of the component $\left\{\overset{\approx(2)}{\mathbf{F}_{\boldsymbol{\beta}}}\right\}$

It successively results the relations

$$\begin{bmatrix} \dfrac{\partial T_2}{\partial \dot{\psi}} \\ \dfrac{\partial T_2}{\partial \dot{\theta}} \\ \dfrac{\partial T_2}{\partial \dot{\varphi}} \end{bmatrix} = \begin{bmatrix} \dfrac{\partial \{\boldsymbol{\omega}\}^{\mathrm{T}}}{\partial \dot{\psi}}[\mathbf{S}][\mathbf{A}]^{\mathrm{T}} \\ \dfrac{\partial \{\boldsymbol{\omega}\}^{\mathrm{T}}}{\partial \dot{\theta}}[\mathbf{S}][\mathbf{A}]^{\mathrm{T}} \\ \dfrac{\partial \{\boldsymbol{\omega}\}^{\mathrm{T}}}{\partial \dot{\varphi}}[\mathbf{S}][\mathbf{A}]^{\mathrm{T}} \end{bmatrix} \{\dot{\mathbf{s}}\} \tag{4.41}$$

or

$$\frac{\partial T_2}{\partial \dot{\boldsymbol{\beta}}} = [\mathbf{Q}]^{\mathrm{T}}[\mathbf{S}][\mathbf{A}]^{\mathrm{T}}\{\dot{\mathbf{s}}\}; \tag{4.42}$$

$$\frac{\mathrm{d}}{\mathrm{d}t}\left\{\frac{\partial T_2}{\partial \dot{\boldsymbol{\beta}}}\right\} = [\mathbf{Q}]^{\mathrm{T}}[\mathbf{S}][\mathbf{A}]^{\mathrm{T}}\{\ddot{\mathbf{s}}\} + \left[[\dot{\mathbf{Q}}][\mathbf{S}][\mathbf{A}]^{\mathrm{T}} + [\mathbf{Q}]^{\mathrm{T}}[\mathbf{S}][\dot{\mathbf{A}}]^{\mathrm{T}}\right]\{\dot{\mathbf{s}}\}; \tag{4.43}$$

$$\begin{bmatrix} \dfrac{\partial T_2}{\partial \psi} \\ \dfrac{\partial T_2}{\partial \theta} \\ \dfrac{\partial T_2}{\partial \varphi} \end{bmatrix} = \begin{bmatrix} \dfrac{\partial \{\boldsymbol{\omega}\}^{\mathrm{T}}}{\partial \psi}[\mathbf{S}][\mathbf{A}]^{\mathrm{T}} \\ \dfrac{\partial \{\boldsymbol{\omega}\}^{\mathrm{T}}}{\partial \theta}[\mathbf{S}][\mathbf{A}]^{\mathrm{T}} \\ \dfrac{\partial \{\boldsymbol{\omega}\}^{\mathrm{T}}}{\partial \varphi}[\mathbf{S}][\mathbf{A}]^{\mathrm{T}} \end{bmatrix} \{\dot{\mathbf{s}}\} + \begin{bmatrix} \{\boldsymbol{\omega}\}^{\mathrm{T}}[\mathbf{S}]\left[\mathbf{A}_{\psi}\right]^{\mathrm{T}} \\ \{\boldsymbol{\omega}\}^{\mathrm{T}}[\mathbf{S}][\mathbf{A}_{\theta}]^{\mathrm{T}} \\ \{\boldsymbol{\omega}\}^{\mathrm{T}}[\mathbf{S}]\left[\mathbf{A}_{\varphi}\right]^{\mathrm{T}} \end{bmatrix} \{\dot{\mathbf{s}}\}, \tag{4.44}$$

or

$$\left\{\frac{\partial T_2}{\partial \boldsymbol{\beta}}\right\} = [\tilde{\boldsymbol{\omega}}]^{\mathrm{T}}[\mathbf{S}][\mathbf{A}]^{\mathrm{T}}\{\dot{\mathbf{s}}\} + [\mathbf{D}]\{\dot{\mathbf{s}}\}, \tag{4.45}$$

where

$$[\mathbf{D}] = \begin{bmatrix} \{\boldsymbol{\omega}\}^{\mathrm{T}}[\mathbf{S}]\left[\mathbf{A}_{\psi}\right]^{\mathrm{T}} \\ \{\boldsymbol{\omega}\}^{\mathrm{T}}[\mathbf{S}][\mathbf{A}_{\theta}]^{\mathrm{T}} \\ \{\boldsymbol{\omega}\}^{\mathrm{T}}[\mathbf{S}]\left[\mathbf{A}_{\varphi}\right]^{\mathrm{T}} \end{bmatrix}. \tag{4.46}$$

Making the replacement

$$[\tilde{\boldsymbol{\omega}}]^{\mathrm{T}} = \left[\dot{\mathbf{Q}}\right]^{\mathrm{T}} - [\mathbf{Q}]^{\mathrm{T}}[\boldsymbol{\omega}], \tag{4.47}$$

it results

$$\left\{\frac{\partial T_2}{\partial \boldsymbol{\beta}}\right\} = \left[\dot{\mathbf{Q}}\right]^{\mathrm{T}}[\mathbf{S}][\mathbf{A}]^{\mathrm{T}}\{\dot{\mathbf{s}}\} - [\mathbf{Q}]^{\mathrm{T}}[\boldsymbol{\omega}][\mathbf{S}][\mathbf{A}]^{\mathrm{T}}\{\dot{\mathbf{s}}\} + [\mathbf{D}]\{\dot{\mathbf{s}}\}; \tag{4.48}$$

hence

$$\frac{\mathrm{d}}{\mathrm{d}t}\left\{\frac{\partial T_2}{\partial \dot{\boldsymbol{\beta}}}\right\} - \left\{\frac{\partial T_2}{\partial \boldsymbol{\beta}}\right\} = [\mathbf{Q}]^{\mathrm{T}}[\mathbf{S}][\mathbf{A}]^{\mathrm{T}}\{\ddot{\mathbf{s}}\} + \left\{\tilde{\tilde{\mathbf{F}}}_{\boldsymbol{\beta}}^{(2)}\right\}, \tag{4.49}$$

where

$$\left\{\tilde{\tilde{\mathbf{F}}}_{\boldsymbol{\beta}}^{(2)}\right\} = \left[[\mathbf{Q}]^{\mathrm{T}}[\mathbf{S}]\left[\dot{\mathbf{A}}\right]^{\mathrm{T}} - [\mathbf{Q}]^{\mathrm{T}}[\boldsymbol{\omega}][\mathbf{S}][\mathbf{A}]^{\mathrm{T}}\{\dot{\mathbf{s}}\} - [\mathbf{D}]\right]\{\dot{\mathbf{s}}\}. \tag{4.50}$$

Transposing the relation (4.50) and replacing the matrix $\left[\dot{\mathbf{A}}\right]$ by $[\mathbf{A}][\boldsymbol{\omega}]$, we get

$$\left\{\tilde{\tilde{\mathbf{F}}}_{\boldsymbol{\beta}}^{(2)}\right\}^{\mathrm{T}} = \{\dot{\mathbf{s}}\}^{\mathrm{T}}[\mathbf{A}]\left[[\mathbf{S}][\boldsymbol{\omega}] - [\boldsymbol{\omega}][\mathbf{S}][\mathbf{Q}] - [\mathbf{A}]^{\mathrm{T}}[\mathbf{D}]^{\mathrm{T}}\right]. \tag{4.51}$$

If we consider the vector product $\mathbf{s} \times \boldsymbol{\omega} = \mathbf{b}$, where $\mathbf{s}$ and $\boldsymbol{\omega}$ are the vectors associated to the matrices $[\mathbf{S}]$ and $[\boldsymbol{\omega}]$, respectively, and if we denote by $[\mathbf{B}]$ the skew matrix associated to the vector $\mathbf{b}$, then the following equality holds true

$$[\mathbf{B}] = [\mathbf{S}][\boldsymbol{\omega}] - [\boldsymbol{\omega}][\mathbf{S}]. \tag{4.52}$$

It also results the successive relations

$$\begin{aligned}[\mathbf{A}]^{\mathrm{T}}[\mathbf{D}]^{\mathrm{T}} &= -\left[[\mathbf{A}]^{\mathrm{T}}\left[\mathbf{A}_{\psi}\right][\mathbf{S}]\{\boldsymbol{\omega}\} \;\; [\mathbf{A}]^{\mathrm{T}}[\mathbf{A}_{\theta}][\mathbf{S}]\{\boldsymbol{\omega}\} \;\; [\mathbf{A}]^{\mathrm{T}}\left[\mathbf{A}_{\varphi}\right][\mathbf{S}]\{\boldsymbol{\omega}\}\right] \\ &= -\left[[\varphi]^{\mathrm{T}}[\boldsymbol{\theta}]^{\mathrm{T}}\left[\mathbf{U}_{\psi}\right][\boldsymbol{\theta}][\varphi][\mathbf{S}]\{\boldsymbol{\omega}\} \;\; [\varphi]^{\mathrm{T}}[\mathbf{U}_{\theta}][\varphi][\mathbf{S}]\{\boldsymbol{\omega}\} \;\; \left[\mathbf{U}_{\varphi}\right][\mathbf{S}]\{\boldsymbol{\omega}\}\right];\end{aligned} \tag{4.53}$$

recalling the properties of the skew symmetric matrices, given by the relations (1.80) and (1.82), one obtains

$$[\mathbf{A}]^{\mathrm{T}}[\mathbf{D}]^{\mathrm{T}} = [\mathbf{B}]\left[[\boldsymbol{\varphi}]^{\mathrm{T}}[\boldsymbol{\theta}]^{\mathrm{T}}\{\mathbf{U}_{\psi}\} \quad [\boldsymbol{\varphi}]^{\mathrm{T}}\{\mathbf{U}_{\theta}\} \quad \{\mathbf{U}_{\varphi}\} \right]. \tag{4.54}$$

Taking into account the equality

$$[\mathbf{Q}] = \left[[\boldsymbol{\varphi}]^{\mathrm{T}}[\boldsymbol{\theta}]^{\mathrm{T}}\{\mathbf{U}_{\psi}\} \quad [\boldsymbol{\varphi}]^{\mathrm{T}}\{\mathbf{U}_{\theta}\} \quad \{\mathbf{U}_{\varphi}\} \right], \tag{4.55}$$

it results

$$[\mathbf{A}]^{\mathrm{T}}[\mathbf{D}]^{\mathrm{T}} = [\mathbf{B}][\mathbf{Q}]; \tag{4.56}$$

consequently, the relation (4.51) becomes

$$\left\{ \overset{\approx(2)}{\mathbf{F}_{\boldsymbol{\beta}}} \right\}^{\mathrm{T}} = \{\dot{\mathbf{s}}\}^{\mathrm{T}}[\mathbf{A}][[\mathbf{B}][\mathbf{Q}] - [\mathbf{B}][\mathbf{Q}]], \tag{4.57}$$

that is

$$\left\{ \overset{\approx(2)}{\mathbf{F}_{\boldsymbol{\beta}}} \right\}^{\mathrm{T}} = \{\mathbf{0}\}. \tag{4.58}$$

B. Calculation of the component $\left\{ \overset{\approx(3)}{\mathbf{F}_{\boldsymbol{\beta}}} \right\}$

Analogically, in this case, one may successively write the relations

$$\begin{bmatrix} \dfrac{\partial T_3}{\partial \dot{\psi}} \\[2ex] \dfrac{\partial T_3}{\partial \dot{\theta}} \\[2ex] \dfrac{\partial T_3}{\partial \dot{\varphi}} \end{bmatrix} = \begin{bmatrix} \dfrac{\partial\{\boldsymbol{\omega}\}^{\mathrm{T}}}{\partial \dot{\psi}}[\mathbf{J}_O]\{\boldsymbol{\omega}\} + \{\boldsymbol{\omega}\}^{\mathrm{T}}[\mathbf{J}_O]\dfrac{\partial\{\boldsymbol{\omega}\}}{\partial \dot{\psi}} \\[2ex] \dfrac{\partial\{\boldsymbol{\omega}\}^{\mathrm{T}}}{\partial \dot{\theta}}[\mathbf{J}_O]\{\boldsymbol{\omega}\} + \{\boldsymbol{\omega}\}^{\mathrm{T}}[\mathbf{J}_O]\dfrac{\partial\{\boldsymbol{\omega}\}}{\partial \dot{\theta}} \\[2ex] \dfrac{\partial\{\boldsymbol{\omega}\}^{\mathrm{T}}}{\partial \dot{\varphi}}[\mathbf{J}_O]\{\boldsymbol{\omega}\} + \{\boldsymbol{\omega}\}^{\mathrm{T}}[\mathbf{J}_O]\dfrac{\partial\{\boldsymbol{\omega}\}}{\partial \dot{\varphi}} \end{bmatrix} = \begin{bmatrix} \dfrac{\partial\{\boldsymbol{\omega}\}^{\mathrm{T}}}{\partial \dot{\psi}}[\mathbf{J}_O]\{\boldsymbol{\omega}\} \\[2ex] \dfrac{\partial\{\boldsymbol{\omega}\}^{\mathrm{T}}}{\partial \dot{\theta}}[\mathbf{J}_O]\{\boldsymbol{\omega}\} \\[2ex] \dfrac{\partial\{\boldsymbol{\omega}\}^{\mathrm{T}}}{\partial \dot{\varphi}}[\mathbf{J}_O]\{\boldsymbol{\omega}\} \end{bmatrix} \tag{4.59}$$

or

$$\left\{\frac{\partial T_3}{\partial \boldsymbol{\beta}}\right\} = [\mathbf{Q}]^{\mathrm{T}}[\mathbf{J}_O][\mathbf{Q}]\{\dot{\mathbf{q}}_2\}; \tag{4.60}$$

$$\frac{\mathrm{d}}{\mathrm{d}t}\left\{\frac{\partial T_3}{\partial \dot{\boldsymbol{\beta}}}\right\} = [\mathbf{Q}]^{\mathrm{T}}[\mathbf{J}_O][\mathbf{Q}]\{\ddot{\boldsymbol{\beta}}\} + \left[[\dot{\mathbf{Q}}][\mathbf{J}_O][\mathbf{Q}] + [\mathbf{Q}]^{\mathrm{T}}[\mathbf{J}_O][\dot{\mathbf{Q}}]\right]\{\dot{\boldsymbol{\beta}}\}; \tag{4.61}$$

$$\begin{bmatrix} \dfrac{\partial T_3}{\partial \psi} \\ \dfrac{\partial T_3}{\partial \theta} \\ \dfrac{\partial T_3}{\partial \varphi} \end{bmatrix} = \frac{1}{2}\begin{bmatrix} \dfrac{\partial\{\boldsymbol{\omega}\}^{\mathrm{T}}}{\partial \psi}[\mathbf{J}_O]\{\boldsymbol{\omega}\} + \{\boldsymbol{\omega}\}^{\mathrm{T}}[\mathbf{J}_O]\dfrac{\partial\{\boldsymbol{\omega}\}}{\partial \psi} \\ \dfrac{\partial\{\boldsymbol{\omega}\}^{\mathrm{T}}}{\partial \theta}[\mathbf{J}_O]\{\boldsymbol{\omega}\} + \{\boldsymbol{\omega}\}^{\mathrm{T}}[\mathbf{J}_O]\dfrac{\partial\{\boldsymbol{\omega}\}}{\partial \theta} \\ \dfrac{\partial\{\boldsymbol{\omega}\}^{\mathrm{T}}}{\partial \varphi}[\mathbf{J}_O]\{\boldsymbol{\omega}\} + \{\boldsymbol{\omega}\}^{\mathrm{T}}[\mathbf{J}_O]\dfrac{\partial\{\boldsymbol{\omega}\}}{\partial \varphi} \end{bmatrix} = \frac{1}{2}\begin{bmatrix} \dfrac{\partial\{\boldsymbol{\omega}\}^{\mathrm{T}}}{\partial \psi}[\mathbf{J}_O]\{\boldsymbol{\omega}\} \\ \dfrac{\partial\{\boldsymbol{\omega}\}^{\mathrm{T}}}{\partial \theta}[\mathbf{J}_O]\{\boldsymbol{\omega}\} \\ \dfrac{\partial\{\boldsymbol{\omega}\}^{\mathrm{T}}}{\partial \varphi}[\mathbf{J}_O]\{\boldsymbol{\omega}\} \end{bmatrix} \tag{4.62}$$

or

$$\left\{\frac{\partial T_3}{\partial \boldsymbol{\beta}}\right\} = [\widetilde{\boldsymbol{\omega}}]^{\mathrm{T}}[\mathbf{J}_O][\mathbf{Q}]\{\dot{\boldsymbol{\beta}}\}. \tag{4.63}$$

In the conditions of the relation (4.47) the expression (4.63) becomes

$$\left\{\frac{\partial T_3}{\partial \boldsymbol{\beta}}\right\} = [\dot{\mathbf{Q}}]^{\mathrm{T}}[\mathbf{J}_O][\mathbf{Q}]\{\dot{\boldsymbol{\beta}}\} - [\mathbf{Q}]^{\mathrm{T}}[\boldsymbol{\omega}][\mathbf{J}_O][\mathbf{Q}]\{\dot{\boldsymbol{\beta}}\}, \tag{4.64}$$

and one deduces the equality

$$\frac{\mathrm{d}}{\mathrm{d}t}\left\{\frac{\partial T_3}{\partial \dot{\boldsymbol{\beta}}}\right\} - \left\{\frac{\partial T_3}{\partial \boldsymbol{\beta}}\right\} = [\mathbf{Q}]^{\mathrm{T}}[\mathbf{J}_O][\mathbf{Q}]\{\ddot{\boldsymbol{\beta}}\} + \left[[\mathbf{Q}]^{\mathrm{T}}[\mathbf{J}_O][\dot{\mathbf{Q}}] + [\mathbf{Q}]^{\mathrm{T}}[\boldsymbol{\omega}][\mathbf{J}_O][\mathbf{Q}]\right]\{\dot{\boldsymbol{\beta}}\}, \tag{4.65}$$

from which it results

$$\left\{\widetilde{\widetilde{\mathbf{F}}}_{\boldsymbol{\beta}}^{(3)}\right\} = \left\{\widetilde{\widetilde{\mathbf{F}}}_{\boldsymbol{\beta}}\right\} = -\left[[\mathbf{Q}]^{\mathrm{T}}[\mathbf{J}_O][\dot{\mathbf{Q}}] + [\mathbf{Q}]^{\mathrm{T}}[\boldsymbol{\omega}][\mathbf{J}_O][\mathbf{Q}]\right]\{\dot{\boldsymbol{\beta}}\} = \left\{\widetilde{\mathbf{F}}_{\boldsymbol{\beta}}\right\}; \tag{4.66}$$

hence, the second equality (4.31) is proved too.

4.4 The Matrix Differential Equations for the Motion of the Constrained Rigid Solid

4.4.1 The Matrix of Constraints

Generally speaking, if the rigid solid has *p constraints*, then between the kinematic parameters there exist *p* functions of constraint.

In the case of the *holonomic constraints* these functions may be written in the form

$$f_i(X_O, Y_O, Z_O, \psi, \theta, \varphi) = 0,\ i = 1, 2, \ldots, p, \tag{4.67}$$

or

$$f_i(q_1, q_2, q_3, q_4, q_5, q_6) = 0,\ i = 1, 2, \ldots, p. \tag{4.68}$$

By derivation the equations (4.68) with respect to time, it results

$$\sum_{k=1}^{6} \frac{\partial f_i}{\partial q_k} \dot{q}_k = 0,\ i = 1, 2, \ldots, p; \tag{4.69}$$

if one uses the notation

$$[\mathbf{B}] = \frac{D(f_1, f_2, \ldots, f_p)}{D(q_1, q_2, \ldots, q_6)} = \begin{bmatrix} \frac{\partial f_1}{\partial q_1} & \frac{\partial f_1}{\partial q_2} & \cdots & \frac{\partial f_1}{\partial q_6} \\ \frac{\partial f_2}{\partial q_1} & \frac{\partial f_2}{\partial q_2} & \cdots & \frac{\partial f_2}{\partial q_6} \\ \cdots & \cdots & \cdots & \cdots \\ \frac{\partial f_p}{\partial q_1} & \frac{\partial f_p}{\partial q_2} & \cdots & \frac{\partial f_p}{\partial q_6} \end{bmatrix}, \tag{4.70}$$

for the *Jacobi matrix*, [**B**] being called *the matrix of constraints*, then one obtains the equality

$$[\mathbf{B}]\{\dot{\mathbf{q}}\} = \{\mathbf{0}\}. \tag{4.71}$$

In the more general case, the matrix relation of constraints reads

$$[\mathbf{B}]\{\dot{\mathbf{q}}\} = \{\mathbf{C}\}, \tag{4.72}$$

where the column matrix $\{\mathbf{C}\}$ depends on the coordinates q_i, $i = 1, 2, \ldots, 6$.

The case of *the non-holonomic constraints* directly leads to the existence of the matrix relation (4.72), relation which is not *integrable.*

Example 4.4.1 Determine the matrix of constraints for a rigid solid knowing that its point A ($x_A = l$, $y_A = 0$, $z_A = 0$) is situated in the plan $Z = 0$, while its point B ($x_B = -l$, $y_B = z_B = 0$) is situated in the plan $X = 0$.

Solution: Choosing the Euler angles as parameters, it results

$$[\mathbf{A}] = \begin{bmatrix} \mathrm{c}\psi\mathrm{c}\varphi - \mathrm{s}\psi\mathrm{c}\theta\mathrm{s}\varphi & -\mathrm{c}\psi\mathrm{s}\varphi - \mathrm{s}\psi\mathrm{c}\theta\mathrm{c}\varphi & \mathrm{s}\psi\mathrm{s}\theta \\ \mathrm{s}\psi\mathrm{c}\varphi + \mathrm{c}\psi\mathrm{c}\theta\mathrm{s}\varphi & -\mathrm{s}\psi\mathrm{s}\varphi + \mathrm{c}\psi\mathrm{c}\theta\mathrm{c}\varphi & -\mathrm{c}\psi\mathrm{s}\theta \\ \mathrm{s}\theta\mathrm{s}\varphi & \mathrm{s}\theta\mathrm{c}\varphi & \mathrm{c}\theta \end{bmatrix};$$

from the matrix relations

$$\begin{bmatrix} X_A \\ Y_A \\ 0 \end{bmatrix} = \begin{bmatrix} X_O \\ Y_O \\ Z_O \end{bmatrix} + [\mathbf{A}]\begin{bmatrix} l \\ 0 \\ 0 \end{bmatrix}, \begin{bmatrix} 0 \\ Y_B \\ Z_B \end{bmatrix} = \begin{bmatrix} X_O \\ Y_O \\ Z_O \end{bmatrix} + [\mathbf{A}]\begin{bmatrix} -l \\ 0 \\ 0 \end{bmatrix}$$

one obtains the constraint functions

$$Z_O + l\sin\theta\sin\varphi = 0, X_O - l(\cos\psi\cos\varphi - \sin\psi\cos\theta\sin\varphi) = 0,$$

while the matrix of constraints reads

$$[\mathbf{B}] = \begin{bmatrix} 0 & 0 & 1 & 0 & l\mathrm{c}\theta\mathrm{s}\varphi & l\mathrm{s}\theta\mathrm{c}\varphi \\ 1 & 0 & 0 & l(\mathrm{s}\psi\mathrm{c}\varphi + \mathrm{c}\psi\mathrm{c}\theta\mathrm{s}\varphi) & -l\mathrm{s}\psi\mathrm{s}\theta\mathrm{s}\varphi & l(\mathrm{c}\psi\mathrm{s}\varphi + \mathrm{s}\psi\mathrm{c}\theta\mathrm{c}\varphi) \end{bmatrix}.$$

Example 4.4.2 Determine the matrix of constraints and the equation of constraints for the planar bar of length $2l$ in Fig. 4.1, knowing that the point A moves with constant velocity u_0 on the circle of radius R, while the point B moves on the O_0X axis; the origin of the mobile system is situated at the middle of the bar.

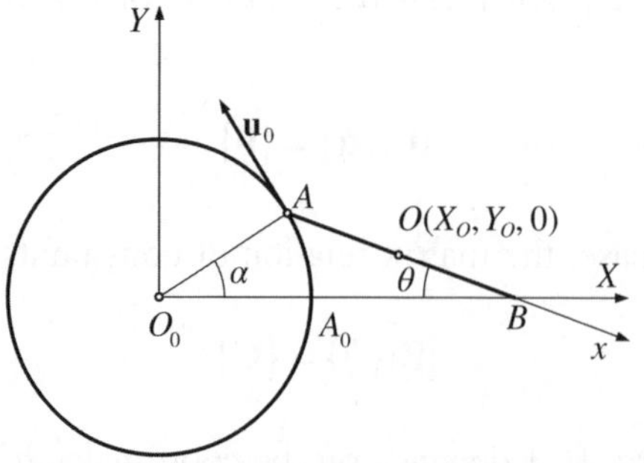

Figure 4.1 Determination the matrix of constraints and the equation of constraints for the planar bar in Example 4.4.2.

Solution: The kinematic parameters are X_O, Y_O, θ. Using the notation $\omega = \frac{u_0}{R}$, it results the relations

$$X_O - l\cos\theta = R\cos\omega t,\ Y_O + l\sin\theta = R\sin\omega t,\ Y_O - l\sin\theta = 0,$$

by derivation of which one obtains the matrix equation of constraints

$$\begin{bmatrix} 1 & 0 & l\sin\theta \\ 0 & 1 & l\cos\theta \\ 0 & 1 & -l\cos\theta \end{bmatrix} \begin{bmatrix} \dot{X}_O \\ \dot{Y}_O \\ \dot{\theta} \end{bmatrix} = \begin{bmatrix} -R\omega\sin\omega t \\ R\omega\cos\omega t \\ 0 \end{bmatrix};$$

hence

$$[\mathbf{B}] = \begin{bmatrix} 1 & 0 & l\sin\theta \\ 0 & 1 & l\cos\theta \\ 0 & 1 & -l\cos\theta \end{bmatrix},\ \{\mathbf{C}\} = \begin{bmatrix} -R\omega\sin\omega t \\ R\omega\cos\omega t \\ 0 \end{bmatrix}.$$

4.4.2 The Lagrange Equations for Mechanical Systems with Constraints

We consider the same notations: q_k, $k = 1, 2, \ldots, n$, for the generalized coordinates, T for the kinetic energy, and F_k, $k = 1, 2, \ldots, n$, for the generalized forces corresponding to *the given forces effectively acting upon the system* (we do not include the constraint forces). If the constraints are defined by the equation

$$[\mathbf{B}]\{\dot{\mathbf{q}}\} = \{\mathbf{C}\}, \tag{4.73}$$

where

$$[\mathbf{B}] = \begin{bmatrix} B_{11} & B_{12} & \ldots & B_{1n} \\ B_{21} & B_{22} & \ldots & B_{2n} \\ \ldots & \ldots & \ldots & \ldots \\ B_{p1} & B_{p2} & \ldots & B_{pn} \end{bmatrix} \tag{4.74}$$

is *the matrix of constraints*, then the Lagrange equations for mechanical systems with constraints read

$$\frac{\mathrm{d}}{\mathrm{dt}}\left(\frac{\partial T}{\partial \dot{q}_k}\right) - \frac{\partial T}{\partial q_k} = F_{q_k} + \sum_{i=1}^{p} B_{ik}\lambda_i,\ k = 1, 2, \ldots, n, \tag{4.75}$$

λ_1, λ_2, …, λ_p being *the Lagrange multipliers*. Using the notations

$$\{\boldsymbol{\lambda}\} = [\lambda_1\ \ \lambda_2\ \ \ldots\ \ \lambda_p]^{\mathrm{T}} \tag{4.76}$$

for the matrix of the Lagrange multipliers, and the notations (4.20), from the expression (4.75) one deduces *the matrix form of the Lagrange equations*

$$\frac{\mathrm{d}}{\mathrm{dt}}\left\{\frac{\partial T}{\partial \dot{\mathbf{q}}_k}\right\}-\left\{\frac{\partial T}{\partial \mathbf{q}_k}\right\}=\{\mathbf{F}_\mathbf{q}\}+[\mathbf{B}]^\mathrm{T}\{\boldsymbol{\lambda}\}. \tag{4.77}$$

In the most general case, of *the rheonomic constraints*, the vector of position $\mathbf{r}_i$ of the particle A_i writes in the form

$$\mathbf{r}_i=\mathbf{r}_i(q_1,q_2,\ldots,q_n,t), \tag{4.78}$$

while the velocity $\mathbf{v}_i$ is

$$\mathbf{v}_i=\sum_{k=1}^{n}\frac{\partial \mathbf{r}_i}{\partial q_k}\dot{q}_k+\frac{\partial \mathbf{r}_i}{\partial t}. \tag{4.79}$$

With the notations

$$\left[\mathbf{M}^{(ii)}\right]=m_i\begin{bmatrix}\left(\dfrac{\partial \mathbf{r}_i}{\partial q_1}\right)^2 & \dfrac{\partial \mathbf{r}_i}{\partial q_1}\dfrac{\partial \mathbf{r}_i}{\partial q_2} & \cdots & \dfrac{\partial \mathbf{r}_i}{\partial q_1}\dfrac{\partial \mathbf{r}_i}{\partial q_n}\\ \dfrac{\partial \mathbf{r}_i}{\partial q_2}\dfrac{\partial \mathbf{r}_i}{\partial q_1} & \left(\dfrac{\partial \mathbf{r}_i}{\partial q_2}\right)^2 & \cdots & \dfrac{\partial \mathbf{r}_i}{\partial q_2}\dfrac{\partial \mathbf{r}_i}{\partial q_n}\\ \cdots & \cdots & \cdots & \cdots\\ \dfrac{\partial \mathbf{r}_i}{\partial q_n}\dfrac{\partial \mathbf{r}_i}{\partial q_1} & \dfrac{\partial \mathbf{r}_i}{\partial q_n}\dfrac{\partial \mathbf{r}_i}{\partial q_2} & \cdots & \left(\dfrac{\partial \mathbf{r}_i}{\partial q_n}\right)^2\end{bmatrix}, \tag{4.80}$$

$$\left\{\mathbf{M}^{(i1)}\right\}=m_i\begin{bmatrix}\dfrac{\partial \mathbf{r}_i}{\partial t}\dfrac{\partial \mathbf{r}_i}{\partial q_1} & \dfrac{\partial \mathbf{r}_i}{\partial t}\dfrac{\partial \mathbf{r}_i}{\partial q_2} & \cdots & \dfrac{\partial \mathbf{r}_i}{\partial t}\dfrac{\partial \mathbf{r}_i}{\partial q_n}\end{bmatrix}^\mathrm{T}, \tag{4.81}$$

$$[\mathbf{M}]=\sum_{i=1}^{n}\left[\mathbf{M}^{(ii)}\right],\ \left\{\mathbf{M}^{(1)}\right\}^\mathrm{T}=\sum_{i=1}^{n}\left\{\mathbf{M}^{(i1)}\right\}^\mathrm{T},\ M^{(0)}=\frac{1}{2}\sum_{i=1}^{n}m_i\left(\frac{\partial \mathbf{r}_i}{\partial t}\right)^2, \tag{4.82}$$

one obtains the expression of the kinetic energy of the system

$$T=\frac{1}{2}\{\dot{\mathbf{q}}\}^\mathrm{T}[\mathbf{M}]\{\dot{\mathbf{q}}\}+\left\{\mathbf{M}^{(1)}\right\}^\mathrm{T}\{\dot{\mathbf{q}}\}+M^{(0)}. \tag{4.83}$$

It successively results the relations

$$\left\{\frac{\partial T}{\partial \dot{\mathbf{q}}}\right\}=[\mathbf{M}]\{\dot{\mathbf{q}}\}+\left\{\mathbf{M}^{(1)}\right\}, \tag{4.84}$$

$$\frac{\mathrm{d}}{\mathrm{dt}}\left\{\frac{\partial T}{\partial \dot{\mathbf{q}}}\right\} = [\mathbf{M}]\{\ddot{\mathbf{q}}\} + [\dot{\mathbf{M}}]\{\dot{\mathbf{q}}\} + \left\{\dot{\mathbf{M}}^{(1)}\right\}^{\mathrm{T}}, \tag{4.85}$$

$$\left\{\frac{\partial T}{\partial \mathbf{q}}\right\} = \frac{1}{2}\begin{bmatrix} \{\dot{\mathbf{q}}\}^{\mathrm{T}}\dfrac{\partial[\mathbf{M}]}{\partial q_1}\{\dot{\mathbf{q}}\} \\ \{\dot{\mathbf{q}}\}^{\mathrm{T}}\dfrac{\partial[\mathbf{M}]}{\partial q_2}\{\dot{\mathbf{q}}\} \\ \cdots \\ \{\dot{\mathbf{q}}\}^{\mathrm{T}}\dfrac{\partial[\mathbf{M}]}{\partial q_n}\{\dot{\mathbf{q}}\} \end{bmatrix} + \begin{bmatrix} \dfrac{\partial\{\mathbf{M}^{(1)}\}}{\partial q_1}\{\dot{\mathbf{q}}\} \\ \dfrac{\partial\{\mathbf{M}^{(1)}\}}{\partial q_2}\{\dot{\mathbf{q}}\} \\ \cdots \\ \dfrac{\partial\{\mathbf{M}^{(1)}\}}{\partial q_n}\{\dot{\mathbf{q}}\} \end{bmatrix} + \begin{bmatrix} \dfrac{\partial M^{(0)}}{\partial q_1} \\ \dfrac{\partial M^{(0)}}{\partial q_2} \\ \cdots \\ \dfrac{\partial M^{(0)}}{\partial q_n} \end{bmatrix}; \tag{4.86}$$

if one makes the notation

$$\left\{\tilde{\mathbf{F}}_{\mathbf{q}}\right\} = -[\dot{\mathbf{M}}]\{\dot{\mathbf{q}}\} - \left\{\dot{\mathbf{M}}^{(1)}\right\} + \left\{\frac{\partial T}{\partial \mathbf{q}}\right\}, \tag{4.87}$$

then from the relation (4.77) one deduces the matrix differential equation

$$[\mathbf{M}]\{\ddot{\mathbf{q}}\} = \{\mathbf{F}_{\mathbf{q}}\} + \left\{\tilde{\mathbf{F}}_{\mathbf{q}}\right\} + [\mathbf{B}]^{\mathrm{T}}\{\boldsymbol{\lambda}\}. \tag{4.88}$$

If we associate the derivative with respect to time of the equation (4.73) to the equation (4.88),

$$[\mathbf{B}]\{\ddot{\mathbf{q}}\} + [\dot{\mathbf{B}}]\{\dot{\mathbf{q}}\} = \{\dot{\mathbf{C}}\}, \tag{4.89}$$

then from the relations (4.88) and (4.89) we obtain the final matrix equation written in the general form for every mechanical system

$$\begin{bmatrix} [\mathbf{M}] & -[\mathbf{B}]^{\mathrm{T}} \\ [\mathbf{B}] & [\mathbf{0}] \end{bmatrix}\begin{bmatrix} \{\ddot{\mathbf{q}}\} \\ \{\boldsymbol{\lambda}\} \end{bmatrix} = \begin{bmatrix} \{\mathbf{F}_{\mathbf{q}}\} + \left\{\tilde{\mathbf{F}}_{\mathbf{q}}\right\} \\ \{\dot{\mathbf{C}}\} - [\dot{\mathbf{B}}]\{\dot{\mathbf{q}}\} \end{bmatrix}. \tag{4.90}$$

In the case of the *scleronomic* systems, the relations (4.83)–(4.87) become simpler because $\left\{\mathbf{M}^{(1)}\right\} = \{\mathbf{0}\}$ and $M^{(0)} = 0$.

4.4.3 The Mathematical Model of the Motion of Rigid Solid with Constraints

Since the Lagrange equations are equivalent to those obtained from the general theorems, for the study of the motion of the rigid solid we may use the equation (4.16) in which we separate the vectors of the generalized forces, and of constraint forces, respectively; we thus replace in the right-hand term the global vector $\{\mathbf{F_q}\}$ of the given generalized forces and of the constraint forces, with the sum of the vector of *the given generalized forces denoted by* $\{\mathbf{F_q}\}$ and of the vector of *the constraint forces which is given in the relation* (4.77) *in the form* $[\mathbf{B}]^{\mathrm{T}}\{\boldsymbol{\lambda}\}$.

Hence, it results the matrix equation

$$\begin{bmatrix} [\mathbf{M}] & -[\mathbf{B}]^{\mathrm{T}} \\ [\mathbf{B}] & [\mathbf{0}] \end{bmatrix} \begin{bmatrix} \{\ddot{\mathbf{q}}\} \\ \{\boldsymbol{\lambda}\} \end{bmatrix} = \begin{bmatrix} \{\mathbf{F_q}\} + \{\widetilde{\mathbf{F}}_\mathbf{q}\} \\ \{\dot{\mathbf{C}}\} - [\dot{\mathbf{B}}]\{\dot{\mathbf{q}}\} \end{bmatrix}; \tag{4.91}$$

in the previous equation one uses the following notations and equalities

$$\{\mathbf{s}\} = [X_O \ \ Y_O \ \ Z_O]^{\mathrm{T}}, \{\boldsymbol{\beta}\} = [\psi \ \ \theta \ \ \varphi]^{\mathrm{T}}, \{\mathbf{q}\} = [X_O \ \ Y_O \ \ Z_O \ \ \psi \ \ \theta \ \ \varphi]^{\mathrm{T}}, \tag{4.92}$$

$$[\mathbf{M}] = \begin{bmatrix} [\mathbf{m}] & [\mathbf{A}][\mathbf{S}]^{\mathrm{T}}[\mathbf{Q}] \\ [\mathbf{Q}]^{\mathrm{T}}[\mathbf{S}][\mathbf{A}]^{\mathrm{T}} & [\mathbf{Q}]^{\mathrm{T}}[\mathbf{J}_O][\mathbf{Q}] \end{bmatrix}, \tag{4.93}$$

$$\{\widetilde{\mathbf{F}}_\mathbf{s}\} = -\left[[\mathbf{A}][\mathbf{S}]^{\mathrm{T}}[\dot{\mathbf{Q}}] + [\dot{\mathbf{A}}][\mathbf{S}]^{\mathrm{T}}[\mathbf{Q}]\right]\{\dot{\boldsymbol{\beta}}\}, \tag{4.94}$$

$$\{\widetilde{\mathbf{F}}_{\boldsymbol{\beta}}\} = -\left[[\mathbf{Q}]^{\mathrm{T}}[\mathbf{J}_O][\dot{\mathbf{Q}}] + [\mathbf{Q}]^{\mathrm{T}}[\boldsymbol{\omega}][\mathbf{J}_O][\mathbf{Q}]\right]\{\dot{\boldsymbol{\beta}}\}, \tag{4.95}$$

$$\{\widetilde{\mathbf{F}}_\mathbf{q}\} = \begin{bmatrix} \{\widetilde{\mathbf{F}}_\mathbf{s}\} \\ \{\widetilde{\mathbf{F}}_{\boldsymbol{\beta}}\} \end{bmatrix}. \tag{4.96}$$

The matrix equation (4.91) shows that the general equation of the rigid solid *differs from one case to other only by the constraints matrices*; hence, the determination of this matrix has a special role in the study of the motion for a rigid solid.

4.4.4 General Algorithm of Calculation

One consider as known:

- the mechanical and geometric constants

$$m, [\mathbf{J}_O] = \begin{bmatrix} J_x & -J_{xy} & -J_{xz} \\ -J_{yx} & J_y & -J_{yz} \\ -J_{zx} & -J_{zy} & J_z \end{bmatrix}, [\mathbf{S}] = \begin{bmatrix} 0 & -mz_C & my_C \\ mz_C & 0 & -mx_C \\ -my_C & mx_C & 0 \end{bmatrix}; \tag{4.97}$$

- the matrix of constraints [**B**];
- the variant of rotation about the axes;
- the projections F_X, F_Y, and F_Z of the resultant of forces that act upon the rigid solid onto *the fixed axes* O_0X, O_0Y, and O_0Z, respectively;
- the projections M_x, M_y, and M_z of the resultant moment at the point O of the given forces that act upon the rigid solid onto *the mobile axes* Ox, Oy, and Oz, respectively;
- the initial values (at $t=0$) of the parameters $X_O, Y_O, Z_O, \psi, \theta, \varphi, \dot{X}_O, \dot{Y}_O, \dot{Z}_O, \dot{\psi}, \dot{\theta}, \dot{\varphi}$, that is, the initial values of the column matrices $\{\mathbf{q}\}$ and $\{\dot{\mathbf{q}}\}$.

One successively calculates:

- the matrices [$\boldsymbol{\psi}$], [$\boldsymbol{\theta}$], [$\boldsymbol{\varphi}$] defined by *the chosen variant* of rotations about the axes, and the derivatives with respect to time of the matrices $[\dot{\mathbf{B}}]$, and $\{\dot{\mathbf{C}}\}$,

$$[\mathbf{A}] = [\boldsymbol{\psi}][\boldsymbol{\theta}][\boldsymbol{\varphi}], \tag{4.98}$$

$$[\boldsymbol{\psi}_p] = [\mathbf{U}_\psi][\boldsymbol{\psi}],\ [\boldsymbol{\theta}_p] = [\mathbf{U}_\theta][\boldsymbol{\theta}],\ [\boldsymbol{\varphi}_p] = [\mathbf{U}_\varphi][\boldsymbol{\varphi}], \tag{4.99}$$

$$[\mathbf{A}_\psi] = [\boldsymbol{\psi}_p][\boldsymbol{\theta}][\boldsymbol{\varphi}],\ [\mathbf{A}_\theta] = [\boldsymbol{\psi}][\boldsymbol{\theta}_p][\boldsymbol{\varphi}],\ [\mathbf{A}_\varphi] = [\boldsymbol{\psi}][\boldsymbol{\theta}][\boldsymbol{\varphi}_p], \tag{4.100}$$

$$[\dot{\mathbf{A}}] = \dot{\psi}[\mathbf{A}_\psi] + \dot{\theta}[\mathbf{A}_\theta] + \dot{\varphi}[\mathbf{A}_\varphi], \tag{4.101}$$

$$[\boldsymbol{\omega}] = \dot{\psi}[\boldsymbol{\varphi}]^{\mathrm{T}}[\boldsymbol{\theta}]^{\mathrm{T}}[\mathbf{U}_\psi][\boldsymbol{\theta}][\boldsymbol{\varphi}] + \dot{\theta}[\boldsymbol{\varphi}]^{\mathrm{T}}[\mathbf{U}_\theta][\boldsymbol{\varphi}] + \dot{\varphi}[\mathbf{U}_\varphi], \tag{4.102}$$

$$[\mathbf{Q}] = [\boldsymbol{\varphi}]^{\mathrm{T}}\left[[\boldsymbol{\theta}]^{\mathrm{T}}\{\mathbf{u}_\psi\}\ \ \{\mathbf{u}_\theta\}\ \ \{\mathbf{u}_\varphi\}\right], \tag{4.103}$$

$$[\mathbf{Q}_\varphi] = [\boldsymbol{\varphi}_p]^{\mathrm{T}}\left[[\boldsymbol{\theta}]^{\mathrm{T}}\{\mathbf{u}_\psi\}\ \ \{\mathbf{u}_\theta\}\ \ \{\mathbf{u}_\varphi\}\right], [\mathbf{Q}_\theta] = [\boldsymbol{\varphi}]^{\mathrm{T}}\left[[\boldsymbol{\theta}_p]^{\mathrm{T}}\{\mathbf{u}_\psi\}\ \ \{\mathbf{0}\}\ \ \{\mathbf{0}\}\right], \tag{4.104}$$

$$[\dot{\mathbf{Q}}] = \dot{\varphi}[\mathbf{Q}_\varphi] + \dot{\theta}[\mathbf{Q}_\theta], \tag{4.105}$$

$$[\mathbf{M}] = \begin{bmatrix} [\mathbf{m}] & [\mathbf{A}][\mathbf{S}]^{\mathrm{T}}[\mathbf{Q}] \\ [\mathbf{Q}]^{\mathrm{T}}[\mathbf{S}][\mathbf{A}]^{\mathrm{T}} & [\mathbf{Q}]^{\mathrm{T}}[\mathbf{J}_O][\mathbf{Q}] \end{bmatrix}, \tag{4.106}$$

$$\{\mathbf{F}_\mathbf{s}\} = [F_x\ \ F_y\ \ F_z]^{\mathrm{T}},\ \{\mathbf{F}_{\boldsymbol{\beta}}\} = [\mathbf{Q}]^{\mathrm{T}}[M_x\ \ M_y\ \ M_z]^{\mathrm{T}}, \tag{4.107}$$

$$\{\mathbf{F}_\mathbf{q}\} = \begin{bmatrix} \{\mathbf{F}_\mathbf{s}\} \\ \{\mathbf{F}_{\boldsymbol{\beta}}\} \end{bmatrix}, \tag{4.108}$$

$$\{\widetilde{\mathbf{F}}_\mathbf{s}\} = -\left[[\mathbf{A}][\mathbf{S}]^{\mathrm{T}}[\dot{\mathbf{Q}}] + [\dot{\mathbf{A}}][\mathbf{S}]^{\mathrm{T}}[\mathbf{Q}]\right]\{\dot{\boldsymbol{\beta}}\}, \tag{4.109}$$

$$\{\widetilde{\mathbf{F}}_{\boldsymbol{\beta}}\} = -\left[[\mathbf{Q}]^{\mathrm{T}}[\mathbf{J}_O][\dot{\mathbf{Q}}] + [\mathbf{Q}]^{\mathrm{T}}[\boldsymbol{\omega}][\mathbf{J}_O][\mathbf{Q}]\right]\{\dot{\boldsymbol{\beta}}\}, \tag{4.110}$$

$$\left\{\widetilde{\mathbf{F}}_{\mathbf{q}}\right\} = \begin{bmatrix} \left\{\widetilde{\mathbf{F}}_{\mathbf{s}}\right\} \\ \left\{\widetilde{\mathbf{F}}_{\boldsymbol{\beta}}\right\} \end{bmatrix}. \tag{4.111}$$

With these values known, from the relation (4.90) one determines the values of the matrices $\{\ddot{\mathbf{q}}\}$, $\{\boldsymbol{\lambda}\}$; by solving the first order system

$$\frac{\mathrm{d}\{\mathbf{w}\}}{\mathrm{d}t} = \begin{bmatrix} \{\dot{\mathbf{q}}\} \\ \{\ddot{\mathbf{q}}\} \end{bmatrix}, \tag{4.112}$$

where

$$\{\mathbf{w}\} = \begin{bmatrix} q_1 & q_2 & \dots & q_6 & \dot{q}_1 & \dot{q}_2 & \dots & \dot{q}_6 \end{bmatrix}^{\mathrm{T}}, \tag{4.113}$$

with the aid of the one-step fourth order Runge-Kutta method, one determines the solution $\{\mathbf{q}\} = \{\mathbf{q}(t)\}$.

If the origin O of the mobile system $Oxyz$ coincides to the center of weight, then the calculation method simplifies because $[\mathbf{S}] = [\mathbf{0}]$, and $\left\{\widetilde{\mathbf{F}}_{\mathbf{s}}\right\} = \{\mathbf{0}\}$.

4.4.5 The Calculation of the Forces of Constraints

Writing the matrix of constraints in the form

$$[\mathbf{B}] = \begin{bmatrix} B_{11} & B_{12} & \dots & B_{16} \\ B_{21} & B_{22} & \dots & B_{26} \\ \dots & \dots & \dots & \dots \\ B_{p1} & B_{p2} & \dots & B_{p6} \end{bmatrix}, \tag{4.114}$$

it results that the generalized force corresponding to the constraint defined by the index i has the components

$$\left\{\mathbf{F}_{G\mathbf{s}}^{(i)}\right\} = \lambda_i \begin{bmatrix} B_{i1} \\ B_{i2} \\ B_{i3} \end{bmatrix}, \left\{\mathbf{F}_{G\boldsymbol{\beta}}^{(i)}\right\} = \lambda_i \begin{bmatrix} B_{i4} \\ B_{i5} \\ B_{i6} \end{bmatrix}, \tag{4.115}$$

where $\lambda_i B_{i1}$, $\lambda_i B_{i2}$, and $\lambda_i B_{i3}$ represent the projections of the constraint force onto the axes O_0X, O_0Y, and O_0Z, respectively; the projections onto the axes Ox, Oy, and Oz of the moment of the constraint force relative to the point O are obtained from the matrix expression $\left[\mathbf{Q}^{-1}\right]^{\mathrm{T}}\left\{\mathbf{F}_{G\boldsymbol{\beta}}^{(i)}\right\}$.

Example 4.4.3 Write the mathematical model for the motion of a heavy solid rigid of mass m, knowing the matrices [**S**], [$\mathbf{J}_O$]; in addition, the point $A(l,0,0)$ remains situated in the plan $Z=0$, while the point $B(-l,0,0)$ remains situated in the plan $X=0$. Choose as parameters the Euler angles, and the initial conditions are: $t=0$; $X_O=X_O^0$; $Y_O=Y_O^0$, $Z_O=Z_O^0$, $\psi=\psi_0$, $\theta=\theta_0$, $\varphi=\varphi_0$, $\dot{X}_O=\dot{X}_O^0$, $\dot{Y}_O=\dot{Y}_O^0$, $\dot{Z}_O=\dot{Z}_O^0$, $\dot{\psi}=\dot{\psi}_0$, $\dot{\theta}=\dot{\theta}_0$, $\dot{\varphi}=\dot{\varphi}_0$. Consider that the O_0Z axis is vertical.

Solution: The matrix of constraints was determined in the paragraph 4.4.1 (example 4.4.1) and it has the elements

$$B_{11}=B_{12}=0, B_{13}=1, B_{14}=0, B_{15}=\cos\theta\sin\varphi, B_{16}=\sin\theta\cos\varphi, B_{21}=1,$$

$$B_{22}=B_{23}=0, B_{24}=\sin\psi\cos\varphi+\cos\psi\cos\theta\sin\varphi, B_{25}=-\sin\psi\sin\theta\sin\varphi,$$

$$B_{26}=\cos\psi\sin\varphi+\sin\psi\cos\theta\cos\varphi;$$

the matrix $\{\mathbf{C}\}$ is null.

The derivative of the matrix of constraints has the elements

$$\dot{B}_{11}=\dot{B}_{12}=\dot{B}_{13}=\dot{B}_{14}=0, \dot{B}_{15}=-\dot{\theta}\sin\theta\sin\varphi+\dot{\varphi}\cos\theta\cos\varphi,$$

$$\dot{B}_{16}=\dot{\theta}\cos\theta\cos\varphi-\dot{\varphi}\sin\theta\sin\varphi, \dot{B}_{21}=\dot{B}_{22}=\dot{B}_{23}=0,$$

$$\dot{B}_{24}=\dot{\psi}(\cos\psi\cos\varphi-\sin\psi\cos\theta\sin\varphi)-\dot{\theta}\cos\psi\sin\theta\sin\varphi$$
$$+\dot{\varphi}(-\sin\psi\sin\varphi-\cos\psi\cos\theta\cos\varphi),$$

$$\dot{B}_{25}=-\dot{\psi}\cos\psi\sin\theta\sin\varphi-\dot{\theta}\sin\psi\cos\theta\sin\varphi-\dot{\varphi}\sin\psi\sin\theta\cos\varphi,$$

$$\dot{B}_{26}=\dot{\psi}(-\sin\psi\sin\varphi+\cos\psi\cos\theta\cos\varphi)-\dot{\theta}\sin\psi\sin\theta\cos\varphi$$
$$+\dot{\varphi}(\cos\psi\cos\varphi-\sin\psi\cos\theta\sin\varphi).$$

For the variant of Euler's angles, the rotational matrices are

$$[\boldsymbol{\psi}]=\begin{bmatrix}\cos\psi & -\sin\psi & 0\\ \sin\psi & \cos\psi & 0\\ 0 & 0 & 1\end{bmatrix}, [\boldsymbol{\theta}]=\begin{bmatrix}1 & 0 & 0\\ 0 & \cos\theta & -\sin\theta\\ 0 & \sin\theta & \cos\theta\end{bmatrix}, [\boldsymbol{\varphi}]=\begin{bmatrix}\cos\varphi & -\sin\varphi & 0\\ \sin\varphi & \cos\varphi & 0\\ 0 & 0 & 1\end{bmatrix}$$

and it results

$$[\mathbf{U}_\psi]=[\mathbf{U}_\varphi]=\begin{bmatrix}0 & -1 & 0\\ 1 & 0 & 0\\ 0 & 0 & 0\end{bmatrix}, [\mathbf{U}_\theta]=\begin{bmatrix}0 & 0 & 0\\ 0 & 0 & -1\\ 0 & 1 & 0\end{bmatrix}.$$

The matrices [**A**], and [**ω**] are calculated with the aid of the relations (4.98), and (4.102), respectively, while the matrices [**Q**] and $[\dot{\mathbf{Q}}]$ are

$$[\mathbf{Q}] = \begin{bmatrix} \sin\varphi\sin\theta & \cos\varphi & 0 \\ \cos\varphi\sin\theta & -\sin\varphi & 0 \\ \cos\theta & 0 & 1 \end{bmatrix},$$

$$[\dot{\mathbf{Q}}] = \dot{\theta}\begin{bmatrix} \sin\varphi\cos\theta & 0 & 0 \\ \cos\varphi\cos\theta & 0 & 0 \\ -\sin\theta & 0 & 0 \end{bmatrix} + \dot{\varphi}\begin{bmatrix} \cos\varphi\sin\theta & -\sin\varphi & 0 \\ -\sin\varphi\sin\theta & -\cos\varphi & 0 \\ 0 & 0 & 0 \end{bmatrix}.$$

The matrix [**M**] is given by the relation (4.106), while the generalized forces read

$$\{\mathbf{F_s}\} = [0 \;\; 0 \;\; -mg]^{\mathrm{T}}, \{\mathbf{F_\beta}\} = [\mathbf{Q}]^{\mathrm{T}}[\mathbf{S}][\mathbf{A}]^{\mathrm{T}}\{\mathbf{F}_s\}.$$

One calculates the components $\left\{\tilde{\mathbf{F}}_\mathbf{s}\right\}$, and $\left\{\tilde{\mathbf{F}}_\boldsymbol{\beta}\right\}$ with the relations (4.108), and (4.109), respectively.

The constraint force (the normal reaction at the point *A*), which corresponds to the first constraint is $\left\{\mathbf{F}_{\mathbf{Gs}}^{(1)}\right\} = \lambda_1[0 \;\; 0 \;\; 1]^{\mathrm{T}}$, while the constraint force (the normal reaction at the point *B*), which corresponds to the second constraint is $\left\{\mathbf{F}_{\mathbf{Gs}}^{(2)}\right\} = \lambda_2[1 \;\; 0 \;\; 0]^{\mathrm{T}}$.

4.4.6 The Elimination of the Matrix of the Lagrange Multipliers

A. The case when the matrix [M] is nonsingular

From the matrix general equation of the motion of the rigid solid (4.90) one obtains the equalities

$$[\mathbf{M}]\{\ddot{\mathbf{q}}\} - [\mathbf{B}]^{\mathrm{T}}\{\boldsymbol{\lambda}\} = \{\mathbf{F_q}\} + \left\{\tilde{\mathbf{F}}_\mathbf{q}\right\}, \tag{4.116}$$

$$[\mathbf{B}]\{\ddot{\mathbf{q}}\} = \{\dot{\mathbf{C}}\} - [\dot{\mathbf{B}}]\{\dot{\mathbf{q}}\}. \tag{4.117}$$

If the matrix [**M**] is an invertible one, then from the relation (4.116) one deduces the expression

$$\{\ddot{\mathbf{q}}\} = [\mathbf{M}]^{-1}\left\{\{\mathbf{F_q}\} + \left\{\tilde{\mathbf{F}}_\mathbf{q}\right\}\right\} + [\mathbf{M}]^{-1}[\mathbf{B}]^{\mathrm{T}}\{\boldsymbol{\lambda}\}. \tag{4.118}$$

Further on, from the relation (4.117) one obtains

$$\{\boldsymbol{\lambda}\} = \left[[\mathbf{B}][\mathbf{M}]^{-1}[\mathbf{B}]^{\mathrm{T}}\right]^{-1}\left\{\{\dot{\mathbf{C}}\} - [\dot{\mathbf{B}}]\{\dot{\mathbf{q}}\} - [\mathbf{B}][\mathbf{M}]^{-1}\left\{\{\mathbf{F_q}\} + \left\{\tilde{\mathbf{F}}_\mathbf{q}\right\}\right\}\right\}; \tag{4.119}$$

hence, the system of matrix differential equations (4.116) and (4.117) reduces to the equation (4.75) in which the matrix $\{\boldsymbol{\lambda}\}$ has the expression (4.119).

B. The case when the matrix [M] is singular

Since the constraints of the rigid solid are independent, it results that the matrix of constraints [**B**] has the rank equal to p; consequently, there exists a nonsingular square matrix of order equal to p, inside the matrix [**B**].

For instance, if $p = 3$, and the nonsingular matrix is formed with the second, third and sixth columns, then, by the substitutions $q_1^* = q_2$. $q_2^* = q_3$, $q_3^* = q_6$, $q_4^* = q_1$, $q_5^* = q_4$, $q_6^* = q_5$, that is, by the matrix substitution

$$\{\mathbf{q}\} = [\mathbf{D}]\{\mathbf{q}^*\}, \tag{4.120}$$

with

$$[\mathbf{D}] = \begin{bmatrix} 0 & 0 & 0 & 1 & 0 & 0 \\ 1 & 0 & 0 & 0 & 0 & 0 \\ 0 & 1 & 0 & 0 & 0 & 0 \\ 0 & 0 & 0 & 0 & 1 & 0 \\ 0 & 0 & 0 & 0 & 0 & 1 \\ 0 & 0 & 1 & 0 & 0 & 0 \end{bmatrix}, \tag{4.121}$$

one obtains a new matrix of constraints $[\mathbf{B}^*] = [\mathbf{B}][\mathbf{D}]$, in which the matrix formed with the three columns is an invertible one.

In the general case, if we make a new rearrangement of the variables q_k, $k = 1, 2, \ldots, 6$, then, by the substitution (4.120), the matrices [**B**], and [**M**] transform into the matrices $[\mathbf{B}^*]$, and $[\mathbf{M}^*]$, respectively, defined by the relations

$$[\mathbf{B}^*] = [\mathbf{B}][\mathbf{D}],\ [\mathbf{M}^*] = [\mathbf{M}][\mathbf{D}]. \tag{4.122}$$

In these conditions, denoting by $[\mathbf{B}_1^*]$ the matrix consisting in the first p columns of the matrix $[\mathbf{B}^*]$, then one may make the partition

$$[\mathbf{B}^*] = \left[[\mathbf{B}_1^*] \quad [\mathbf{B}_2^*] \right] \tag{4.123}$$

and denoting by $[\mathbf{M}_1^*]$ the matrix formed with the first p rows of the matrix $[\mathbf{M}^*]$, then one may write

$$[\mathbf{M}^*] = \begin{bmatrix} [\mathbf{M}_1^*] \\ [\mathbf{M}_2^*] \end{bmatrix}. \tag{4.124}$$

Analogically, making the partitions after the first p rows for the matrices $\left\{\mathbf{F}_{\mathbf{q}}^{*}\right\}$, and $\left\{\widetilde{\mathbf{F}}_{\mathbf{q}}^{*}\right\}$,

$$\left\{\mathbf{F}_{\mathbf{q}}^{*}\right\}=\begin{bmatrix}\left\{\mathbf{F}_{\mathbf{q}_1}^{*}\right\}\\ \left\{\mathbf{F}_{\mathbf{q}_2}^{*}\right\}\end{bmatrix},\ \left\{\widetilde{\mathbf{F}}_{\mathbf{q}}^{*}\right\}=\begin{bmatrix}\left\{\widetilde{\mathbf{F}}_{\mathbf{q}_1}^{*}\right\}\\ \left\{\widetilde{\mathbf{F}}_{\mathbf{q}_2}^{*}\right\}\end{bmatrix}, \tag{4.125}$$

from the relation (4.116) one obtains the system

$$[\mathbf{M}_1^*]\{\ddot{\mathbf{q}}^*\}-[\mathbf{B}_1^*]^{\mathrm{T}}\{\boldsymbol{\lambda}\}=\left\{\mathbf{F}_{\mathbf{q}_1}^{*}\right\}+\left\{\widetilde{\mathbf{F}}_{\mathbf{q}_1}^{*}\right\},\ [\mathbf{M}_2^*]\{\ddot{\mathbf{q}}^*\}-[\mathbf{B}_2^*]^{\mathrm{T}}\{\boldsymbol{\lambda}\}=\left\{\mathbf{F}_{\mathbf{q}_2}^{*}\right\}+\left\{\widetilde{\mathbf{F}}_{\mathbf{q}_2}^{*}\right\}. \tag{4.126}$$

From the first equation (4.126) we deduce

$$\{\boldsymbol{\lambda}\}=\left[[\mathbf{B}_1^*]^{-1}\right]^{\mathrm{T}}\left\{[\mathbf{M}_1^*]\{\ddot{\mathbf{q}}^*\}-\left\{\mathbf{F}_{\mathbf{q}_1}^{*}\right\}-\left\{\widetilde{\mathbf{F}}_{\mathbf{q}_1}^{*}\right\}\right\}; \tag{4.127}$$

using the notations

$$\begin{aligned}[\mathbf{M}_2^{**}]&=[\mathbf{M}_2^*]-[\mathbf{B}_2^*]^{\mathrm{T}}\left[[\mathbf{B}_1^*]^{-1}\right]^{\mathrm{T}}[\mathbf{M}_1^*],\\ \{\mathbf{F}_2^{**}\}&=\left\{\mathbf{F}_{\mathbf{q}_2}^{*}\right\}+\left\{\widetilde{\mathbf{F}}_{\mathbf{q}_2}^{*}\right\}-[\mathbf{B}_2^*]^{\mathrm{T}}\left[[\mathbf{B}_1^*]^{-1}\right]^{\mathrm{T}}\left\{\left\{\mathbf{F}_{\mathbf{q}_1}^{*}\right\}+\left\{\widetilde{\mathbf{F}}_{\mathbf{q}_1}^{*}\right\}\right\}\end{aligned} \tag{4.128}$$

and replacing in the second equation (4.126), it results

$$[\mathbf{M}_2^{**}]\{\ddot{\mathbf{q}}^*\}=\{\mathbf{F}_2^{**}\}. \tag{4.129}$$

The system (4.116), (4.117) reduces to the equation

$$[\mathbf{M}^{**}]\{\ddot{\mathbf{q}}^*\}=\{\mathbf{F}^{**}\}, \tag{4.130}$$

where

$$[\mathbf{M}^{**}]=\begin{bmatrix}[\mathbf{M}_2^{**}]\\ [\mathbf{B}^*]\end{bmatrix},\ \{\mathbf{F}^{**}\}=\begin{bmatrix}\{\mathbf{F}^*\}\\ \{\dot{\mathbf{C}}\}-[\dot{\mathbf{B}}]\{\dot{\mathbf{q}}\}\end{bmatrix}; \tag{4.131}$$

the column matrix of the Lagrange multipliers is given by the relation

$$\{\boldsymbol{\lambda}\}=\left[[\mathbf{B}_1^*]^{-1}\right]^{\mathrm{T}}\left\{[\mathbf{M}_1^*][\mathbf{M}_2^{**}]^{-1}\{\mathbf{F}_2^{**}\}-\left\{\mathbf{F}_{\mathbf{q}_1}^{*}\right\}-\left\{\widetilde{\mathbf{F}}_{\mathbf{q}_1}^{*}\right\}\right\}. \tag{4.132}$$

Example 4.4.4 Write the calculation algorithm for the motion of a heavy rigid solid knowing that the mobile axes are the principal central axes of inertia ($O \equiv C$, $J_{xy} = J_{yz} = J_{zx} = 0$); $J_x = 0$; the point $A(l, 0, 0)$ remains situated in the plan $Z = 0$; the rotational angles are the Euler angles. One knows the initial conditions: $t = 0$, $X_O^0 = \frac{2l}{5}$, $Y_O^0 = -\frac{3l}{10}$, $Z_O^0 = \frac{l\sqrt{3}}{2}$, $\psi_0 = 0$, $\theta_0 = \arctan\frac{5\sqrt{3}}{3}$, $\varphi_0 = -\arctan\frac{\sqrt{21}}{2}$, $\dot{X}_O^0 = 0$, $\dot{Y}_O^0 = 0$, $\dot{Z}_O^0 = 0$, $\dot{\psi}_0 = 0$, $\dot{\theta}_0 = 0$, $\dot{\varphi}_0 = 0$.

Solution: Since the point $A(l, 0, 0)$ is situated in the plan $Z = 0$ one gets the condition

$$Z_O + l\sin\theta\sin\varphi = 0,$$

the derivation of which with respect to time being

$$\dot{Z}_O + l\dot{\theta}\cos\theta\sin\varphi + l\dot{\varphi}\sin\theta\cos\varphi = 0$$

or

$$\begin{bmatrix}1 & 0 & l\cos\theta\sin\varphi & l\sin\theta\cos\varphi\end{bmatrix}\begin{bmatrix}\dot{Z}_O & \dot{\psi} & \dot{\theta} & \dot{\varphi}\end{bmatrix}^{\mathrm{T}} = 0.$$

On the other hand, the coincidence $O \equiv C$ implies that the matrix $[\mathbf{S}]$ is null, and the relation (4.94) leads to $\left\{\widetilde{\mathbf{F}}_{\mathbf{s}}\right\} = \{\mathbf{0}\}$.

The matrix of the generalized forces reads

$$\{\mathbf{F_q}\} = [0 \;\; 0 \;\; -mg \;\; 0 \;\; 0 \;\; 0]^{\mathrm{T}},$$

so that from the relation (4.91) one obtains the equations

$$[\mathbf{m}]\begin{bmatrix}\ddot{X}_O \\ \ddot{Y}_O \\ \ddot{Z}_O\end{bmatrix} = \begin{bmatrix}0 \\ 0 \\ -mg\end{bmatrix} + \lambda\begin{bmatrix}0 \\ 0 \\ 1\end{bmatrix},\; [\mathbf{J}^*]\begin{bmatrix}\ddot{\psi} \\ \ddot{\theta} \\ \ddot{\varphi}\end{bmatrix} = \lambda\begin{bmatrix}0 \\ B_{15} \\ B_{16}\end{bmatrix} + \left\{\widetilde{\mathbf{F}}_{\boldsymbol{\beta}}\right\},$$

where

$$B_{15} = l\cos\theta\sin\varphi, B_{16} = l\sin\theta\cos\varphi, [\mathbf{Q}] = \begin{bmatrix}\sin\varphi\sin\theta & \cos\varphi & 0 \\ \cos\varphi\sin\theta & -\sin\varphi & 0 \\ \cos\theta & 0 & 1\end{bmatrix},$$

$$[\mathbf{J}_O] = \begin{bmatrix}0 & 0 & 0 \\ 0 & J_y & 0 \\ 0 & 0 & J_z\end{bmatrix},\; [\mathbf{J}^*] = [\mathbf{Q}]^{\mathrm{T}}[\mathbf{J}_O][\mathbf{Q}],$$

and the matrix $\left\{\widetilde{\mathbf{F}}_{\boldsymbol{\beta}}\right\}$ is calculated with the relation (4.95).

For the initial conditions $t=0$, $X_O=X_O^0$, $Y_O=Y_O^0$, $\dot{X}_O=0$, $\dot{Y}_O=0$, one obtains the equations $m\ddot{X}_O=0$, $m\ddot{Y}_O=0$ and it results that the point O moves on the vertical straight line $X=X_O^0$, $Y=Y_O^0$; hence, the system of differential equations becomes

$$m\ddot{Z}_O=-mg+\lambda,\ [\mathbf{J}^*]\begin{bmatrix}\ddot{\psi}\\ \ddot{\theta}\\ \ddot{\varphi}\end{bmatrix}=\lambda\begin{bmatrix}0\\ B_{15}\\ B_{16}\end{bmatrix}+\left\{\widetilde{\mathbf{F}}_{\boldsymbol{\beta}}\right\}.$$

With the aid of the notations

$$\{\mathbf{0}\}=[0\ \ 0\ \ 0]^{\mathrm{T}},\ [\mathbf{M}^*]=\begin{bmatrix}m & \{\mathbf{0}\}^{\mathrm{T}}\\ \{\mathbf{0}\} & [\mathbf{J}^*]\end{bmatrix},\ \{\mathbf{q}\}=[Z_O\ \ \psi\ \ \theta\ \ \varphi]^{\mathrm{T}},$$

one obtains

$$[\mathbf{M}^*]\{\ddot{\mathbf{q}}\}=\begin{bmatrix}-mg\\ 0\\ 0\\ 0\end{bmatrix}+\lambda\begin{bmatrix}1\\ 0\\ B_{15}\\ B_{16}\end{bmatrix}+\begin{bmatrix}0\\ \left\{\widetilde{\mathbf{F}}_{\boldsymbol{\beta}}\right\}\end{bmatrix}.$$

Calculating the determinant of the matrix $[\mathbf{M}^*]$ and keeping into account that $\det[\mathbf{J}_O]=0$, it results

$$\det[\mathbf{M}^*]=m\det[\mathbf{J}^*]=m\det[\mathbf{Q}]^{\mathrm{T}}\det[\mathbf{J}_O]\det[\mathbf{Q}]=0;$$

consequently, the matrix $[\mathbf{M}^*]$ is a non-invertible one.

Applying the method described above, one obtains the relations

$$\lambda=mg+[m\ \ 0\ \ 0\ \ 0]\{\ddot{\mathbf{q}}\},$$

$$\begin{bmatrix}\begin{bmatrix}0\\0\\0\end{bmatrix} & [\mathbf{J}^*]\end{bmatrix}\{\ddot{\mathbf{q}}\}=mg\begin{bmatrix}0\\ B_{15}\\ B_{16}\end{bmatrix}+\begin{bmatrix}0\\ B_{15}\\ B_{16}\end{bmatrix}[m\ \ 0\ \ 0\ \ 0]\{\ddot{\mathbf{q}}\}+\left\{\widetilde{\mathbf{F}}_{\boldsymbol{\beta}}\right\}$$

or

$$\begin{bmatrix}\begin{bmatrix}0\\0\\0\end{bmatrix} & [\mathbf{J}^*]\end{bmatrix}\{\ddot{\mathbf{q}}\}-\begin{bmatrix}0 & 0 & 0 & 0\\ mB_{15} & 0 & 0 & 0\\ mB_{16} & 0 & 0 & 0\end{bmatrix}\{\ddot{\mathbf{q}}\}=mg\begin{bmatrix}0\\ B_{15}\\ B_{16}\end{bmatrix}+\left\{\widetilde{\mathbf{F}}_{\boldsymbol{\beta}}\right\};$$

it results the expression

$$\begin{bmatrix} \begin{bmatrix} 0 \\ mB_{15} \\ mB_{16} \end{bmatrix} & [\mathbf{J}^*] \end{bmatrix} \{\ddot{\mathbf{q}}\} = mg \begin{bmatrix} 0 \\ B_{15} \\ B_{16} \end{bmatrix} + \left\{ \widetilde{\mathbf{F}}_{\boldsymbol{\beta}} \right\},$$

at which we attach the relation

$$\begin{bmatrix} 1 & 0 & B_{15} & B_{16} \end{bmatrix} \{\ddot{\mathbf{q}}\} = -\begin{bmatrix} 0 & 0 & \dot{B}_{15} & \dot{B}_{16} \end{bmatrix} \{\dot{\mathbf{q}}\}.$$

Finally, one obtains the matrix equation

$$[\mathbf{M}^{**}]\{\ddot{\mathbf{q}}\} = \begin{bmatrix} mg \begin{bmatrix} 0 \\ B_{15} \\ B_{16} \end{bmatrix} + \left\{ \widetilde{\mathbf{F}}_{\boldsymbol{\beta}} \right\} \\ -\begin{bmatrix} 0 & 0 & \dot{B}_{15} & \dot{B}_{16} \end{bmatrix} \{\dot{\mathbf{q}}\} \end{bmatrix},$$

where the matrix

$$[\mathbf{M}^{**}] = \begin{bmatrix} \begin{bmatrix} 0 \\ B_{15} \\ B_{16} \end{bmatrix} & [\mathbf{J}^*] \\ 1 & \begin{bmatrix} 0 & B_{15} & B_{16} \end{bmatrix} \end{bmatrix}$$

is nonsingular.

Further Reading

Amirouche F (2005). Fundamentals of Multibody Dynamics. Theory and Applications. Boston: Birkhäuser.

Arnold VI (1997). Mathematical Methods of Classical Mechanics (Graduate Texts in Mathematics, Vol. 60). 2nd ed. New York: Springer.

Ball RS (1998). A Treatise on the Theory of Screws. Cambridge: Cambridge University Press.

Bloch AM, Baillieul J, Crouch P, Marsden J (2007). Nonholonomic Mechanics and Control (Interdisciplinary Applied Mathematics). Berlin: Springer.

Blundell M, Harty D (2004). The Multibody Systems Approach to Vehicle Dynamics. Amsterdam: Elsevier Butterworth-Heinemann.

Chaudhary H, Saha SH (2008). Dynamics and Balancing of Multibody Systems (Lecture Notes in Applied and Computational Mechanics). Berlin: Springer.

Cortes J (2002). Geometric, Control and Numerical Aspects of Nonholonomic Systems. Berlin: Springer.

Coutinho MG (2001). Dynamic Simulations of Multibody Systems. New York: Springer.

Dankowicz HJ (2004). Multibody Mechanics and Visualization. London: Springer.
de Jalón JG, Bayo E (2011). Kinematic and Dynamic Simulation of Multibody Systems: The Real-Time Challenge. Berlin: Springer Verlag.
Den Hartog JP (1961). Mechanics. New York: Dover Publications Inc.
Douglas Gregory R (2006). Classical Mechanics. Cambridge: Cambridge University Press.
Dresig H, Holzweißig F (2010). Dynamics of Machinery: Theory and Applications. Heidelberg: Springer.
Edelen DG (2010). Lagrangian Mechanics of Nonconservative Nonholonomic Systems. Leyden: Kluwer.
Fasano A, Marmi S, Pelloni B (2006). Analytical Mechanics: An Introduction. Oxford: Oxford University Press.
Gattringer H, Gerstmayr J (2013). Multibody Systems Dynamics, Robotics and Control. Wien: Springer.
Ginsberg JH (1998). Advanced Engineering Dynamics. 2nd ed. Cambridge: Cambridge University Press.
Ginsberg J (2007). Engineering Dynamics. Cambridge: Cambridge University Press.
Goldstein H, Poole CP Jr, Safko JL (2001). Classical Mechanics. 3rd ed. Edinburg: Pearson.
Greenwood DT (2006). Advanced Dynamics. Cambridge: Cambridge University Press.
Greiner W (2009). Classical Mechanics: Systems of Particles and Hamiltonian Dynamics. 2nd ed. Heidelberg: Springer.
Hand LN, Finch JD (1998). Analytical Mechanics. Cambridge: Cambridge University Press.
Jain A (2010). Robot and Multibody Dynamics: Analysis and Algorithms. New York: Springer.
Jazar RN (2011). Advanced Dynamics: Rigid Body, Multibody, and Aerospace Applications. Hoboken: John Wiley & Sons.
Jazar RN (2010). Theory of Applied Robotics: Kinematics, Dynamics, and Control. 2nd ed. New York: Springer.
Karnopp DC, Margolis DL, Rosenberg RC (2012). System Dynamics: Modeling, Simulation, and Control of Mechatronic Systems. Hoboken: John Wiley & Sons.
Kibble TWB, Berkshire FH (2004). Classical Mechanics. 5th ed. London: Imperial College Press.
Kleppner D, Kolenkow R (2013). An Introduction to Mechanics. 2nd ed. Cambridge: Cambridge University Press.
Lagrange JL (1996). Analytical Mechanics. Dordrecht: Kluwer Academic Publishers,
Lurie AI (2002). Analytical Mechanics. Berlin: Springer.
Meriam JL, Kraige LG (2012). Engineering Mechanics: Dynamics. Hoboken: John Wiley & Sons.
Moon FC (1998). Applied Dynamics: With Applications to Multibody and Mechatronic Systems. New York: John Wiley & Sons, Inc.
Morin D (2008). Introduction to Classical Mechanics: With Problems and Solution. Cambridge: Cambridge University Press.
Murray RM, Li Z, Shankar Sastry S (1994). A Mathematical Introduction to Robotic Manipulation. Boca Raton: CRC Press.
Neimark IuI, Fufaev NA (1972). Dynamics of Nonholonomic Systems (Translations of Mathematical Monographs, V. 33). Providence, Rhode Island: American Mathematical Society.
O'Reilly OM (2008). Intermediate Dynamics for Engineers: A unified Treatment of Newton-Euler and Lagrangian Mechanics. Cambridge: Cambridge University Press
Pandrea N (2000). Elemente de mecanica solidelor în coordonate plückeriene. Bucureşti: Editura Academiei Române (in Romanian).
Pandrea N, Stănescu ND (2002). Mecanica. Bucureşti: Editura Didactică şi Pedagogică (in Romanian).
Papastavridis JG (2002). Analytical Mechanics: A Comprehensive Treatise on the Dynamics of Constrained Systems; For Engineers, Physicists, and Mathematicians. Oxford: Oxford University Press.

Pfeiffer F (2008). Mechanical System Dynamics (Lecture Notes in Applied and Computational Mechanics). Berlin: Springer.
Pfeiffer F, Glocker C (1996). Multibody Dynamics with Unilateral Contacts. New York: John Wiley & Sons, Inc.
Roberson RE, Schwertassek R (1988). Dynamics of Multibody Systems. Berlin: Springer Verlag.
Seifried R (2013). Dynamics of Underactuated Multibody Systems: Modeling, Control and Optimal Design. Heidelberg: Springer.
Shabana AA (2008). Computational Dynamics. New York: John Wiley & Sons.
Shabana AA (2013). Dynamics of Multibody Systems. Cambridge: Cambridge University Press.
Soltakhanov ShKh, Yushkov MPP Zegzhda SA (2009). Mechanics of non-nolonomic systems: A New Class of control systems (Foundations of Engineering Mechanics). Berlin: Springer.
Spong MW, Hutchinson S, Vidyasagar M (2005). Robot Modeling and Control. New York: John Wiley & Sons.
Stănescu ND (2013). Mecanica sistemelor. Bucureşti: Editura Didactică şi Pedagogică (in Romanian).
Taylor JR (2005). Classical Mechanics. Sausalito: University Science Books.
Teodorescu PP (2009). Mechanical Systems, Classical Models: Volume 3: Analytical Mechanics (Mathematical and Analytical Techniques with Applications to Engineering). Dordrecht: Springer.
Udwadia FE, Kalaba RE (2007). Analytical Dynamics: A New Approach. Cambridge: Cambridge University Press.
Uicker JJ, Ravani B, Sheth PN (2013). Matrix Methods in the Design Analysis of Mechanisms and Multibody Systems. Cambridge: Cambridge University Press.
Wittenburg J (2007). Dynamics of Multibody Systems. Berlin: Springer.
Woodhouse N (2010). Introduction to Analytical Dynamics. London: Springer.

5

Generalized Forces: The Equilibrium of the Rigid Solid

In the present chapter we discuss the equilibrium of the rigid solid; we introduce the generalized forces and their expressions; as a particular case we study the equilibrium of a rigid solid hanged by springs.

5.1 The Generalized Forces in the Case of a Mechanical System

We consider the most general case of a mechanical system having the generalized coordinates q_k, $k=\overline{1,n}$. If at the points N_i, $i=\overline{1,n}$, with the vectors of position $\mathbf{r}_i=\mathbf{O}_0\mathbf{N}_i$, act the given forces $\mathbf{P}_i$, $i=\overline{1,n}$, then the generalized forces are given by the relations

$$F_k=\sum_{i=1}^{n}\mathbf{P}_i\cdot\frac{\partial \mathbf{r}_i}{\partial q_k}, k=1, 2, \ldots, n, \tag{5.1}$$

or

$$F_k=\sum_{i=1}^{n}\left(P_{iX}\frac{\partial X_i}{\partial q_k}+P_{iY}\frac{\partial Y_i}{\partial q_k}+P_{iZ}\frac{\partial Z_i}{\partial q_k}\right), \tag{5.2}$$

where P_{iX}, P_{iY}, P_{iZ}, and X_i, Y_i, Z_i are the projections of the vectors $\mathbf{P}_i$, and $\mathbf{r}_i$, respectively, onto *the fixed axes* O_0X, O_0Y, O_0Z.

Dynamics of the Rigid Solid with General Constraints by a Multibody Approach, First Edition.
Nicolae Pandrea and Nicolae-Doru Stănescu.

Companion website: www.wiley.com/go/pandrea

Analogically, if upon the rigid bodies *indexed by the index j and having the angular velocities* $\boldsymbol{\omega}_j$ act the torques of moments $\mathbf{M}_j$, $j=\overline{1,n}$, then the generalized forces are given by the relations

$$F_k=\sum_{j=1}^{n}\mathbf{M}_j\cdot\frac{\partial\boldsymbol{\omega}_j}{\partial\dot{q}_k},\ k=\overline{1,n}, \tag{5.3}$$

or

$$F_k=\sum_{i=1}^{n}\left(M_{jx}\frac{\partial\omega_{jx}}{\partial\dot{q}_k}+M_{jy}\frac{\partial\omega_{jy}}{\partial\dot{q}_k}+M_{jz}\frac{\partial\omega_{jz}}{\partial\dot{q}_k}\right), \tag{5.4}$$

where M_{jx}, M_{jy}, M_{jz}, and ω_{jx}, ω_{jy}, ω_{jz} are the projections of the vectors $\mathbf{M}_j$, and $\boldsymbol{\omega}_j$, respectively, onto *the mobile axes.*

5.2 The General Expressions of the Generalized Forces in the Case of Rigid Solid

5.2.1 The Case When at a Point Acts a Given Force

One considers that at the point A of the rigid solid, of coordinates x_A, y_A, z_A relative to the reference system $Oxyz$ jointed to the rigid solid, acts the force $\mathbf{P}$, which has the projections P_X, P_Y, P_Z onto the fixed axes O_0X, O_0Y, O_0Z.

The coordinates X_A, Y_A, Z_A of the point A relative to the fixed reference system O_0XYZ are given by the equality

$$\begin{bmatrix}X_A\\Y_A\\Z_A\end{bmatrix}=\begin{bmatrix}X_O\\Y_O\\Z_O\end{bmatrix}+[\mathbf{A}]\begin{bmatrix}x_A\\y_A\\z_A\end{bmatrix}. \tag{5.5}$$

The generalized coordinates being X_O, Y_O, Z_O, ψ, θ, φ, it results the relations

$$\frac{\partial X_A}{\partial X_O}=1,\ \frac{\partial Y_A}{\partial X_O}=\frac{\partial Z_A}{\partial X_O}=0 \tag{5.6}$$

and similar; hence, the generalized forces which correspond to the variables X_O, Y_O, Z_O are

$$F_{X_O}=P_X, F_{Y_O}=P_Y, F_{Z_O}=P_Z. \tag{5.7}$$

The generalized force that corresponds to the coordinate ψ deduces from the relation (5.2), that is

$$F_\psi=P_X\frac{\partial X_A}{\partial\psi}+P_Y\frac{\partial Y_A}{\partial\psi}+P_Z\frac{\partial Z_A}{\partial\psi} \tag{5.8}$$

or

$$F_{\psi} = [P_X \ \ P_Y \ \ P_Z]\left[\frac{\partial X_A}{\partial \psi} \ \ \frac{\partial Y_A}{\partial \psi} \ \ \frac{\partial Z_A}{\partial \psi}\right]^{\mathrm{T}}; \tag{5.9}$$

keeping into account the equality (5.5), it results

$$F_{\psi} = [P_X \ \ P_Y \ \ P_Z][\mathbf{A}_{\psi}]\{\mathbf{r}_A\}, \tag{5.10}$$

where

$$\{\mathbf{r}_A\} = [x_A \ \ y_A \ \ z_A]^{\mathrm{T}}. \tag{5.11}$$

The expression (5.10) may be also written in the form

$$F_{\psi} = \{\mathbf{r}_A\}^{\mathrm{T}}[\mathbf{A}_{\psi}]^{\mathrm{T}}\begin{bmatrix} P_X \\ P_Y \\ P_Z \end{bmatrix}. \tag{5.12}$$

Analogically, one obtains for the rest of the generalized forces the expressions

$$F_{\theta} = \{\mathbf{r}_A\}^{\mathrm{T}}[\mathbf{A}_{\theta}]^{\mathrm{T}}\begin{bmatrix} P_X \\ P_Y \\ P_Z \end{bmatrix}, F_{\varphi} = \{\mathbf{r}_A\}^{\mathrm{T}}[\mathbf{A}_{\varphi}]^{\mathrm{T}}\begin{bmatrix} P_X \\ P_Y \\ P_Z \end{bmatrix} \tag{5.13}$$

and it results

$$\begin{bmatrix} F_{\psi} \\ F_{\theta} \\ F_{\varphi} \end{bmatrix} = \begin{bmatrix} \{\mathbf{r}_A\}^{\mathrm{T}}[\mathbf{A}_{\psi}]^{\mathrm{T}} \\ \{\mathbf{r}_A\}^{\mathrm{T}}[\mathbf{A}_{\theta}]^{\mathrm{T}} \\ \{\mathbf{r}_A\}^{\mathrm{T}}[\mathbf{A}_{\varphi}]^{\mathrm{T}} \end{bmatrix}\begin{bmatrix} P_X \\ P_Y \\ P_Z \end{bmatrix}; \tag{5.14}$$

with the notation

$$[\mathbf{QRA}] = \begin{bmatrix} \{\mathbf{r}_A\}^{\mathrm{T}}[\mathbf{A}_{\psi}]^{\mathrm{T}} \\ \{\mathbf{r}_A\}^{\mathrm{T}}[\mathbf{A}_{\theta}]^{\mathrm{T}} \\ \{\mathbf{r}_A\}^{\mathrm{T}}[\mathbf{A}_{\varphi}]^{\mathrm{T}} \end{bmatrix}, \tag{5.15}$$

the equation (5.14) becomes

$$\begin{bmatrix} F_{\psi} \\ F_{\theta} \\ F_{\varphi} \end{bmatrix} = [\mathbf{QRA}]\begin{bmatrix} P_X \\ P_Y \\ P_Z \end{bmatrix}. \tag{5.16}$$

The equalities (5.7) and (5.16) reunite in the matrix expression

$$\{\mathbf{F}\} = \begin{bmatrix} [\mathbf{I}] \\ [\mathbf{QRA}] \end{bmatrix} \begin{bmatrix} P_X \\ P_Y \\ P_Z \end{bmatrix}; \tag{5.17}$$

hence, the column matrix of the generalized forces is given by the relation (5.17).

One may write the successive equalities

$$[\mathbf{QRA}] = \begin{bmatrix} \{\mathbf{r}_A\}^{\mathrm{T}} [\mathbf{A}_\psi]^{\mathrm{T}} [\mathbf{A}] \\ \{\mathbf{r}_A\}^{\mathrm{T}} [\mathbf{A}_\theta]^{\mathrm{T}} [\mathbf{A}] \\ \{\mathbf{r}_A\}^{\mathrm{T}} [\mathbf{A}_\varphi]^{\mathrm{T}} [\mathbf{A}] \end{bmatrix} [\mathbf{A}]^{\mathrm{T}} = \begin{bmatrix} [\mathbf{A}]^{\mathrm{T}} [\mathbf{A}_\psi] \{\mathbf{r}_A\} & [\mathbf{A}]^{\mathrm{T}} [\mathbf{A}_\theta] \{\mathbf{r}_A\} & [\mathbf{A}]^{\mathrm{T}} [\mathbf{A}_\varphi] \{\mathbf{r}_A\} \end{bmatrix} [\mathbf{A}]^{\mathrm{T}}. \tag{5.18}$$

Since the matrices $[\mathbf{A}]^{\mathrm{T}}[\mathbf{A}_\psi]$, $[\mathbf{A}]^{\mathrm{T}}[\mathbf{A}_\theta]$, $[\mathbf{A}]^{\mathrm{T}}[\mathbf{A}_\varphi]$ are skew symmetric and have as associated the column matrices $\{\mathbf{A}_\psi\}$, $\{\mathbf{A}_\theta\}$, $\{\mathbf{A}_\varphi\}$, and taking into account the relations

$$[\mathbf{A}]^{\mathrm{T}} [\mathbf{A}_\psi] \{\mathbf{r}_A\} = -[\mathbf{r}_A] \{\mathbf{A}_\psi\}, [\mathbf{A}]^{\mathrm{T}} [\mathbf{A}_\psi] \{\mathbf{r}_A\} = -[\mathbf{r}_A] \{\mathbf{A}_\psi\}, [\mathbf{A}]^{\mathrm{T}} [\mathbf{A}_\psi] \{\mathbf{r}_A\} = -[\mathbf{r}_A] \{\mathbf{A}_\psi\}, \tag{5.19}$$

where

$$[\mathbf{r}_A] = \begin{bmatrix} 0 & -z_A & y_A \\ z_A & 0 & -x_A \\ -y_A & x_A & 0 \end{bmatrix}, \tag{5.20}$$

one obtains the successive relations

$$\begin{aligned} [\mathbf{QRA}] &= -\begin{bmatrix} [\mathbf{r}_A]\{\mathbf{A}_\psi\} & [\mathbf{r}_A]\{\mathbf{A}_\theta\} & [\mathbf{r}_A]\{\mathbf{A}_\varphi\} \end{bmatrix}^{\mathrm{T}} [\mathbf{A}]^{\mathrm{T}} \\ &= -\begin{bmatrix} [\mathbf{r}_A] \begin{bmatrix} \{\mathbf{A}_\psi\} & \{\mathbf{A}_\theta\} & \{\mathbf{A}_\varphi\} \end{bmatrix} \end{bmatrix}^{\mathrm{T}} [\mathbf{A}]^{\mathrm{T}} = -[[\mathbf{r}_A][\mathbf{Q}]]^{\mathrm{T}} [\mathbf{A}]^{\mathrm{T}} = [\mathbf{Q}]^{\mathrm{T}} [\mathbf{r}_A] [\mathbf{A}]^{\mathrm{T}}, \end{aligned} \tag{5.21}$$

that is

$$[\mathbf{QRA}] = [\mathbf{Q}]^{\mathrm{T}} [\mathbf{r}_A] [\mathbf{A}]^{\mathrm{T}}. \tag{5.22}$$

In these conditions, from the relation (5.17) one obtains

$$\{\mathbf{F}\} = \begin{bmatrix} [\mathbf{I}] \\ [\mathbf{Q}]^{\mathrm{T}}[\mathbf{r}_A][\mathbf{A}]^{\mathrm{T}} \end{bmatrix} \begin{bmatrix} P_X \\ P_Y \\ P_Z \end{bmatrix}. \tag{5.23}$$

If the force $\mathbf{P}$ is given by its projections P_x, P_y, P_z onto the mobile axes Ox, Oy, Oz, then the relation (5.23) becomes

$$\{\mathbf{F}\} = \begin{bmatrix} [\mathbf{A}] \\ [\mathbf{Q}]^{\mathrm{T}}[\mathbf{r}_A] \end{bmatrix} \begin{bmatrix} P_x \\ P_y \\ P_z \end{bmatrix}. \tag{5.24}$$

5.2.2 *The Case When the Rigid Solid is Acted by a Torque of Given Moment*

If upon the rigid solid acts a torque of moment $\mathbf{M}$, and if M_x, M_y, M_z are its projections onto *the mobile axes Ox, Oy, Oz*, then, with the aid of the relation (5.4), it results

$$F_\psi = M_x \frac{\partial \omega_x}{\partial \dot{\psi}} + M_y \frac{\partial \omega_y}{\partial \dot{\psi}} + M_z \frac{\partial \omega_z}{\partial \dot{\psi}} \tag{5.25}$$

or

$$F_\psi = \frac{\partial \{\boldsymbol{\omega}\}^{\mathrm{T}}}{\partial \dot{\psi}} \{\mathbf{M}\}, \tag{5.26}$$

where

$$\{\mathbf{M}\} = [M_x \ \ M_y \ \ M_z]^{\mathrm{T}}. \tag{5.27}$$

Writing the analogous expressions for F_θ, F_φ and knowing that

$$[\mathbf{Q}] = \begin{bmatrix} \dfrac{\partial \{\boldsymbol{\omega}\}}{\partial \dot{\psi}} & \dfrac{\partial \{\boldsymbol{\omega}\}}{\partial \dot{\theta}} & \dfrac{\partial \{\boldsymbol{\omega}\}}{\partial \dot{\varphi}} \end{bmatrix}, \tag{5.28}$$

we get

$$\begin{bmatrix} F_\psi \\ F_\theta \\ F_\varphi \end{bmatrix} = [\mathbf{Q}]^{\mathrm{T}} \{\mathbf{M}\}. \tag{5.29}$$

The relation (5.29) permits an easy proof for the equality (5.22).

Indeed, the moment of the force **P** relative to the point *O*, written in the reference system *Oxyz* is

$$\{\mathbf{M}\} = [\mathbf{r}_A][\mathbf{A}]^{\mathrm{T}} \begin{bmatrix} P_X \\ P_Y \\ P_Z \end{bmatrix}; \tag{5.30}$$

using the relation (5.29), it results the equality

$$\begin{bmatrix} F_\psi \\ F_\theta \\ F_\varphi \end{bmatrix} = [\mathbf{Q}]^{\mathrm{T}}[\mathbf{r}_A][\mathbf{A}]^{\mathrm{T}} \begin{bmatrix} P_X \\ P_Y \\ P_Z \end{bmatrix}. \tag{5.31}$$

Comparing now the relations (5.31) and (5.16), we obtain the equality (5.22).

5.3 Conservative Forces

5.3.1 General Aspects

A force **F** with the projections F_X, F_Y, F_Z onto the axes of the fixed reference system and acting at the point *N*(*X*, *Y*, *Z*) is called *conservative* if there exists a function *V*(*X*, *Y*, *Z*) named *potential* or *potential energy* such that

$$F_X = -\frac{\partial V}{\partial X}, F_Y = -\frac{\partial V}{\partial Y}, F_Z = -\frac{\partial V}{\partial Z}. \tag{5.32}$$

From the mathematical equalities

$$\frac{\partial^2 V}{\partial X \partial Y} = \frac{\partial^2 V}{\partial Y \partial X}, \frac{\partial^2 V}{\partial X \partial Z} = \frac{\partial^2 V}{\partial Z \partial X}, \frac{\partial^2 V}{\partial Y \partial Z} = \frac{\partial^2 V}{\partial Z \partial Y} \tag{5.33}$$

it results that the force $\mathbf{F}(F_X, F_Y, F_Z)$ is conservative if

$$\frac{\partial F_X}{\partial Y} = \frac{\partial F_Y}{\partial X}, \frac{\partial F_X}{\partial Z} = \frac{\partial F_Z}{\partial X}, \frac{\partial F_Y}{\partial Z} = \frac{\partial F_Z}{\partial Y}. \tag{5.34}$$

If a force is conservative (it respects the conditions (5.34)), then the potential deduces from the equality

$$V(X,Y,Z) = -\int_{X_0}^{X} F_X(X,Y_0,Z_0)\mathrm{d}X - \int_{Y_0}^{Y} F_Y(X,Y,Z_0)\mathrm{d}Y - \int_{Z_0}^{Z} F_Z(X,Y,Z)\mathrm{d}Z + C, \tag{5.35}$$

where *C* is an arbitrary constant.

For instance, if

$$F_X = -2XYZ, F_Y = -X^2Z, F_Z = -X^2Y, \tag{5.36}$$

then the conditions (5.34) hold true, and from the relation (5.35) one obtains

$$V = X^2YZ + \widetilde{C}, \tag{5.37}$$

where $\widetilde{C}$ is an arbitrary constant.

If the conservative force $\mathbf{F}(F_X, F_Y, F_Z)$ acts at the point $N(x, y, z)$ of the rigid solid, the coordinates x, y, z being constant, then, keeping into account the equality

$$\begin{bmatrix} X \\ Y \\ Z \end{bmatrix} = \begin{bmatrix} X_O \\ Y_O \\ Z_O \end{bmatrix} + [\mathbf{A}]\{\mathbf{r}\}, \tag{5.38}$$

where

$$\{\mathbf{r}\} = [x \ \ y \ \ z]^{\mathrm{T}}, \tag{5.39}$$

it results that the potential V is a function of the generalized coordinates (X_O, Y_O, Z_O, ψ, θ, φ) by means of the functions X, Y, Z.

By virtue of the relations

$$\frac{\partial X}{\partial X_O} = 1, \frac{\partial V}{\partial X_O} = \frac{\partial V}{\partial X}\frac{\partial X}{\partial X_O} \tag{5.40}$$

and similar, it results that the generalized forces corresponding to the generalized coordinates X_O, Y_O, Z_O read

$$F_{X_O} = F_X = -\frac{\partial V}{\partial X_O}, F_{Y_O} = F_Y = -\frac{\partial V}{\partial Y_O}, F_{Z_O} = F_Z = -\frac{\partial V}{\partial Z_O}. \tag{5.41}$$

Further on, it results for the coordinate ψ

$$\frac{\partial V}{\partial \psi} = \frac{\partial V}{\partial X}\frac{\partial X}{\partial \psi} + \frac{\partial V}{\partial Y}\frac{\partial Y}{\partial \psi} + \frac{\partial V}{\partial Z}\frac{\partial Z}{\partial \psi}, \tag{5.42}$$

and by virtue of the relation (5.38) one deduces

$$\frac{\partial V}{\partial \psi} = -\{\mathbf{r}\}^{\mathrm{T}}[\mathbf{A}_\psi]^{\mathrm{T}} \begin{bmatrix} F_{X_O} \\ F_{Y_O} \\ F_{Z_O} \end{bmatrix}. \tag{5.43}$$

Comparing this result to the relation (5.12), it results

$$F_{\psi} = -\frac{\partial V}{\partial \psi}; \tag{5.44}$$

hence, the column matrix of the generalized forces reads

$$\{\mathbf{F}\} = -\begin{bmatrix} \frac{\partial V}{\partial X_O} & \frac{\partial V}{\partial Y_O} & \frac{\partial V}{\partial Z_O} & \frac{\partial V}{\partial \psi} & \frac{\partial V}{\partial \theta} & \frac{\partial V}{\partial \varphi} \end{bmatrix}^{\mathrm{T}} \tag{5.45}$$

or

$$\{\mathbf{F}\} = \begin{bmatrix} [\mathbf{I}] \\ [\mathbf{Q}]^{\mathrm{T}}[\mathbf{r}][\mathbf{A}]^{\mathrm{T}} \end{bmatrix} \begin{bmatrix} F_{X_O} \\ F_{Y_O} \\ F_{Z_O} \end{bmatrix}. \tag{5.46}$$

5.3.2 *The Weight*

If we denote by **u** the unit vector of the vertical direction (orientated to the surface of Earth), then the weight has the vector form

$$\mathbf{G} = mg\mathbf{u}; \tag{5.47}$$

if we denote by a, b, and c the projections of the unit vector **u** onto the fixed axes O_0X, O_0Y, and O_0Z, respectively, it results

$$G_x = mga,\ G_y = mgb,\ G_z = mgc. \tag{5.48}$$

Further on, from the relation (5.35) one deduces

$$V = -mg(aX + bY + cZ). \tag{5.49}$$

Since the force **G** acts at the center of weight C defined by the coordinates X_C, Y_C, Z_C relative to the fixed system of coordinates O_0XYZ and by the constant coordinates x_C, y_C, z_C relative to the mobile system of coordinates $Oxyz$, it results

$$V = -mg(aX_C + bY_C + cZ_C), \tag{5.50}$$

and by virtue of the expression (5.23) one deduces the column matrix of the generalized coordinates

$$\{\mathbf{F}\} = mg\begin{bmatrix} [\mathbf{I}] \\ [\mathbf{Q}]^{\mathrm{T}}[\mathbf{r}_C][\mathbf{A}]^{\mathrm{T}} \end{bmatrix}\begin{bmatrix} a \\ b \\ c \end{bmatrix}. \tag{5.51}$$

5.3.3 The Elastic Force of a Spring

We consider the rigid solid in Fig. 5.1 related to the fixed reference system of coordinates O_0XYZ and to the mobile reference system of coordinates $Oxyz$ jointed to the rigid solid. Let $A(x_A, y_A, z_A)$ be a point of the rigid solid at which is linked the spring AB of stiffness k.

Denoted by l_0 the not deformed length of the spring, one obtains the force in the spring

$$\mathbf{F} = k(AB - l_0)\frac{\mathbf{AB}}{AB} \tag{5.52}$$

or, on the components

$$F_X = k(AB - l_0)\frac{X_B - X_A}{AB},\ F_Y = k(AB - l_0)\frac{Y_B - Y_A}{AB},\ F_Z = k(AB - l_0)\frac{Z_B - Z_A}{AB}, \tag{5.53}$$

where

$$AB = \sqrt{(X_B - X_A)^2 + (Y_B - Y_A)^2 + (Z_B - Z_A)^2}. \tag{5.54}$$

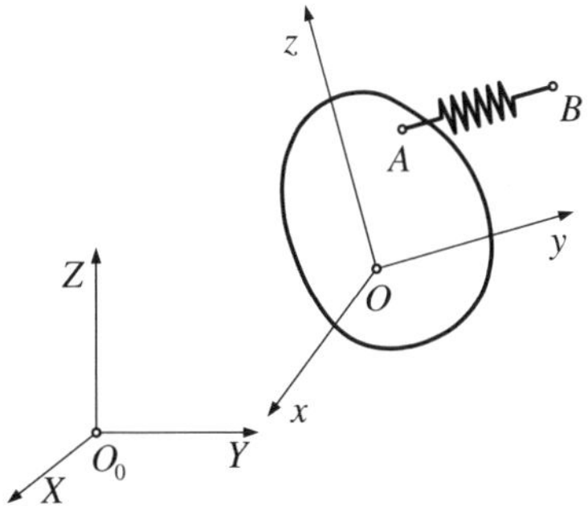

Figure 5.1 The elastic force of a spring.

It is easy to prove that the potential function reads

$$V = -\frac{k}{2}(AB - l_0)^2. \tag{5.55}$$

Indeed, if one derivates the expression (5.55) with respect to X_A, Y_A, Z_A, and takes into account the equalities

$$\frac{\partial AB}{\partial X_A} = -\frac{X_B - X_A}{AB}, \frac{\partial AB}{\partial Y_A} = -\frac{Y_B - Y_A}{AB}, \frac{\partial AB}{\partial Z_A} = -\frac{Z_B - Z_A}{AB}, \tag{5.56}$$

then one obtains the components defined by the relations (5.53).

The column matrix of the generalized forces is obtained from the relation (5.46), that is,

$$\{\mathbf{F}\} = \frac{k(AB - l_0)}{AB}\begin{bmatrix} [\mathbf{I}] \\ [\mathbf{Q}]^{\mathrm{T}}[\mathbf{r}][\mathbf{A}]^{\mathrm{T}} \end{bmatrix}\begin{bmatrix} X_B - X_A \\ Y_B - Y_A \\ Z_B - Z_A \end{bmatrix}, \tag{5.57}$$

where

$$\begin{bmatrix} X_A \\ Y_A \\ Z_A \end{bmatrix} = \begin{bmatrix} X_O \\ Y_O \\ Z_O \end{bmatrix} + [\mathbf{A}]\{\mathbf{r}_A\}. \tag{5.58}$$

Observation 5.3.1. The point B may be either fixed or it may belong to another rigid for which the motion *is known*; in this case one uses the matrix relation

$$\begin{bmatrix} X_B \\ Y_B \\ Z_B \end{bmatrix} = \begin{bmatrix} X_{O^*} \\ Y_{O^*} \\ Z_{O^*} \end{bmatrix} + [\mathbf{A}^*]\{\mathbf{r}_B^*\}, \tag{5.59}$$

the parameters X_{O^*}, Y_{O^*}, Z_{O^*}, $[\mathbf{A}^*]$, and $\{\mathbf{r}_B^*\}$ being related to the reference system $O^*x^*y^*z^*$ jointed to the rigid solid that contains the point B.

5.4 The Equilibrium of the Constrained Rigid Solid

5.4.1 The Equations of Equilibrium: Numerical Solution

In the case of the mechanical systems with *holonomic constraints* the kinetic energy is null. From the Lagrange equations one obtains the matrix equation

$$\{\mathbf{F_q}\} + [\mathbf{B}]^{\mathrm{T}}\{\boldsymbol{\lambda}\} = \{\mathbf{0}\}, \tag{5.60}$$

at which one adds the matrix function of constraint

$$\{\mathbf{f}\}=\begin{bmatrix} f_1(q_1,q_2,\ldots,q_n) \\ f_2(q_1,q_2,\ldots,q_n) \\ \cdots\cdots\cdots\cdots \\ f_p(q_1,q_2,\ldots,q_n) \end{bmatrix}=\begin{bmatrix} 0 \\ 0 \\ \ldots \\ 0 \end{bmatrix}, \tag{5.61}$$

where $\{\mathbf{F_q}\}$ is the column matrix of *the generalized forces*, $\{\mathbf{q}\}$ is the column matrix of *the generalized coordinates*, $[\mathbf{B}]$ is *the matrix of constraints*, that is *the Jacobi matrix* of the function $\{\mathbf{f}\}$, while $\{\boldsymbol{\lambda}\}$ is the column matrix of *the Lagrange multipliers*.

In the case of the equilibrium of the rigid solid one obtains the same equations in which

$$\{\mathbf{q}\}=[X_O \;\; Y_O \;\; Z_O \;\; \psi \;\; \theta \;\; \varphi]^{\mathrm{T}}. \tag{5.62}$$

To be out to numerically solve the systems of equations (5.60) and (5.61), one applies the Newton-Raphson method and makes the development into series around the approximate solution X_O^0, Y_O^0, Z_O^0, ψ^0, θ^0, φ^0, λ_1^0, …, λ_p^0. If one makes the notations

$$\{\mathbf{F}_{X_O}\}=\frac{\partial\{\mathbf{F}_q\}}{\partial X_O},\{\mathbf{F}_{Y_O}\}=\frac{\partial\{\mathbf{F}_q\}}{\partial Y_O},\ldots,\{\mathbf{F}_{\varphi}\}=\frac{\partial\{\mathbf{F}_q\}}{\partial \varphi}, \tag{5.63}$$

$$[\mathbf{JF}]=\left[\{\mathbf{F}_{X_O}\}\;\;\{\mathbf{F}_{Y_O}\}\;\;\{\mathbf{F}_{Z_O}\}\;\;\{\mathbf{F}_{\psi}\}\;\;\{\mathbf{F}_{\theta}\}\;\;\{\mathbf{F}_{\varphi}\}\right], \tag{5.64}$$

$$[\mathbf{B}_{X_O}]=\frac{\partial[\mathbf{B}]^{\mathrm{T}}}{\partial X_O},[\mathbf{B}_{Y_O}]=\frac{\partial[\mathbf{B}]^{\mathrm{T}}}{\partial Y_O},\ldots,[\mathbf{B}_{\varphi}]=\frac{\partial[\mathbf{B}]^{\mathrm{T}}}{\partial \varphi}, \tag{5.65}$$

$$[\mathbf{JB}]=\left[[\mathbf{B}_{X_O}]\{\boldsymbol{\lambda}\}\;\;[\mathbf{B}_{Y_O}]\{\boldsymbol{\lambda}\}\;\;[\mathbf{B}_{Z_O}]\{\boldsymbol{\lambda}\}\;\;[\mathbf{B}_{\psi}]\{\boldsymbol{\lambda}\}\;\;[\mathbf{B}_{\theta}]\{\boldsymbol{\lambda}\}\;\;[\mathbf{B}_{\varphi}]\{\boldsymbol{\lambda}\}\right], \tag{5.66}$$

then it results the matrix equation

$$\begin{bmatrix} [\mathbf{JF}]+[\mathbf{JB}] & [\mathbf{B}]^{\mathrm{T}} \\ [\mathbf{B}] & [\mathbf{0}] \end{bmatrix}\begin{bmatrix} \{\Delta\mathbf{q}\} \\ \{\Delta\boldsymbol{\lambda}\} \end{bmatrix}=\begin{bmatrix} -\{\mathbf{F}_q\}-[\mathbf{B}]^{\mathrm{T}}\{\boldsymbol{\lambda}\} \\ -\{\mathbf{f}\} \end{bmatrix} \tag{5.67}$$

from which one determines the variations $\{\Delta\mathbf{q}\}$, $\{\Delta\boldsymbol{\lambda}\}$.

The calculation process is an iterative one, that is, one makes the replacements $q_i^0 \to q_i^0+\Delta q_i$, $\lambda_i^0 \to \lambda_i^0+\Delta\lambda_i$ and recalculates the values $\{\Delta\mathbf{q}\}$, $\{\Delta\boldsymbol{\lambda}\}$ until the chosen precision is reached ($|\Delta q_i|<\varepsilon_1$, $|\Delta\lambda_i|<\varepsilon_2$) for any $i=\overline{1,n}$, and $j=\overline{1,p}$.

5.4.2 The Case When the Functions of Constraints Introduce Auxiliary Coordinates (Pseudo-Coordinates)

We consider the case in which the p functions of constraint depend on the basic generalized coordinates q_1, q_2, ..., q_n and on *the generalized pseudo-coordinates* ξ_1, ξ_2, ..., ξ_k, that is

$$f_i(q_1, q_2, \ldots, q_n, \xi_1, \xi_2, \ldots, \xi_k) = 0, i = \overline{1, p}. \quad (5.68)$$

Deriving with respect to time, denoting the Jacobi matrix by [**B**], and using the notations

$$\{\mathbf{q}\} = [X_O \ Y_O \ Z_O \ \psi \ \theta \ \varphi]^{\mathrm{T}}, \{\boldsymbol{\xi}\} = [\xi_1, \xi_2, \ldots, \xi_k]^{\mathrm{T}}, \quad (5.69)$$

$$[\mathbf{B}] = \begin{bmatrix} [\mathbf{B}_{11}] & [\mathbf{B}_{12}] \\ [\mathbf{B}_{21}] & [\mathbf{B}_{22}] \end{bmatrix}, \quad (5.70)$$

where $[\mathbf{B}_{11}]$ has $p-k$ rows and 6 columns, $[\mathbf{B}_{12}]$ has $p-k$ rows and k columns, $[\mathbf{B}_{21}]$ has k rows and 6 columns, while $[\mathbf{B}_{22}]$ has k rows and k columns, one obtains the equality

$$\begin{bmatrix} [\mathbf{B}_{11}] & [\mathbf{B}_{12}] \\ [\mathbf{B}_{21}] & [\mathbf{B}_{22}] \end{bmatrix} \begin{bmatrix} \{\dot{\mathbf{q}}\} \\ \{\dot{\boldsymbol{\xi}}\} \end{bmatrix} = \{\mathbf{0}\}, \quad (5.71)$$

which separates into the equations

$$[\mathbf{B}_{11}]\{\dot{\mathbf{q}}\} + [\mathbf{B}_{12}]\{\dot{\boldsymbol{\xi}}\} = \{\mathbf{0}\}, [\mathbf{B}_{21}]\{\dot{\mathbf{q}}\} + [\mathbf{B}_{22}]\{\dot{\boldsymbol{\xi}}\} = \{\mathbf{0}\}. \quad (5.72)$$

Considering that the matrix $[\mathbf{B}_{22}]$ is invertible, with the aid of the notation

$$[\mathbf{B}^*] = [\mathbf{B}_{11}] - [\mathbf{B}_{12}][\mathbf{B}_{22}]^{-1}[\mathbf{B}_{21}], \quad (5.73)$$

from the relation (5.72) one deduces the equality

$$[\mathbf{B}^*]\{\dot{\mathbf{q}}\} = \{\mathbf{0}\}; \quad (5.74)$$

hence the pseudo-coordinates are eliminated and the matrix equation of equilibrium (5.60) becomes

$$\{\mathbf{F}_q\} + [\mathbf{B}^*]^{\mathrm{T}}\{\boldsymbol{\lambda}\} = \{\mathbf{0}\}, \quad (5.75)$$

where the matrix $\{\boldsymbol{\lambda}\}$ has $p-k$ elements.

One may obtain the matrix equation of equilibrium (5.75) using another argumentation: considering as generalized coordinates the all $6+k$ coordinates X_O, Y_O, Z_O, ψ, θ, φ, ξ_1, ξ_2, ..., ξ_k, it results that the generalized forces F_{ξ_k} vanish and, consequently, the equation of equilibrium (5.60) reads

$$\begin{bmatrix} \{\mathbf{F}_q\} \\ \{\mathbf{0}\} \end{bmatrix} + \begin{bmatrix} [\mathbf{B}_{11}]^{\mathrm{T}} & [\mathbf{B}_{12}]^{\mathrm{T}} \\ [\mathbf{B}_{21}]^{\mathrm{T}} & [\mathbf{B}_{22}]^{\mathrm{T}} \end{bmatrix} \begin{bmatrix} \{\boldsymbol{\lambda}\} \\ \{\boldsymbol{\lambda}^*\} \end{bmatrix} = \{\mathbf{0}\} \tag{5.76}$$

or

$$\{\mathbf{F}_q\} + [\mathbf{B}_{11}]^{\mathrm{T}}\{\boldsymbol{\lambda}\} + [\mathbf{B}_{21}]^{\mathrm{T}}\{\boldsymbol{\lambda}^*\} = \{\mathbf{0}\}, [\mathbf{B}_{12}]^{\mathrm{T}}\{\boldsymbol{\lambda}\} + [\mathbf{B}_{22}]^{\mathrm{T}}\{\boldsymbol{\lambda}^*\} = \{\mathbf{0}\}; \tag{5.77}$$

by elimination of the matrix $\{\boldsymbol{\lambda}^*\}$, one obtains the equation (5.75).

In this way, it is more advantageous to consider as conditions of equilibrium the equations

$$\{\mathbf{F}_q\} + [\mathbf{B}_{11}]^{\mathrm{T}}\{\boldsymbol{\lambda}\} + [\mathbf{B}_{21}]^{\mathrm{T}}\{\boldsymbol{\lambda}^*\} = \{\mathbf{0}\}, [\mathbf{B}_{12}]^{\mathrm{T}}\{\boldsymbol{\lambda}\} + [\mathbf{B}_{22}]^{\mathrm{T}}\{\boldsymbol{\lambda}^*\} = \{\mathbf{0}\}, \{\mathbf{f}\} = \{\mathbf{0}\}, \tag{5.78}$$

where

$$\{\mathbf{f}\} = [f_1 \ f_2 \ \dots \ f_p]^{\mathrm{T}}, \tag{5.79}$$

f_i, $i=\overline{1,p}$ being the functions (5.67).

The parameters which appear in the $6+k+p$ scalar equations (5.78) are also in number of de $6+k+p$, that is, X_O, Y_O, Z_O, ψ, θ, φ, ξ_1, ξ_2, ..., ξ_k, λ_1, λ_2, ..., λ_{p-k}, λ_1^*, λ_2^*, ..., λ_k^*.

Using notations similar to (5.63)–(5.70), the numerical solution of the system (5.79) by the Newton-Raphson method leads to the system

$$\begin{bmatrix} [\mathbf{JF}] + [\mathbf{JB}_{11}] + [\mathbf{JB}_{12}] & \left[\widetilde{\mathbf{JB}}_{11}\right] + \left[\widetilde{\mathbf{JB}}_{21}\right] & [\mathbf{B}_{11}]^{\mathrm{T}} & [\mathbf{B}_{21}]^{\mathrm{T}} \\ [\mathbf{JB}_{12}] + [\mathbf{JB}_{22}] & \left[\widetilde{\mathbf{JB}}_{12}\right] + \left[\widetilde{\mathbf{JB}}_{22}\right] & [\mathbf{B}_{12}]^{\mathrm{T}} & [\mathbf{B}_{22}]^{\mathrm{T}} \\ [\mathbf{B}_{11}] & [\mathbf{B}_{12}] & [\mathbf{0}] & [\mathbf{0}] \\ [\mathbf{B}_{21}] & [\mathbf{B}_{22}] & [\mathbf{0}] & [\mathbf{0}] \end{bmatrix} \begin{bmatrix} \{\Delta\mathbf{q}\} \\ \{\Delta\boldsymbol{\xi}\} \\ \{\Delta\boldsymbol{\lambda}\} \\ \{\Delta\boldsymbol{\lambda}^*\} \end{bmatrix} \tag{5.80}$$

$$= -\begin{bmatrix} \{\mathbf{F}_q\} + [\mathbf{B}_{11}]^{\mathrm{T}}\{\boldsymbol{\lambda}\} + [\mathbf{B}_{21}]^{\mathrm{T}}\{\boldsymbol{\lambda}^*\} \\ [\mathbf{B}_{11}]^{\mathrm{T}}\{\boldsymbol{\lambda}\} + [\mathbf{B}_{22}]^{\mathrm{T}}\{\boldsymbol{\lambda}^*\} \\ \{\mathbf{f}\} \end{bmatrix}.$$

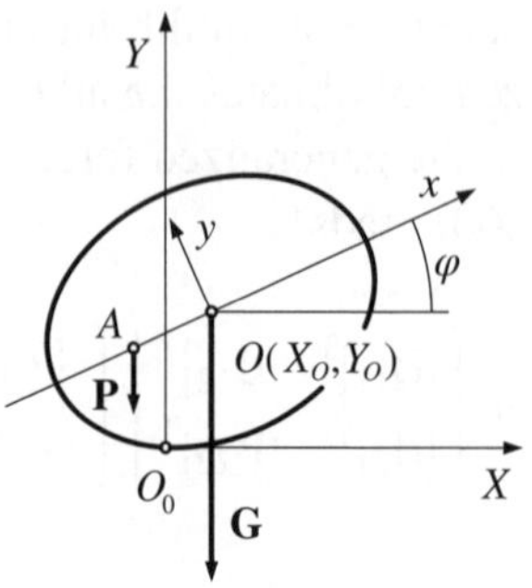

Figure 5.2 Determination of the equilibrium for the elliptical shell in Example 5.4.1.

Example 5.4.1 Determine the equilibrium position for the elliptical shell in Fig. 5.2, of semi-axes a, b and weight G, supported at the point O_0 and acted at the point A ($OA = d$) by another force P.

Solution: One considers the parametric equations of the ellipse $x = a\cos\xi$, $y = b\sin\xi$. The basic generalized coordinates are X_O, Y_O, φ, while ξ is a *pseudo-coordinate.*

The contact at the point O_0 leads to the equations

$$X_O + a\cos\varphi\cos\xi - b\sin\varphi\sin\xi = 0, Y_O + a\sin\varphi\cos\xi + b\cos\varphi\cos\xi = 0,$$

while the tangency at the same point offers the equation

$$-a\sin\varphi\sin\xi + b\cos\varphi\cos\xi = 0.$$

One obtains the matrices

$$[\mathbf{B}_{11}] = \begin{bmatrix} 1 & 0 & -a\sin\varphi\cos\xi - b\cos\varphi\sin\xi \\ 0 & 1 & a\cos\varphi\cos\xi - b\sin\varphi\sin\xi \end{bmatrix}, [\mathbf{B}_{12}] = \begin{bmatrix} -a\cos\varphi\sin\xi - b\sin\varphi\cos\xi \\ -a\sin\varphi\sin\xi + b\cos\varphi\cos\xi \end{bmatrix},$$

$$[\mathbf{B}_{21}] = [0 \;\; 0 \;\; -a\cos\varphi\sin\xi - b\sin\varphi\cos\xi], [\mathbf{B}_{22}] = [-a\sin\varphi\cos\xi - b\cos\varphi\sin\xi].$$

The potential energy reads

$$V = GY_O + P(Y_O - d\sin\varphi)$$

and leads to the matrix of generalized forces

$$\{\mathbf{F}_q\} = \begin{bmatrix} 0 \\ -G-P \\ Pd\cos\varphi \end{bmatrix}.$$

The system (5.78) takes the form

$$\begin{bmatrix} 0 \\ -G-P \\ Pd\cos\varphi \end{bmatrix} \begin{bmatrix} 1 & 0 \\ 0 & 1 \\ -a\sin\varphi\cos\xi - b\cos\varphi\sin\xi & a\cos\varphi\cos\xi - b\sin\varphi\sin\xi \end{bmatrix} \begin{bmatrix} \lambda_1 \\ \lambda_2 \end{bmatrix}$$

$$+ \begin{bmatrix} 0 \\ 0 \\ -a\cos\varphi\sin\xi - b\sin\varphi\cos\xi \end{bmatrix} \lambda^* = \begin{bmatrix} 0 \\ 0 \\ 0 \end{bmatrix},$$

$$[-a\cos\varphi\sin\xi - b\sin\varphi\cos\xi \quad 0] \begin{bmatrix} \lambda_1 \\ \lambda_2 \end{bmatrix} + [-a\sin\varphi\cos\xi - b\cos\varphi\sin\xi]\{\boldsymbol{\lambda}^*\} = 0,$$

$$X_O + a\cos\varphi\cos\xi - b\sin\varphi\sin\xi = 0, Y_O + a\sin\varphi\cos\xi + b\cos\varphi\sin\xi = 0,$$

$$-a\sin\varphi\sin\xi + b\cos\varphi\cos\xi = 0.$$

First of all, one deduces the equalities

$$\lambda_1 = 0, \lambda_2 = G + P, \lambda^* = 0, Pd\cos\varphi + (a\cos\varphi\cos\xi - b\sin\varphi\sin\xi)(G + P) = 0$$

resulting the expression

$$X_O = \frac{Pd\cos\varphi}{P+G} = \frac{(a^2-b^2)\cos\varphi\sin\varphi}{\sqrt{a^2\sin^2\varphi + b^2\cos^2\varphi}},$$

from which one obtains

$$\tan\varphi = \frac{Pdb}{\sqrt{(P+G)^2(a^2-b^2)^2 - P^2a^2d^2}}.$$

It also results

$$Y_O = \sqrt{a^2\sin^2\varphi + b^2\cos^2\varphi},$$

$$\cot\xi = \frac{a}{b}\tan\varphi.$$

If the ellipse becomes a circle, $a = b = R$, then the previous formulae offer

$$X_O = 0, \cos\varphi = 0, \varphi = \frac{\pi}{2} \text{ or } \varphi = -\frac{\pi}{2}, Y_O = R, \xi = \pi \text{ or } \xi = 0.$$

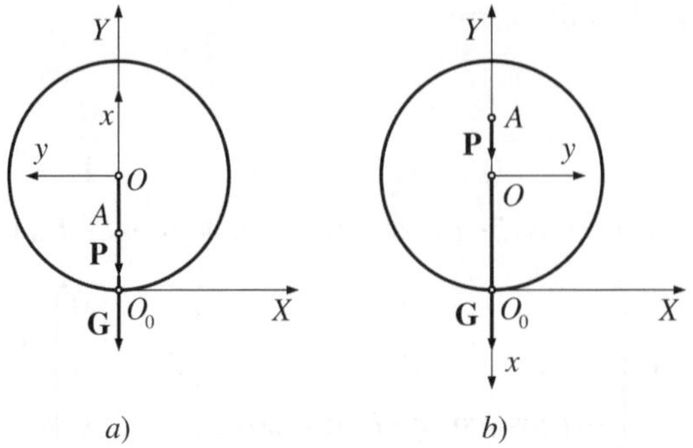

Figure 5.3 The equilibrium positions for the circle.

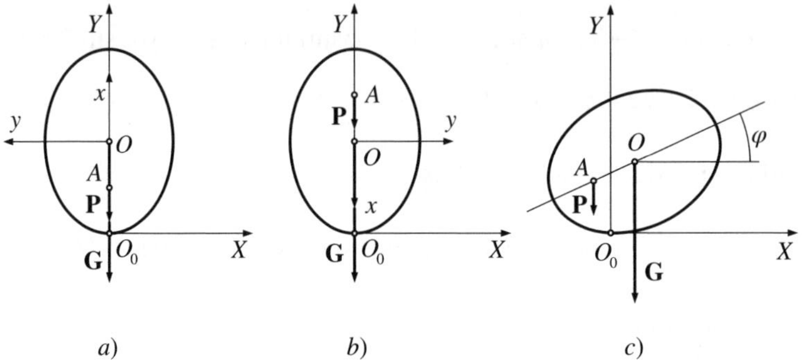

Figure 5.4 The equilibrium positions for the ellipse.

The two equilibrium positions are represented in Fig. 5.3.

In the case of the ellipse may exist two or three equilibrium positions as follows:

- if $P < \dfrac{G(a^2-b^2)}{da-(a^2-b^2)}$, $P>0$, then there exist three equilibrium positions drawn in Fig. 5.4 *a*), *b*) and *c*);
- if $P \geq \dfrac{G(a^2-b^2)}{da-(a^2-b^2)}$, $P>0$, there exist only two equilibrium positions drawn in Fig 5.4 *a*) and *b*).

5.5 The Equilibrium of the Heavy Rigid Solid Hanged by Springs

5.5.1 The Matrix Equation of Equilibrium

One considers (Fig. 5.5) the rigid solid of mass m, hanged by the springs A_iB_i, $i=\overline{1,n}$, of stiffness k_i, lengths l_{0i} in the non-deformed statuses, and lengths L_i at the equilibrium status.

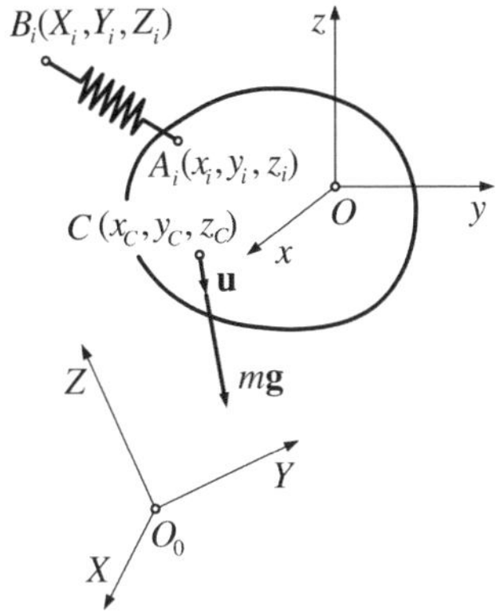

Figure 5.5 The equilibrium of the heavy rigid solid hanged by springs.

One uses the following notations:

- $\mathbf{u}$ – the unit vector of the vertical direction orientated to the surface of Earth;
- a, b, c – the projections of the unit vector $\mathbf{u}$ onto the axes of the reference system O_0XYZ;
- $\{\mathbf{u}\}$ – the column matrix

$$\{\mathbf{u}\} = [a \ \ b \ \ c]^{\mathrm{T}}; \tag{5.81}$$

- x_i, y_i, z_i – the coordinates of the points A_i relative to the local reference system $Oxyz$, $i = \overline{1,n}$;
- $\{\mathbf{r}_i\}$ – the column matrix

$$\{\mathbf{r}_i\} = [x_i \ \ y_i \ \ z_i]^{\mathrm{T}}; \tag{5.82}$$

- $[\mathbf{r}_i]$ – the skew matrix

$$[\mathbf{r}_i] = \begin{bmatrix} 0 & -z_i & y_i \\ z_i & 0 & -x_i \\ -y_i & x_i & 0 \end{bmatrix}; \tag{5.83}$$

- x_C, y_C, z_C – the coordinates of the center of weight relative to the local reference system $Oxyz$;
- $[\mathbf{r}_C]$ – the skew matrix

$$[\mathbf{r}_C] = \begin{bmatrix} 0 & -z_C & y_C \\ z_C & 0 & -x_C \\ -y_C & x_C & 0 \end{bmatrix}; \tag{5.84}$$

- $[\mathbf{QR}_i\mathbf{A}]$, and $[\mathbf{QR}_C\mathbf{A}]$ – the products of matrices

$$[\mathbf{QR}_i\mathbf{A}] = [\mathbf{Q}]^{\mathrm{T}}[\mathbf{r}_i][\mathbf{A}]^{\mathrm{T}}, \tag{5.85}$$

and

$$[\mathbf{QR}_C\mathbf{A}] = [\mathbf{Q}]^{\mathrm{T}}[\mathbf{r}_C][\mathbf{A}]^{\mathrm{T}}, \tag{5.86}$$

respectively;
- X_i, Y_i, Z_i – the coordinates of the fixed points B_i relative to the reference system O_0XYZ;
- $\{\mathbf{R}_i\}$ – the column matrices

$$\{\mathbf{R}_i\} = [X_i \;\; Y_i \;\; Z_i]^{\mathrm{T}}; \tag{5.87}$$

- $\{\mathbf{R}_O\}$ – the column matrix

$$\{\mathbf{R}_O\} = [X_O \;\; Y_O \;\; Z_O]^{\mathrm{T}}, \tag{5.88}$$

- s_i – the elongations of the springs at the rigid solid's equilibrium position

$$s_i = L_i - l_{0i}, i = \overline{1,n}. \tag{5.89}$$

With these notations, by virtue of the relations (5.51) and (5.57), one writes *the matrix equation of equilibrium*

$$\sum \frac{k_i s_i}{L_i}\begin{bmatrix} [\mathbf{I}] \\ [\mathbf{QR}_i\mathbf{A}] \end{bmatrix}\{\{\mathbf{R}_i\} - \{\mathbf{R}_O\} - [\mathbf{A}]\{\mathbf{r}_i\}\} + mg\begin{bmatrix} [\mathbf{I}] \\ [\mathbf{QR}_C\mathbf{A}] \end{bmatrix}\{\mathbf{u}\} = \{\mathbf{0}\}, \tag{5.90}$$

which splits into two matrix equations of equilibrium

$$\{\mathbf{F}_\mathbf{s}\} = \sum \frac{k_i s_i}{L_i}\{\{\mathbf{R}_i\} - \{\mathbf{R}_O\} - [\mathbf{A}]\{\mathbf{r}_i\}\} + mg\{\mathbf{u}\} = \{\mathbf{0}\}, \tag{5.91}$$

$$\{\mathbf{F}_{\boldsymbol{\beta}}\} = \sum \frac{k_i s_i}{L_i}[\mathbf{QR}_i\mathbf{A}]\{\{\mathbf{R}_i\} - \{\mathbf{R}_O\}\} + mg[\mathbf{QR}_C\mathbf{A}]\{\mathbf{u}\} = \{\mathbf{0}\}, \tag{5.92}$$

5.5.2 Numerical Solution

We consider the approximate solution

$$\left\{\mathbf{q}^{(0)}\right\} = \left[\Delta X_O^{(0)} \;\; \Delta Y_O^{(0)} \;\; \Delta Z_O^{(0)} \;\; \psi^{(0)} \;\; \theta^{(0)} \;\; \varphi^{(0)}\right]^{\mathrm{T}} \tag{5.93}$$

and the matrices $\{\mathbf{F}_\mathbf{s}^{(0)}\}$, and $\left\{\mathbf{F}_\boldsymbol{\beta}^{(0)}\right\}$ which represent the matrices $\{\mathbf{F}_\mathbf{s}\}$, and $\{\mathbf{F}_\boldsymbol{\beta}\}$, respectively, calculated for this solution.

Developing into series the functions $\{\mathbf{F}_\mathbf{s}\}$, and $\{\mathbf{F}_\boldsymbol{\beta}\}$ around the solutions $\{\mathbf{q}^{(0)}\}$ and neglecting the nonlinear terms, one obtains the expressions

$$-\sum\frac{k_i s_i}{L_i}\left\{\begin{bmatrix}\Delta X_0\\ \Delta Y_0\\ \Delta Z_0\end{bmatrix}-[\mathbf{A}_\psi\{\mathbf{r}_i\}\ \ \mathbf{A}_\theta\{\mathbf{r}_i\}\ \ \mathbf{A}_\varphi\{\mathbf{r}_i\}]\begin{bmatrix}\Delta\psi\\ \Delta\theta\\ \Delta\varphi\end{bmatrix}\right\}+\left\{\mathbf{F}_\mathbf{s}^{(0)}\right\}=\{\mathbf{0}\},$$

$$-\sum\frac{k_i s_i}{L_i}[\mathbf{QR_iA}]\begin{bmatrix}\Delta X_O\\ \Delta Y_O\\ \Delta Z_O\end{bmatrix}+\left[\sum\frac{k_i s_i}{L_i}\left[\mathbf{Q\widetilde{R}_iA}\right]+\left[\mathbf{Q\widetilde{R}_CA}\right]\right]\begin{bmatrix}\Delta\psi\\ \Delta\theta\\ \Delta\varphi\end{bmatrix}+\left\{F_\boldsymbol{\beta}^{(0)}\right\}=\{\mathbf{0}\},\tag{5.94}$$

where

$$\begin{aligned}
&\left[\mathbf{Q\widetilde{R}}_i\mathbf{A}\right]=\left[[\mathbf{QR}_i\mathbf{A}_\psi]\{\{\mathbf{R}_i\}-\{\mathbf{R}_O\}\}\ \ [\mathbf{QR}_i\mathbf{A}_\theta]\{\{\mathbf{R}_i\}-\{\mathbf{R}_O\}\}\ \ [\mathbf{QR}_i\mathbf{A}_\varphi]\{\{\mathbf{R}_i\}-\{\mathbf{R}_O\}\}\right],\\
&\left[\mathbf{Q\widetilde{R}}_C\mathbf{A}\right]=\left[[\mathbf{QR}_C\mathbf{A}_\psi]\{\mathbf{u}\}\ \ [\mathbf{QR}_C\mathbf{A}_\theta]\{\mathbf{u}\}\ \ [\mathbf{QR}_C\mathbf{A}_\varphi]\{\mathbf{u}\}\right],[\mathbf{QR}_i\mathbf{A}_\psi]=[\mathbf{Q}]^\mathrm{T}[\mathbf{r}_i][\mathbf{A}_\psi]^\mathrm{T},\\
&[\mathbf{QR}_i\mathbf{A}_\theta]=[\mathbf{Q}_\theta]^\mathrm{T}[\mathbf{r}_i][\mathbf{A}]^\mathrm{T}+[\mathbf{Q}]^\mathrm{T}[\mathbf{r}_i][\mathbf{A}_\theta]^\mathrm{T},\ [\mathbf{QR}_i\mathbf{A}_\varphi]=[\mathbf{Q}_\varphi]^\mathrm{T}[\mathbf{r}_i][\mathbf{A}]^\mathrm{T}+[\mathbf{Q}]^\mathrm{T}[\mathbf{r}_i][\mathbf{A}_\varphi]^\mathrm{T},\\
&[\mathbf{QR}_C\mathbf{A}_\psi]=[\mathbf{Q}]^\mathrm{T}[\mathbf{r}_C][\mathbf{A}_\psi]^\mathrm{T},[\mathbf{QR}_C\mathbf{A}_\theta]=[\mathbf{Q}_\theta]^\mathrm{T}[\mathbf{r}_C][\mathbf{A}]^\mathrm{T}+[\mathbf{Q}]^\mathrm{T}[\mathbf{r}_C][\mathbf{A}_\theta]^\mathrm{T},\\
&[\mathbf{QR}_C\mathbf{A}_\varphi]=[\mathbf{Q}_\varphi]^\mathrm{T}[\mathbf{r}_C][\mathbf{A}]^\mathrm{T}+[\mathbf{Q}]^\mathrm{T}[\mathbf{r}_C][\mathbf{A}_\varphi]^\mathrm{T}.
\end{aligned}\tag{5.95}$$

Taking into account the equality

$$\left[[\mathbf{A}_\psi]\{\mathbf{r}_i\}\ \ [\mathbf{A}_\theta]\{\mathbf{r}_i\}\ \ [\mathbf{A}_\varphi]\{\mathbf{r}_i\}\right]=[\mathbf{QR}_i\mathbf{A}]^\mathrm{T},\tag{5.96}$$

from the relation (5.94) one deduces the expression

$$\begin{bmatrix}-\sum\frac{k_i s_i}{L_i}[\mathbf{I}] & -\sum\frac{k_i s_i}{L_i}[\mathbf{QR}_i\mathbf{A}]^\mathrm{T}\\ -\sum\frac{k_i s_i}{L_i}[\mathbf{QR}_i\mathbf{A}] & \sum\frac{k_i s_i}{L_i}\left[\mathbf{Q\widetilde{R}}_i\mathbf{A}\right]+\left[\mathbf{Q\widetilde{R}}_C\mathbf{A}\right]\end{bmatrix}\{\Delta\mathbf{q}\}+\begin{bmatrix}\left\{\mathbf{F}_\mathbf{s}^{(0)}\right\}\\ \left\{\mathbf{F}_\boldsymbol{\beta}^{(0)}\right\}\end{bmatrix}=\{\mathbf{0}\};\tag{5.97}$$

from the last equation one may obtain the variation $\{\boldsymbol{\Delta}\mathbf{q}\}$ and then the column matrix

$$\{\mathbf{q}\}=\{\mathbf{q}^0\}+\{\Delta\mathbf{q}\},\tag{5.98}$$

which is the approximate solution for a new iteration.

5.5.3 *The Case When the Fixed Reference System Coincides to the Local Reference System at the Equilibrium Position*

If we multiply the equation (5.92) at the left by $\left[[\mathbf{Q}]^{\mathrm{T}}\right]^{-1}$, then the equilibrium equations (5.91) and (5.92) become

$$\sum \frac{k_i s_i}{L_i}\{\{\mathbf{R}_i\}-\{\mathbf{R}_O\}-[\mathbf{A}]\{\mathbf{r}_i\}\}+mg\{\mathbf{u}\}=\{\mathbf{0}\}, \tag{5.99}$$

$$\sum \frac{k_i s_i}{L_i}[\mathbf{r}_i][\mathbf{A}]^{\mathrm{T}}\{\{\mathbf{R}_i\}-\{\mathbf{R}_O\}-[\mathbf{A}]\{\mathbf{r}_i\}\}+mg[\mathbf{r}_C][\mathbf{A}]^{\mathrm{T}}\{\mathbf{u}\}=\{\mathbf{0}\}. \tag{5.100}$$

If the fixed reference system coincides to the local reference system at the equilibrium position, then one obtains the relations

$$\{\mathbf{R}_O\}=\{\mathbf{0}\},[\mathbf{A}]=[\mathbf{I}];\{\mathbf{R}_i\}-\{\mathbf{r}_i\}=L_i\{\mathbf{u}_i\}, \tag{5.101}$$

and the equilibrium equations become

$$\sum k_i s_i\{\mathbf{u}_i\}+mg\{\mathbf{u}\}=\{\mathbf{0}\},\sum k_i s_i[\mathbf{r}_i]\{\mathbf{u}_i\}+mg[\mathbf{r}_C]\{\mathbf{u}\}=\{\mathbf{0}\}; \tag{5.102}$$

in the last relation $\{\mathbf{u}_i\}$ represents the column matrix of the projections of the unit vectors of the vectors $\mathbf{A}_i\mathbf{B}_i$ onto the fixed reference system.

Example 5.5.1 A parallelepiped of mass m and lengths of sides l_1, l_2, l_3, (Fig. 5.6) is hanged at the points A_i by the springs of stiffness k_{i1}, k_{i2}, k_{i3}. Knowing that at the position given in Fig. 5.6 the parallelepiped is in equilibrium, write the equations of equilibrium and determine the elongations at the equilibrium in the hypothesis that the elongations s_{11}, s_{12}, and s_{32} are null.

Solution: Denoting by s_{ij} the elongations of the springs of stiffness k_{ij}, one obtains the equilibrium equations

$$k_{11}s_{11}+k_{21}s_{21}+k_{31}s_{31}=0, k_{12}s_{12}+k_{22}s_{22}+k_{32}s_{32}=0, k_{13}s_{13}+k_{23}s_{23}+k_{33}s_{33}=mg,$$
$$l_2k_{23}s_{23}-l_2k_{32}s_{32}=mg\frac{l_2}{2}, -l_1k_{13}s_{13}+l_3k_{31}s_{31}=-mg\frac{l_1}{2}, l_1k_{12}s_{12}-l_2k_{21}s_{21}=0, \tag{5.103}$$

wherefrom it results $s_{21}=s_{31}=s_{22}=s_{33}=0$, $s_{23}=\dfrac{mg}{2k_{23}}$, $s_{13}=\dfrac{mg}{2k_{13}}$.

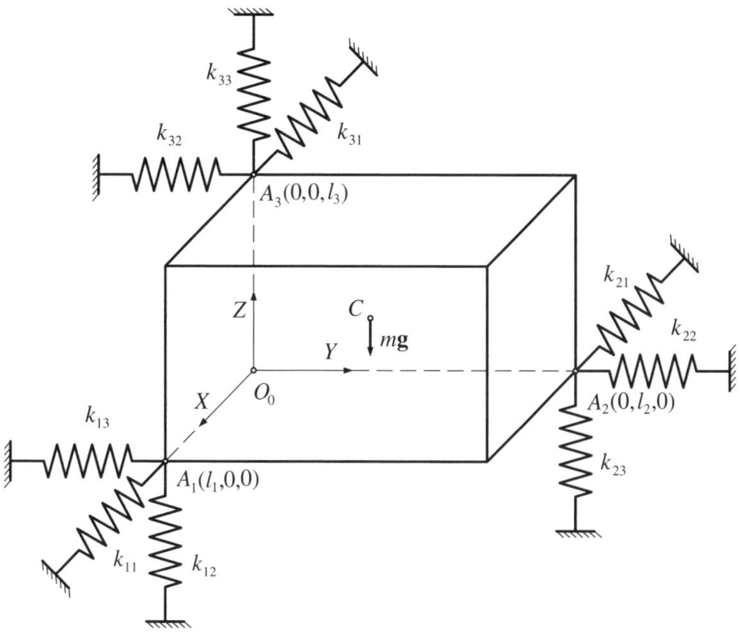

Figure 5.6 Determination of the equilibrium position for the parallelepiped in Example 5.5.1.

Further Reading

Arnold VI (1997). Mathematical Methods of Classical Mechanics (Graduate Texts in Mathematics, Vol. 60). 2nd ed. New York: Springer.

Ball RS (1998). A Treatise on the Theory of Screws. Cambridge: Cambridge University Press.

Bloch AM, Baillieul J, Crouch P, Marsden J (2007). Nonholonomic Mechanics and Control (Interdisciplinary Applied Mathematics). Berlin: Springer.

Bremer H (2008). Elastic Multibody Dynamics: A Direct Ritz Approach. Berlin: Springer.

Chaudhary H, Saha SH (2008). Dynamics and Balancing of Multibody Systems (Lecture Notes in Applied and Computational Mechanics). Berlin: Springer.

Cortes J (2002). Geometric, Control and Numerical Aspects of Nonholonomic Systems. Berlin: Springer.

Coutinho MG (2001). Dynamic Simulations of Multibody Systems. New York: Springer.

de Jalón JG, Bayo E (2011). Kinematic and Dynamic Simulation of Multibody Systems: The Real-Time Challenge. Berlin: Springer Verlag.

Den Hartog JP (1961). Mechanics. New York: Dover Publications Inc.

Douglas Gregory R (2006). Classical Mechanics. Cambridge: Cambridge University Press.

Edelen DG (2010). Lagrangian Mechanics of Nonconservative Nonholonomic Systems. Leyden: Kluwer.

Fasano A, Marmi S, Pelloni B (2006). Analytical Mechanics: An Introduction. Oxford: Oxford University Press.

Fearstone R (2007). Rigid body Dynamics Algorithms. New York: Springer.

Gattringer H, Gerstmayr J (2013). Multibody Systems Dynamics, Robotics and Control. Wien: Springer.

Ginsberg JH (1998). Advanced Engineering Dynamics. 2nd ed. Cambridge: Cambridge University Press.

Greenwood DT (2006). Advanced Dynamics. Cambridge: Cambridge University Press.
Greiner W (2009). Classical Mechanics: Systems of Particles and Hamiltonian Dynamics. 2nd ed. Heidelberg: Springer.
Hahn H (2002). Rigid Body Dynamics of Mechanisms 1: Theoretical Basis. Berlin: Springer.
Hand LN, Finch JD (1998). Analytical Mechanics. Cambridge: Cambridge University Press.
Jain A (2010). Robot and Multibody Dynamics: Analysis and Algorithms. New York: Springer.
Jazar RN (2011). Advanced Dynamics: Rigid Body, Multibody, and Aerospace Applications. Hoboken: John Wiley & Sons.
Jazar RN (2010). Theory of Applied Robotics: Kinematics, Dynamics, and Control. 2nd ed. New York: Springer.
Karnopp DC, Margolis DL, Rosenberg RC (2012). System Dynamics: Modeling, Simulation, and Control of Mechatronic Systems. Hoboken: John Wiley & Sons.
Kibble TWB, Berkshire FH (2004). Classical Mechanics. 5th ed. London: Imperial College Press.
Lagrange JL (1996). Analytical Mechanics. Dordrecht: Kluwer Academic Publishers,
Lurie AI (2002). Analytical Mechanics. Berlin: Springer.
Meriam JL, Kraige LG (2012). Engineering Mechanics: Dynamics. Hoboken: John Wiley & Sons.
Morin D (2008). Introduction to Classical Mechanics: With Problems and Solution. Cambridge: Cambridge University Press.
Murray RM, Li Z, Shankar Sastry S (1994). A Mathematical Introduction to Robotic Manipulation. Boca Raton: CRC Press.
Neimark IuI, Fufaev NA (1972). Dynamics of Nonholonomic Systems (Translations of Mathematical Monographs, V. 33). Providence, Rhode Island: American Mathematical Society.
O'Reilly OM (2008). Intermediate Dynamics for Engineers: A unified Treatment of Newton-Euler and Lagrangian Mechanics. Cambridge: Cambridge University Press
Pandrea N (2000). Elemente de mecanica solidelor în coordonate plückeriene. Bucureşti: Editura Academiei Române (in Romanian).
Pandrea N, Stănescu ND (2002). Mecanica. Bucureşti: Editura Didactică şi Pedagogică (in Romanian).
Papastavridis JG (2002). Analytical Mechanics: A Comprehensive Treatise on the Dynamics of Constrained Systems; For Engineers, Physicists, and Mathematicians. Oxford: Oxford University Press.
Pfeiffer F, Glocker C (1996). Multibody Dynamics with Unilateral Contacts. New York: John Wiley & Sons, Inc.
Roberson RE, Schwertassek R (1988). Dynamics of Multibody Systems. Berlin: Springer Verlag.
Seifried R (2013). Dynamics of Underactuated Multibody Systems: Modeling, Control and Optimal Design. Heidelberg: Springer.
Shabana AA (2013). Dynamics of Multibody Systems. Cambridge: Cambridge University Press.
Soltakhanov ShKh, Yushkov MPP Zegzhda SA (2009). Mechanics of non-nolonomic systems: A New Class of control systems (Foundations of Engineering Mechanics). Berlin: Springer.
Spong MW, Hutchinson S, Vidyasagar M (2005). Robot Modeling and Control. New York: John Wiley & Sons.
Stănescu ND (2013). Mecanica sistemelor. Bucureşti: Editura Didactică şi Pedagogică (in Romanian).
Stănescu ND, Munteanu L, Chiroiu V, Pandrea N (2007). Sisteme dinamice. Teorie şi aplicaţii vol. I. Bucureşti: Editura Academiei Române (in Romanian).
Stănescu ND, Munteanu L, Chiroiu V, Pandrea N (2011). Sisteme dinamice. Teorie şi aplicaţii vol. II. Bucureşti: Editura Academiei Române (in Romanian).
Taylor JR (2005). Classical Mechanics. Sausalito: University Science Books.
Teodorescu PP (2009). Mechanical Systems, Classical Models: Volume 3: Analytical Mechanics (Mathematical and Analytical Techniques with Applications to Engineering). Dordrecht: Springer.

Teodorescu PP, Stănescu ND, Pandrea N (2013). Numerical Analysis with Applications in Mechanics and Engineering. Hoboken: Wiley.
Udwadia FE, Kalaba RE (2007). Analytical Dynamics: A New Approach. Cambridge: Cambridge University Press.
Uicker JJ, Ravani B, Sheth PN (2013). Matrix Methods in the Design Analysis of Mechanisms and Multibody Systems. Cambridge: Cambridge University Press.
Wittenburg J (2007). Dynamics of Multibody Systems. Berlin: Springer.
Woodhouse N (2010). Introduction to Analytical Dynamics. London: Springer.

6

The Motion of the Rigid Solid with Constraints at Given Proper Points

This chapter deals with the motion of the rigid solid having constraints at given proper points; we discuss the rigid body with one fixed point, the rigid body in rotational motion, the rigid body with one or several points situated on given surfaces or curves.

6.1 General Aspects: Classification

The motion of the rigid solid with constraints at given proper points is realized when one or several point of the rigid solid remain either fixed, or situated on fixed surfaces or curves.

This type of motion is characteristic for:

a. the rigid solid with one fixed point;
b. the rigid solid with two fixed points (the rotational motion);
c. the rigid solid with a given point situated on a fixed surface;
d. the rigid solid with several given points situated on fixed surfaces (curves);
e. the rigid solid with one fixed point and another point situated on a fixed surface;
f. the rigid solid with two given points situated on a fixed curve.

Dynamics of the Rigid Solid with General Constraints by a Multibody Approach, First Edition.
Nicolae Pandrea and Nicolae-Doru Stănescu.

Companion website: www.wiley.com/go/pandrea

6.2 Mathematical Aspects: Notations

6.2.1 The Case of the Motion Depending on Only the Generalized Coordinates X_O, Y_O, Z_O, ψ, θ, φ

In chapter 4 we established that the matrix equation of motion for the rigid solid with constraints reads

$$\begin{bmatrix} [\mathbf{M}] & -[\mathbf{B}]^{\mathrm{T}} \\ [\mathbf{B}] & [\mathbf{0}] \end{bmatrix} \begin{bmatrix} \{\ddot{\mathbf{q}}\} \\ \{\boldsymbol{\lambda}\} \end{bmatrix} = \begin{bmatrix} \{\mathbf{F_q}\} + \{\widetilde{\mathbf{F}}_{\mathbf{q}}\} \\ \{\dot{\mathbf{C}}\} - [\dot{\mathbf{B}}]\{\dot{\mathbf{q}}\} \end{bmatrix}, \tag{6.1}$$

the notations being those used in the previous chapters, that is:

$$\{\mathbf{s}\} = [X_O \ \ Y_O \ \ Z_O]^{\mathrm{T}},\ \{\boldsymbol{\beta}\} = [\psi \ \ \theta \ \ \varphi]^{\mathrm{T}},\ \{\mathbf{q}\} = [X_O \ \ Y_O \ \ Z_O \ \ \psi \ \ \theta \ \ \varphi]^{\mathrm{T}}, \tag{6.2}$$

$$[\mathbf{M}] = \begin{bmatrix} [\mathbf{m}] & [\mathbf{A}][\mathbf{S}]^{\mathrm{T}}[\mathbf{Q}] \\ [\mathbf{Q}]^{\mathrm{T}}[\mathbf{S}][\mathbf{A}]^{\mathrm{T}} & [\mathbf{Q}]^{\mathrm{T}}[\mathbf{J}_O][\mathbf{Q}] \end{bmatrix}, \tag{6.3}$$

$$[\boldsymbol{\psi}] = \begin{bmatrix} 1 & 0 & 0 \\ 0 & \cos\psi & -\sin\psi \\ 0 & \sin\psi & \cos\psi \end{bmatrix},\ [\boldsymbol{\theta}] = \begin{bmatrix} \cos\theta & 0 & \sin\theta \\ 0 & 1 & 0 \\ -\sin\theta & 0 & \cos\theta \end{bmatrix},\ [\boldsymbol{\varphi}] = \begin{bmatrix} \cos\varphi & -\sin\varphi & 0 \\ \sin\varphi & \cos\varphi & 0 \\ 0 & 0 & 1 \end{bmatrix}, \tag{6.4}$$

$$[\mathbf{A}] = [\boldsymbol{\psi}][\boldsymbol{\theta}][\boldsymbol{\varphi}], \tag{6.5}$$

$$\{\widetilde{\mathbf{F}}_{\mathbf{s}}\} = -\left[[\mathbf{A}][\mathbf{S}]^{\mathrm{T}}[\dot{\mathbf{Q}}] + [\dot{\mathbf{A}}][\mathbf{S}]^{\mathrm{T}}[\mathbf{Q}]\right]\{\dot{\boldsymbol{\beta}}\}, \tag{6.6}$$

$$\{\widetilde{\mathbf{F}}_{\boldsymbol{\beta}}\} = -\left[[\mathbf{Q}]^{\mathrm{T}}[\mathbf{J}_O][\dot{\mathbf{Q}}] + [\mathbf{Q}]^{\mathrm{T}}[\boldsymbol{\omega}][\mathbf{J}_O][\mathbf{Q}]\right]\{\dot{\boldsymbol{\beta}}\}, \tag{6.7}$$

$$\{\widetilde{\mathbf{F}}_{\mathbf{q}}\} = \left[\{\widetilde{\mathbf{F}}_{\mathbf{s}}\}^{\mathrm{T}} \ \ \{\widetilde{\mathbf{F}}_{\boldsymbol{\beta}}\}^{\mathrm{T}}\right], \tag{6.8}$$

$$[\mathbf{Q}] = \begin{bmatrix} \cos\varphi\cos\theta & \sin\varphi & 0 \\ -\sin\varphi\cos\theta & \cos\varphi & 0 \\ \sin\theta & 0 & 1 \end{bmatrix}. \tag{6.9}$$

We also set out that the angles ψ, θ, φ are the Bryan angles, $\{\mathbf{F_q}\}$ is the column matrix of the external given generalized forces, $\{\boldsymbol{\lambda}\}$ is *the column matrix of the*

Lagrange multipliers, while **[B]** is *the matrix of constraints* which depends on the generalized coordinates (X_O, Y_O, Z_O, ψ, θ, φ) and verifies the equality

$$[\mathbf{B}]\{\dot{\mathbf{q}}\} = \{\mathbf{C}\}. \tag{6.10}$$

6.2.2 *The Case of the Constraints Depending on the Pseudo-Coordinates Too*

In the case when in the constraint equation appear, excepted the coordinates X_O, Y_O, Z_O, ψ, θ, φ, other parameters too (*pseudo-coordinates*) ξ_1, ξ_2, …, ξ_p which, obviously, do not intervene in the expression of the kinetic energy, then these parameters do not appear in the left-hand side of the Lagrange equations, but they appear in the term that contains the Lagrange multipliers; consequently, we have to operate some modifications in the equation (6.1).

For that purpose, one uses the notations

$$\{\boldsymbol{\xi}\} = \begin{bmatrix}\xi_1 & \xi_2 & \cdots & \xi_p\end{bmatrix}^{\mathrm{T}}, \tag{6.11}$$

$$\{\mathbf{q}^*\} = \begin{bmatrix}\{\mathbf{q}\}^{\mathrm{T}} & \{\boldsymbol{\xi}\}^{\mathrm{T}}\end{bmatrix}^{\mathrm{T}}, \tag{6.12}$$

$$[\mathbf{M}^*] = \begin{bmatrix}[\mathbf{M}] & [\mathbf{0}_{6p}] \\ [\mathbf{0}_{p6}] & [\mathbf{0}_{pp}]\end{bmatrix}, \tag{6.13}$$

$$\{\mathbf{F}_{\mathbf{q}^*}\} = \begin{bmatrix}\{\mathbf{F}_{\mathbf{q}}\}^{\mathrm{T}} & \{\mathbf{0}_{p1}\}^{\mathrm{T}}\end{bmatrix}^{\mathrm{T}}, \left\{\widetilde{\mathbf{F}}_{\mathbf{q}^*}\right\} = \begin{bmatrix}\left\{\widetilde{\mathbf{F}}_{\mathbf{q}}\right\}^{\mathrm{T}} & \{\mathbf{0}_{p1}\}^{\mathrm{T}}\end{bmatrix}^{\mathrm{T}}, \tag{6.14}$$

where $[\mathbf{0}_{np}]$ is the null matrix with n rows and p columns; hence, the equation (6.1) becomes

$$\begin{bmatrix}[\mathbf{M}^*] & -[\mathbf{B}]^{\mathrm{T}} \\ [\mathbf{B}] & [\mathbf{0}_{n,n}]\end{bmatrix}\begin{bmatrix}\{\ddot{\mathbf{q}}^*\} \\ \{\boldsymbol{\lambda}\}\end{bmatrix} = \begin{bmatrix}\{\mathbf{F}_{\mathbf{q}^*}\} + \left\{\widetilde{\mathbf{F}}_{\mathbf{q}^*}\right\} \\ \{\dot{\mathbf{C}}\} - [\dot{\mathbf{B}}]\{\dot{\mathbf{q}}^*\}\end{bmatrix}. \tag{6.15}$$

6.2.3 *Relations of Calculation Necessary for the Numerical Algorithm*

For that purpose, one attaches the equalities

$$[\mathbf{S}] = \begin{bmatrix}0 & -mz_C & my_C \\ mz_C & 0 & -mx_C \\ -my_C & mx_C & 0\end{bmatrix}, \tag{6.16}$$

x_C, y_C, z_C being the coordinates of the center of weight C relative to the local reference system $Oxyz$,

$$[\mathbf{U}_\psi]=\begin{bmatrix}0&0&0\\0&0&-1\\0&1&0\end{bmatrix},[\mathbf{U}_\theta]=\begin{bmatrix}0&0&1\\0&0&0\\-1&0&0\end{bmatrix},[\mathbf{U}_\varphi]=\begin{bmatrix}0&-1&0\\1&0&0\\0&0&0\end{bmatrix}, \tag{6.17}$$

$$[\mathbf{A}_\psi]=[\mathbf{U}_\psi][\mathbf{A}],[\mathbf{A}_\theta]=[\mathbf{A}][\boldsymbol{\varphi}]^{\mathrm{T}}[\mathbf{U}_\theta][\boldsymbol{\varphi}],[\mathbf{A}_\varphi]=[\mathbf{A}][\mathbf{U}_\varphi], \tag{6.18}$$

$$[\dot{\mathbf{A}}]=\dot{\psi}[\mathbf{A}_\psi]+\dot{\theta}[\mathbf{A}_\theta]+\dot{\varphi}[\mathbf{A}_\varphi], \tag{6.19}$$

$$[\mathbf{Q}_\theta]=\begin{bmatrix}-\cos\varphi\sin\theta&0&0\\\sin\varphi\sin\theta&0&0\\\cos\theta&0&0\end{bmatrix},[\mathbf{Q}_\varphi]=\begin{bmatrix}-\sin\varphi\cos\theta&\cos\varphi&0\\-\cos\varphi\cos\theta&-\sin\varphi&0\\0&0&0\end{bmatrix},[\dot{\mathbf{Q}}]=\dot{\theta}[\mathbf{Q}_\theta]+\dot{\varphi}[\mathbf{Q}_\varphi], \tag{6.20}$$

$$[\boldsymbol{\omega}]=[\mathbf{A}]^{\mathrm{T}}[\dot{\mathbf{A}}]. \tag{6.21}$$

To be out to study the types of motions, classified in the first paragraph, we will deduce below the matrices of constraints and the equations of motion.

6.3 The Study of the Rigid Solid with a Fixed Point

If one considers that the point of the rigid solid, defined by the local coordinates x, y, z coincides to the fixed point of coordinates X, Y, Z, and uses the notations

$$\{\mathbf{r}\}=[x\ \ y\ \ z]^{\mathrm{T}},[\mathbf{r}]=\begin{bmatrix}0&-z&y\\z&0&-x\\-y&x&0\end{bmatrix}, \tag{6.22}$$

$$\{\mathbf{R}\}=[X\ \ Y\ \ Z]^{\mathrm{T}},\{\mathbf{R}_O\}=[X_O\ \ Y_O\ \ Z_O]^{\mathrm{T}}, \tag{6.23}$$

then one obtains the equality

$$\{\mathbf{R}\}=\{\mathbf{R}_O\}+[\mathbf{A}]\{\mathbf{r}\}; \tag{6.24}$$

by derivation with respect to time, the last relation becomes

$$\{\dot{\mathbf{R}}_O\}+[\dot{\mathbf{A}}]\{\mathbf{r}\}=\{\mathbf{0}\}. \tag{6.25}$$

Further on, based on the relations

$$[\dot{\mathbf{A}}] = [\mathbf{A}][\boldsymbol{\omega}], [\boldsymbol{\omega}]\{\mathbf{r}\} = [\mathbf{r}]^{\mathrm{T}}\{\boldsymbol{\omega}\}, \{\boldsymbol{\omega}\} = [\mathbf{Q}]\{\dot{\boldsymbol{\beta}}\} \tag{6.26}$$

and using the notations

$$[\boldsymbol{\Delta}] = [\mathbf{A}][\mathbf{r}]^{\mathrm{T}}[\mathbf{Q}], \tag{6.27}$$

$$[\boldsymbol{\Gamma}] = [\,[\mathbf{I}] \ \ [\boldsymbol{\Delta}]\,], \tag{6.28}$$

from the expression (6.25) one obtains the equality

$$[\boldsymbol{\Gamma}]\{\dot{\mathbf{q}}\} = \{\mathbf{0}\}; \tag{6.29}$$

hence, *the matrix of constraints* reads

$$[\mathbf{B}] = [\boldsymbol{\Gamma}] = [\,[\mathbf{I}] \ \ [\boldsymbol{\Delta}]\,]. \tag{6.30}$$

If the fixed point is the point O, and it coincides to the point O_0, then one obtains $[\mathbf{r}] = [\mathbf{0}]$, the matrix of constraints becomes

$$[\mathbf{B}] = [\,[\mathbf{I}] \ \ [\mathbf{0}]\,], \tag{6.31}$$

while the column matrix $\{\mathbf{q}\}$ reads

$$\{\mathbf{q}\} = \begin{bmatrix} 0 & 0 & 0 & \{\boldsymbol{\beta}\}^{\mathrm{T}} \end{bmatrix}^{\mathrm{T}}. \tag{6.32}$$

In these conditions, from the equation (6.1) one obtains the expressions

$$[\mathbf{A}][\mathbf{S}]^{\mathrm{T}}[\mathbf{Q}]\{\ddot{\boldsymbol{\beta}}\} = \{\mathbf{F}_{\mathrm{s}}\} + \left\{\widetilde{\mathbf{F}}_{\mathrm{s}}\right\} + \{\boldsymbol{\lambda}\}, \tag{6.33}$$

$$[\mathbf{Q}]^{\mathrm{T}}[\mathbf{J}_O][\mathbf{Q}]\{\ddot{\boldsymbol{\beta}}\} = \{\mathbf{F}_{\boldsymbol{\beta}}\} + \left\{\widetilde{\mathbf{F}}_{\boldsymbol{\beta}}\right\}. \tag{6.34}$$

Keeping into account the third relation (6.26), knowing that

$$\{\mathbf{F}_{\boldsymbol{\beta}}\} = [\mathbf{Q}]^{\mathrm{T}}\{\mathbf{M}_O\}, \tag{6.35}$$

where $\{\mathbf{M}_O\}$ is the column matrix of the components of the external moment that acts upon the rigid solid, then the equation (6.34), multiplied at the left by $\left[[\mathbf{Q}]^{\mathrm{T}}\right]^{-1}$, becomes

$$[\mathbf{J}_O]\{\dot{\boldsymbol{\omega}}\} + [\boldsymbol{\omega}][\mathbf{J}_O]\{\boldsymbol{\omega}\} = \{\mathbf{M}_O\}; \tag{6.36}$$

this matrix equation decomposes in the Euler scalar equations.

6.4 The Rigid Solid with Two Fixed Points (the Rotational Motion of the Rigid Solid)

In the most general case, if the points A_1, and A_2 of the rigid solid, having the local coordinates x_1, y_1, z_1, and x_2, y_2, z_2, respectively, are fixed, that is, they have the constant coordinates X_1, Y_1, Z_1, and X_2, Y_2, Z_2, respectively, relative to the general reference system O_0XYZ, then, with the aid of the notations

$$\{\mathbf{r}_i\} = [x_i \;\; y_i \;\; z_i]^{\mathrm{T}}, [\mathbf{r}_i] = \begin{bmatrix} 0 & -z_i & y_i \\ z_i & 0 & -x_i \\ -y_i & x_i & 0 \end{bmatrix}, i = 1, 2, \tag{6.37}$$

$$\{\mathbf{R}_i\} = [X_i \;\; Y_i \;\; Z_i]^{\mathrm{T}}, i = 1, 2, \tag{6.38}$$

one obtains the equalities

$$\{\mathbf{R}_i\} = \{\mathbf{R}_O\} + [\mathbf{A}]\{\mathbf{r}_i\}, i = 1, 2, \tag{6.39}$$

$$\{\dot{\mathbf{R}}_O\} + [\dot{\mathbf{A}}]\{\mathbf{r}_i\} = \{\mathbf{0}\}, i = 1, 2. \tag{6.40}$$

Based on the relations

$$[\dot{\mathbf{A}}] = [\mathbf{A}][\boldsymbol{\omega}], [\boldsymbol{\omega}]\{\mathbf{r}_i\} = [\mathbf{r}_i]^{\mathrm{T}}\{\boldsymbol{\omega}\}, \{\boldsymbol{\omega}\} = [\mathbf{Q}]\{\dot{\boldsymbol{\beta}}\} \tag{6.41}$$

and using the notations

$$[\boldsymbol{\Delta}_i] = [\mathbf{A}][\mathbf{r}_i]^{\mathrm{T}}[\mathbf{Q}], \; i = 1, 2, \tag{6.42}$$

$$[\boldsymbol{\Gamma}_i] = [[\mathbf{I}] \;\; [\boldsymbol{\Delta}_i]], \; i = 1, 2, \tag{6.43}$$

the expressions (6.40) reunite in the matrix equality

$$\begin{bmatrix} [\boldsymbol{\Gamma}_1] \\ [\boldsymbol{\Gamma}_2] \end{bmatrix} \{\dot{\mathbf{q}}\} = \{\mathbf{0}\}. \tag{6.44}$$

Based on the successive equalities

$$\det\begin{bmatrix}[\boldsymbol{\Gamma}_1]\\ [\boldsymbol{\Gamma}_2]\end{bmatrix}=\det\begin{bmatrix}[\mathbf{I}] & [\boldsymbol{\Delta}_1]\\ [\mathbf{I}] & [\boldsymbol{\Delta}_2]\end{bmatrix}=\det\begin{bmatrix}[\mathbf{I}] & [\boldsymbol{\Delta}_1]\\ [\mathbf{0}] & [\boldsymbol{\Delta}_2]-[\boldsymbol{\Delta}_1]\end{bmatrix}$$

$$=\det[[\boldsymbol{\Delta}_2]-[\boldsymbol{\Delta}_1]]=\det[\mathbf{A}]\det\left[[\mathbf{r}_2]^{\mathrm{T}}-[\mathbf{r}_1]^{\mathrm{T}}\right]\det[\mathbf{Q}] \quad (6.45)$$

and since the matrix $[\mathbf{r}_2]^{\mathrm{T}}-[\mathbf{r}_1]^{\mathrm{T}}$ is of third order skew one, one may prove that the previous determinant is null; consequently, only five of the six scalar equations deduced from the relation (6.44) are independent.

Indeed, denoting by $rank[\cdot]$ the rank function, we may write the relation

$$\operatorname{rank}\begin{bmatrix}[\boldsymbol{\Gamma}_1]\\ [\boldsymbol{\Gamma}_2]\end{bmatrix}=\operatorname{rank}\begin{bmatrix}[\mathbf{I}] & [\boldsymbol{\Delta}_1]\\ [\mathbf{0}] & [\mathbf{A}]\left[[\mathbf{r}_2]^{\mathrm{T}}-[\mathbf{r}_1]^{\mathrm{T}}\right][\mathbf{Q}]\end{bmatrix}; \quad (6.46)$$

since $\operatorname{rank}[\mathbf{I}]=3$, $\operatorname{rank}[\mathbf{A}]=3$, $\operatorname{rank}[\mathbf{Q}]=3$, and

$$\operatorname{rank}\left[[\mathbf{r}_2]^{\mathrm{T}}-[\mathbf{r}_1]^{\mathrm{T}}\right]=\operatorname{rank}\begin{bmatrix}0 & z_2-z_1 & -(y_2-y_1)\\ -(z_2-z_1) & 0 & x_2-x_1\\ y_2-y_1 & -(x_2-x_1) & 0\end{bmatrix}\geq 2, \quad (6.47)$$

because the points A_1 and A_2 are distinct, the statement is proved.

Denoting by $\left[\widetilde{\boldsymbol{\Gamma}}_2\right]$ the matrix consisting in only two rows of the matrix $[\boldsymbol{\Gamma}_2]$, it results that the matrix of constraints has only five rows and writes in the form

$$[\mathbf{B}]=\begin{bmatrix}[\boldsymbol{\Gamma}_1]\\ \left[\widetilde{\boldsymbol{\Gamma}}_2\right]\end{bmatrix}. \quad (6.48)$$

Choosing the references system such that the point O coincides to the points O_0 and A_1, the axes O_0Z, OZ coincide to the straight line A_1A_2, while the center of weight C is situated in the plan Oxz, and denoting by h the distance A_1A_2, one obtains the equalities

$$X_O=Y_O=Z_O=0,\ \psi=\theta=0,\ \{\mathbf{r}_1\}=[0\ \ 0\ \ 0]^{\mathrm{T}},\ \{\mathbf{r}_2\}=[0\ \ 0\ \ h]^{\mathrm{T}}, \quad (6.49)$$

$$[\mathbf{A}]=[\boldsymbol{\varphi}],\ [\mathbf{Q}]=[\boldsymbol{\varphi}]^{\mathrm{T}},\ [\boldsymbol{\Delta}_1]=[\mathbf{0}],\ [\boldsymbol{\Delta}_2]=\begin{bmatrix}0 & h & 0\\ -h & 0 & 0\\ 0 & 0 & 0\end{bmatrix}, \quad (6.50)$$

$$[\mathbf{B}] = \begin{bmatrix} 1 & 0 & 0 & 0 & 0 & 0 \\ 0 & 1 & 0 & 0 & 0 & 0 \\ 0 & 0 & 1 & 0 & 0 & 0 \\ 1 & 0 & 0 & 0 & h & 0 \\ 0 & 1 & 0 & -h & 0 & 0 \end{bmatrix}, \tag{6.51}$$

$$\{\mathbf{q}\} = [0\ \ 0\ \ 0\ \ 0\ \ 0\ \ \varphi]^{\mathrm{T}}. \tag{6.52}$$

In these conditions, from the expression (6.1) one deduces the equations

$$[\boldsymbol{\varphi}][\mathbf{S}]^{\mathrm{T}}[\boldsymbol{\varphi}]^{\mathrm{T}}[0\ \ 0\ \ \ddot{\varphi}]^{\mathrm{T}} = \{\mathbf{F_s}\} + \left\{\widetilde{\mathbf{F}}_{\mathbf{s}}\right\} + [\lambda_1+\lambda_4\ \ \lambda_2+\lambda_5\ \ \lambda_3]^{\mathrm{T}}, \tag{6.53}$$

$$[\boldsymbol{\varphi}][\mathbf{J}_O][\boldsymbol{\varphi}]^{\mathrm{T}}[0\ \ 0\ \ \ddot{\varphi}]^{\mathrm{T}} = \{\mathbf{F}_{\boldsymbol{\beta}}\} + \left\{\widetilde{\mathbf{F}}_{\boldsymbol{\beta}}\right\} + h[-\lambda_5\ \ \lambda_4\ \ 0]^{\mathrm{T}}, \tag{6.54}$$

where

$$\left\{\widetilde{\mathbf{F}}_{\mathbf{s}}\right\} = -\dot{\varphi}\left[[\boldsymbol{\varphi}][\mathbf{S}]^{\mathrm{T}}[\mathbf{U}_{\varphi}][\boldsymbol{\varphi}_p] + [\boldsymbol{\varphi}][\mathbf{U}_{\varphi}][\mathbf{S}]^{\mathrm{T}}[\boldsymbol{\varphi}]\right][0\ \ 0\ \ \dot{\varphi}]^{\mathrm{T}}, \tag{6.55}$$

$$\left\{\widetilde{\mathbf{F}}_{\boldsymbol{\beta}}\right\} = -\dot{\varphi}\left[[\boldsymbol{\varphi}][\mathbf{J}_O][\mathbf{U}_{\varphi}]^{\mathrm{T}}[\boldsymbol{\varphi}]^{\mathrm{T}} + [\boldsymbol{\varphi}][\mathbf{U}_{\varphi}][\mathbf{J}_O][\boldsymbol{\varphi}]^{\mathrm{T}}\right][0\ \ 0\ \ \dot{\varphi}]^{\mathrm{T}}. \tag{6.56}$$

Multiplying the equations (6.53) and (6.54.) at left by $[\boldsymbol{\varphi}]^{\mathrm{T}}$, one obtains the equalities

$$\ddot{\varphi}[\mathbf{S}]^{\mathrm{T}}\begin{bmatrix}0\\0\\1\end{bmatrix} + \dot{\varphi}^2[\mathbf{U}_{\varphi}][\mathbf{S}]^{\mathrm{T}}\begin{bmatrix}0\\0\\1\end{bmatrix} = [\boldsymbol{\varphi}]^{\mathrm{T}}\{\mathbf{F_s}\} + [\boldsymbol{\varphi}]^{\mathrm{T}}\begin{bmatrix}\lambda_1+\lambda_4\\ \lambda_2+\lambda_5\\ \lambda_3\end{bmatrix}, \tag{6.57}$$

$$\ddot{\varphi}[\mathbf{J}_O]\begin{bmatrix}0\\0\\1\end{bmatrix} + \dot{\varphi}^2[\mathbf{U}_{\varphi}][\mathbf{J}_O]\begin{bmatrix}0\\0\\1\end{bmatrix} = [\boldsymbol{\varphi}]^{\mathrm{T}}\{\mathbf{F}_{\boldsymbol{\beta}}\} + h[\boldsymbol{\varphi}]^{\mathrm{T}}\begin{bmatrix}-\lambda_5\\ \lambda_4\\ 0\end{bmatrix}, \tag{6.58}$$

in which

$$[\mathbf{S}]^{\mathrm{T}} = m\begin{bmatrix} 0 & z_C & 0 \\ -z_C & 0 & x_C \\ 0 & -x_C & 0 \end{bmatrix}. \tag{6.59}$$

In equations (6.57) and (6.58) the terms $[\boldsymbol{\varphi}]^{\mathrm{T}}\{\mathbf{F}_{\mathbf{s}}\}$, and $[\boldsymbol{\varphi}]^{\mathrm{T}}\{\mathbf{F}_{\boldsymbol{\beta}}\}$ represents the column matrices of the components of the resultant of external forces, and of the resultant moment of the external forces, respectively, that act upon the rigid solid, relative to the reference system $Oxyz$. We denote these terms by $\begin{bmatrix}R_x & R_y & R_z\end{bmatrix}^{\mathrm{T}}$, and $\begin{bmatrix}M_x & M_y & M_z\end{bmatrix}^{\mathrm{T}}$, respectively.

Analogically, the terms $\begin{bmatrix}\lambda_1+\lambda_4 & \lambda_2+\lambda_5 & \lambda_3\end{bmatrix}^{\mathrm{T}}$, and $[\boldsymbol{\varphi}]^{\mathrm{T}}\begin{bmatrix}-h\lambda_5 & h\lambda_4 & 0\end{bmatrix}$, which represent the column matrices of the components of the resultant of constraint forces, and of the resultant moment of the constraint forces, respectively, relative to the reference system $Oxyz$, are denoted by $\begin{bmatrix}R_x^* & R_y^* & R_z^*\end{bmatrix}^{\mathrm{T}}$, and $\begin{bmatrix}M_x^* & M_y^* & 0\end{bmatrix}^{\mathrm{T}}$, respectively.

From the relations (6.57) and (6.58) one gets the classical scalar equations for the rotational motion of the rigid solid

$$\begin{gathered}-mx_C\dot{\varphi}^2=R_x+R_x^*,-mx_C\ddot{\varphi}=R_y+R_y^*,0=R_z+R_z^*,\\ -J_{xz}\ddot{\varphi}+J_{yz}\dot{\varphi}^2=M_x+M_x^*,-J_{yz}\ddot{\varphi}-J_{xz}\dot{\varphi}^2=M_y+M_y^*,J_z\ddot{\varphi}=M_z.\end{gathered}\tag{6.60}$$

From the last equation one deduces the law of motion $\varphi=\varphi(t)$, while from the first five equations one deduces the five components of the reaction torsor, that is, three reaction forces R_x^*, R_y^*, R_z^*, and two moments M_x^*, M_y^*.

6.5 The Rigid Solid with a Given Point Situated on a Fixed Surface

6.5.1 The Case When the Surface is Defined by an Implicit Equation $F(X,Y,Z)=0$

Considering that the point A, with the local coordinates x, y, z, remains situated on the given surface, and using the notations

$$\{\mathbf{R}\}=\begin{bmatrix}X & Y & Z\end{bmatrix}^{\mathrm{T}},\{\mathbf{r}\}=\begin{bmatrix}x & y & z\end{bmatrix}^{\mathrm{T}},[\mathbf{r}]=\begin{bmatrix}0 & -z & y\\ z & 0 & -x\\ -y & x & 0\end{bmatrix},\tag{6.61}$$

one obtains the relations

$$\{\mathbf{R}\}=\{\mathbf{R}_O\}+[\mathbf{A}]\{\mathbf{r}\},\tag{6.62}$$

$$\{\dot{\mathbf{R}}\}=\{\dot{\mathbf{R}}_O\}+[\dot{\mathbf{A}}]\{\mathbf{r}\};\tag{6.63}$$

since

$$[\dot{\mathbf{A}}]=[\mathbf{A}][\boldsymbol{\omega}],[\boldsymbol{\omega}]\{\mathbf{r}\}=[\mathbf{r}]^{\mathrm{T}}\{\boldsymbol{\omega}\},\{\boldsymbol{\omega}\}=[\mathbf{Q}]\{\dot{\boldsymbol{\beta}}\},\tag{6.64}$$

it results

$$\{\dot{\mathbf{R}}\} = \{\dot{\mathbf{R}}_O\} + [\mathbf{A}][\mathbf{r}]^{\mathrm{T}}[\mathbf{Q}]\{\dot{\boldsymbol{\beta}}\}. \tag{6.65}$$

Deriving the equation

$$F(X,Y,Z) = 0 \tag{6.66}$$

with respect to time and using the notation

$$\{\mathbf{F}_p\} = \begin{bmatrix} \frac{\partial F}{\partial X} & \frac{\partial F}{\partial Y} & \frac{\partial F}{\partial Z} \end{bmatrix}^{\mathrm{T}}, \tag{6.67}$$

one obtains

$$\{\mathbf{F}_p\}^{\mathrm{T}}\{\dot{\mathbf{R}}\} = 0 \tag{6.68}$$

or

$$\{\mathbf{F}_p\}^{\mathrm{T}}\left\{\{\dot{\mathbf{R}}_O\} + [\mathbf{A}][\mathbf{r}]^{\mathrm{T}}[\mathbf{Q}]\{\dot{\boldsymbol{\beta}}\}\right\} = 0. \tag{6.69}$$

Using the notations

$$[\boldsymbol{\Delta}] = [\mathbf{A}][\mathbf{r}]^{\mathrm{T}}[\mathbf{Q}], \tag{6.70}$$

$$[\boldsymbol{\Gamma}] = [[\mathbf{I}] \;\; [\boldsymbol{\Delta}]], \tag{6.71}$$

from the expressions (6.65) and (6.68) one deduces the equality

$$\{\mathbf{F}_p\}^{\mathrm{T}}[\boldsymbol{\Gamma}]\{\dot{\mathbf{q}}\} = 0; \tag{6.72}$$

hence, the matrix of constraints reads

$$[\mathbf{B}] = \{\mathbf{F}_p\}^{\mathrm{T}}[\boldsymbol{\Gamma}]. \tag{6.73}$$

To be out to obtain the matrix equation (6.1) it is necessary to know the derivative of the matrix of constraints

$$[\dot{\mathbf{B}}] = \{\dot{\mathbf{F}}_p\}^{\mathrm{T}}[\boldsymbol{\Gamma}] + \{\mathbf{F}_p\}^{\mathrm{T}}[\dot{\boldsymbol{\Gamma}}]. \tag{6.74}$$

Denoting

$$[\mathbf{F}_s] = \begin{bmatrix} \frac{\partial^2 F}{\partial X^2} & \frac{\partial^2 F}{\partial X \partial Y} & \frac{\partial^2 F}{\partial X \partial Z} \\ \frac{\partial^2 F}{\partial Y \partial X} & \frac{\partial^2 F}{\partial Y^2} & \frac{\partial^2 F}{\partial Y \partial Z} \\ \frac{\partial^2 F}{\partial Z \partial X} & \frac{\partial^2 F}{\partial Z \partial Y} & \frac{\partial^2 F}{\partial Z^2} \end{bmatrix}, \tag{6.75}$$

it results

$$\{\dot{\mathbf{F}}_p\}^{\mathrm{T}} = \{\dot{\mathbf{q}}\}^{\mathrm{T}}[\boldsymbol{\Gamma}][\mathbf{F}_s]; \tag{6.76}$$

since

$$[\dot{\boldsymbol{\Delta}}] = [\dot{\mathbf{A}}][\mathbf{r}]^{\mathrm{T}}[\mathbf{Q}] + [\mathbf{A}][\mathbf{r}]^{\mathrm{T}}[\dot{\mathbf{Q}}], [\dot{\boldsymbol{\Gamma}}] = \left[[\mathbf{0}] \quad [\dot{\boldsymbol{\Delta}}]\right], \tag{6.77}$$

one deduces the result

$$[\dot{\mathbf{B}}] = \{\dot{\mathbf{q}}\}^{\mathrm{T}}[\boldsymbol{\Gamma}][\mathbf{F}_s][\boldsymbol{\Gamma}] + \{\mathbf{F}_p\}^{\mathrm{T}}[\dot{\boldsymbol{\Gamma}}]. \tag{6.78}$$

6.5.2 *The Case When the Surface is Defined by Parametric Equations*

If the surface is defined by the parametric equations

$$X = X(u,v), Y = Y(u,v), Z = Z(u,v), \tag{6.79}$$

then, with the aid of the notations

$$\{\mathbf{R}_u\} = \begin{bmatrix} \frac{\partial X}{\partial u} & \frac{\partial Y}{\partial u} & \frac{\partial Z}{\partial u} \end{bmatrix}^{\mathrm{T}}, \{\mathbf{R}_v\} = \begin{bmatrix} \frac{\partial X}{\partial v} & \frac{\partial Y}{\partial v} & \frac{\partial Z}{\partial v} \end{bmatrix}^{\mathrm{T}}, \tag{6.80}$$

$$[\boldsymbol{\Delta}^*] = -[\{\mathbf{R}_u\} \quad \{\mathbf{R}_v\}], \tag{6.81}$$

from the relation (6.65) one obtains the expression

$$\{\dot{\mathbf{R}}_O\} + [\mathbf{A}][\mathbf{r}]^{\mathrm{T}}[\mathbf{Q}]\{\dot{\boldsymbol{\beta}}\} + [\boldsymbol{\Delta}^*][\dot{u} \quad \dot{v}]^{\mathrm{T}} = \{\mathbf{0}\}. \tag{6.82}$$

Calling *the method of pseudo-coordinates*, and using the notations

$$\{\mathbf{q}^*\}=[X_O\ \ Y_O\ \ Z_O\ \ \psi\ \ \theta\ \ \varphi\ \ u\ \ v]^{\mathrm{T}}, \tag{6.83}$$

$$[\boldsymbol{\Gamma}^*]=[[\mathbf{I}]\ \ [\boldsymbol{\Delta}]\ \ [\boldsymbol{\Delta}^*]], \tag{6.84}$$

one obtains

$$[\boldsymbol{\Gamma}^*]\{\dot{\mathbf{q}}^*\}=\{\mathbf{0}\}; \tag{6.85}$$

hence, the matrix of constraints reads

$$[\mathbf{B}]=[\boldsymbol{\Gamma}^*]. \tag{6.86}$$

To be out to calculate the derivative of the matrix of constraints with respect to time, we use the notations

$$\{\mathbf{R}_{uu}\}=\begin{bmatrix}\dfrac{\partial^2X}{\partial u^2} & \dfrac{\partial^2Y}{\partial u^2} & \dfrac{\partial^2Z}{\partial u^2}\end{bmatrix}^{\mathrm{T}},\{\mathbf{R}_{uv}\}=\{\mathbf{R}_{vu}\}=\begin{bmatrix}\dfrac{\partial^2X}{\partial u\partial v} & \dfrac{\partial^2Y}{\partial u\partial v} & \dfrac{\partial^2Z}{\partial u\partial v}\end{bmatrix}^{\mathrm{T}},$$
$$\{\mathbf{R}_{vv}\}=\begin{bmatrix}\dfrac{\partial^2X}{\partial v^2} & \dfrac{\partial^2Y}{\partial v^2} & \dfrac{\partial^2Z}{\partial v^2}\end{bmatrix}^{\mathrm{T}}, \tag{6.87}$$

$$[\boldsymbol{\Delta}_u^*]=[\{\mathbf{R}_{uu}\}\ \ \{\mathbf{R}_{uv}\}],[\boldsymbol{\Delta}_v^*]=[\{\mathbf{R}_{uv}\}\ \ \{\mathbf{R}_{vv}\}]; \tag{6.88}$$

it results

$$\left[\dot{\boldsymbol{\Delta}}^*\right]=\dot{u}\left[\boldsymbol{\Delta}_u^*\right]+\dot{v}\left[\boldsymbol{\Delta}_v^*\right], \tag{6.89}$$

so that

$$[\dot{\mathbf{B}}]=\left[\dot{\boldsymbol{\Gamma}}^*\right]=\left[[\mathbf{0}]\ \ [\dot{\boldsymbol{\Delta}}]\ \ \left[\dot{\boldsymbol{\Delta}}^*\right]\right]. \tag{6.90}$$

The matrix differential equation of motion is given by the equality (6.15), in which

$$[\mathbf{M}^*]=\begin{bmatrix}[\mathbf{M}] & [\mathbf{0}_{62}]\\ [\mathbf{0}_{26}] & [\mathbf{0}_{22}]\end{bmatrix},\{\mathbf{F}_{\mathbf{q}^*}\}=\left[\{\mathbf{F}_{\mathbf{q}}\}^{\mathrm{T}}\ \ \{\mathbf{0}_{21}\}^{\mathrm{T}}\right]^{\mathrm{T}},$$
$$\left\{\widetilde{\mathbf{F}}_{\mathbf{q}^*}\right\}=\left[\left\{\widetilde{\mathbf{F}}_{\mathbf{q}}\right\}^{\mathrm{T}}\ \ \{\mathbf{0}_{21}\}^{\mathrm{T}}\right]^{\mathrm{T}},\{\mathbf{C}\}=\{\mathbf{0}\}. \tag{6.91}$$

6.6 The Rigid Solid with Several Points Situated on Fixed Surfaces (Curves)

6.6.1 The Case When the Surfaces are Defined by Implicit Equations

We consider the case in which the rigid solid's points A_i, $i = 1, 2, \ldots, n$, $n \leq 5$, described by the local coordinates x_i, y_i, z_i, $i = 1, 2, \ldots, n$, remain situated on the surfaces defined by the implicit equations

$$F_i(X, Y, Z) = 0, i = \overline{1, n}. \tag{6.92}$$

We use the notations

$$\{\mathbf{r}_i\} = [x_i \ \ y_i \ \ z_i]^{\mathrm{T}}, \ [\mathbf{r}_i] = \begin{bmatrix} 0 & -z_i & y_i \\ z_i & 0 & -x_i \\ -y_i & x_i & 0 \end{bmatrix}, \ \{\mathbf{R}_i\} = [X_i \ \ Y_i \ \ Z_i]^{\mathrm{T}}, \tag{6.93}$$

$$\{\mathbf{F}_{ip}\} = \begin{bmatrix} \dfrac{\partial F_i}{\partial X} & \dfrac{\partial F_i}{\partial Y} & \dfrac{\partial F_i}{\partial Z} \end{bmatrix}^{\mathrm{T}}, [\mathbf{F}_{is}] = \begin{bmatrix} \dfrac{\partial^2 F_i}{\partial X^2} & \dfrac{\partial^2 F_i}{\partial X \partial Y} & \dfrac{\partial^2 F_i}{\partial X \partial Z} \\ \dfrac{\partial^2 F_i}{\partial Y \partial X} & \dfrac{\partial^2 F_i}{\partial Y^2} & \dfrac{\partial^2 F_i}{\partial Y \partial Z} \\ \dfrac{\partial^2 F_i}{\partial Z \partial X} & \dfrac{\partial^2 F_i}{\partial Z \partial Y} & \dfrac{\partial^2 F_i}{\partial Z^2} \end{bmatrix}, \tag{6.94}$$

$$[\mathbf{\Delta}_i] = [\mathbf{A}][\mathbf{r}_i]^{\mathrm{T}}[\mathbf{Q}], [\mathbf{\Gamma}_i] = [[\mathbf{I}] \ \ [\mathbf{\Delta}_i]]. \tag{6.95}$$

From the expressions (6.65) and (6.68) it results the relations

$$\{\dot{\mathbf{R}}_i\} = \{\dot{\mathbf{R}}_O\} + [\mathbf{A}][\mathbf{r}_i]^{\mathrm{T}}[\mathbf{Q}]\{\dot{\boldsymbol{\beta}}\}, i = \overline{1, n}, \tag{6.96}$$

$$\{\mathbf{F}_{ip}\}^{\mathrm{T}}\{\dot{\mathbf{R}}_O\} + \{\mathbf{F}_{ip}\}^{\mathrm{T}}[\mathbf{A}][\mathbf{r}_i]^{\mathrm{T}}[\mathbf{Q}]\{\dot{\boldsymbol{\beta}}\} = 0, i = \overline{1, n} \tag{6.97}$$

and thus one deduces the equalities

$$\{\mathbf{F}_{ip}\}^{\mathrm{T}}[\mathbf{\Gamma}_i]\{\dot{\mathbf{q}}\} = 0, i = \overline{1, n}. \tag{6.98}$$

It results from here that the matrix of constraints reads

$$[\mathbf{B}] = \begin{bmatrix} \{\mathbf{F}_{1p}\}^{\mathrm{T}}[\mathbf{\Gamma}_1] \\ \{\mathbf{F}_{2p}\}^{\mathrm{T}}[\mathbf{\Gamma}_2] \\ \cdots \\ \{\mathbf{F}_{np}\}^{\mathrm{T}}[\mathbf{\Gamma}_n] \end{bmatrix}, \tag{6.99}$$

its derivative with respect to time being

$$[\mathbf{B}] = \begin{bmatrix} \{\dot{\mathbf{q}}\}^{\mathrm{T}}[\mathbf{\Gamma}_1][\mathbf{F}_{1s}][\mathbf{\Gamma}_1] + \{\mathbf{F}_{1p}\}^{\mathrm{T}}[\mathbf{\Gamma}_1] \\ \{\dot{\mathbf{q}}\}^{\mathrm{T}}[\mathbf{\Gamma}_2][\mathbf{F}_{2s}][\mathbf{\Gamma}_2] + \{\mathbf{F}_{2p}\}^{\mathrm{T}}[\mathbf{\Gamma}_2] \\ \cdots \\ \{\dot{\mathbf{q}}\}^{\mathrm{T}}[\mathbf{\Gamma}_n][\mathbf{F}_{ns}][\mathbf{\Gamma}_n] + \{\mathbf{F}_{np}\}^{\mathrm{T}}[\mathbf{\Gamma}_n] \end{bmatrix}. \tag{6.100}$$

Due to the fact that a curve can be defined as the intersection of two surfaces, the expressions established above hold true also for the case when a point is situated on a fixed curve.

6.6.2 *The Case When the Surfaces are Defined by Parametric Equations*

If the surfaces are defined by the parametric equations

$$X = X_i(u_i, v_i),\ Y = Y_i(u_i, v_i),\ Z = Z_i(u_i, v_i),\ i = \overline{1, n}, \tag{6.101}$$

then, with the aid of the notations

$$\{\mathbf{R}_{iu}\} = \begin{bmatrix} \frac{\partial X_i}{\partial u_i} & \frac{\partial Y_i}{\partial u_i} & \frac{\partial Z_i}{\partial u_i} \end{bmatrix}^{\mathrm{T}},\ \{\mathbf{R}_{iv}\} = \begin{bmatrix} \frac{\partial X_i}{\partial v_i} & \frac{\partial Y_i}{\partial v_i} & \frac{\partial Z_i}{\partial v_i} \end{bmatrix}^{\mathrm{T}},\ i = \overline{1, n}, \tag{6.102}$$

$$[\mathbf{\Delta}_i^*] = -[\{\mathbf{R}_{iu}\}\ \ \{\mathbf{R}_{iv}\}],\ i = \overline{1, n}, \tag{6.103}$$

$$\{\mathbf{q}^*\} = [X_O\ \ Y_O\ \ Z_O\ \ \psi\ \ \theta\ \ \varphi\ \ u_1\ \ v_1\ \ \ldots\ \ u_n\ \ v_n]^{\mathrm{T}}, \tag{6.104}$$

$$[\mathbf{\Gamma}_i] = \left[[\mathbf{I}]\ \ [\mathbf{\Delta}]\ \ [\mathbf{0}]\ \ \ldots\ \ [\mathbf{\Delta}_i^*]\ \ \ldots\ \ [\mathbf{0}] \right], \tag{6.105}$$

and keeping into account the relations (6.82) and (6.85), one obtains the equalities

$$[\mathbf{\Gamma}_i]\{\dot{\mathbf{q}}^*\} = \{\mathbf{0}\}, i = \overline{1, n}; \tag{6.106}$$

hence, the matrix of constraints is

$$[\mathbf{B}] = \begin{bmatrix} [\mathbf{\Gamma}_1] \\ [\mathbf{\Gamma}_2] \\ \cdots \\ [\mathbf{\Gamma}_n] \end{bmatrix} = \begin{bmatrix} [\mathbf{I}] & [\mathbf{\Delta}] & [\mathbf{\Delta}_1^*] & [\mathbf{0}] & \ldots & [\mathbf{0}] \\ [\mathbf{I}] & [\mathbf{\Delta}] & [\mathbf{0}] & [\mathbf{\Delta}_2^*] & \ldots & [\mathbf{0}] \\ \ldots & \ldots & \ldots & \ldots & \ldots & \ldots \\ [\mathbf{I}] & [\mathbf{\Delta}] & [\mathbf{0}] & [\mathbf{0}] & \ldots & [\mathbf{\Delta}_n^*] \end{bmatrix} \tag{6.107}$$

and it has $3n$ rows and $6 + 2n$ columns.

To be out to calculate the derivative of the matrix of constraints we use the notations

$$\{\mathbf{R}_{iuu}\} = \begin{bmatrix} \frac{\partial^2 X_i}{\partial u_i^2} & \frac{\partial^2 Y_i}{\partial u_i^2} & \frac{\partial^2 Z_i}{\partial u_i^2} \end{bmatrix}^{\mathrm{T}}, \{\mathbf{R}_{iuv}\} = \begin{bmatrix} \frac{\partial^2 X_i}{\partial u_i \partial v_i} & \frac{\partial^2 Y_i}{\partial u_i \partial v_i} & \frac{\partial^2 Z_i}{\partial u_i \partial v_i} \end{bmatrix}^{\mathrm{T}},$$
$$\{\mathbf{R}_{ivv}\} = \begin{bmatrix} \frac{\partial^2 X_i}{\partial v_i^2} & \frac{\partial^2 Y_i}{\partial v_i^2} & \frac{\partial^2 Z_i}{\partial v_i^2} \end{bmatrix}^{\mathrm{T}}, i = \overline{1, n}, \tag{6.108}$$

$$\left[\mathbf{\Delta}_{iu}^*\right] = \left[\{\mathbf{R}_{iuu}\} \quad \{\mathbf{R}_{iuv}\}\right], \left[\mathbf{\Delta}_{iv}^*\right] = \left[\{\mathbf{R}_{iuv}\} \quad \{\mathbf{R}_{ivv}\}\right], i = \overline{1, n}, \tag{6.109}$$

and the relations

$$\left[\dot{\mathbf{\Delta}}_i^*\right] = \dot{u}_i\left[\mathbf{\Delta}_{iu}^*\right] + \dot{v}_i\left[\mathbf{\Delta}_{iv}^*\right], i = \overline{1, n}; \tag{6.110}$$

it results

$$\left[\dot{\mathbf{B}}\right] = = \begin{bmatrix} [\mathbf{0}] & [\dot{\mathbf{\Delta}}] & \left[\dot{\mathbf{\Delta}}_1^*\right] & [\mathbf{0}] & \cdots & [\mathbf{0}] \\ [\mathbf{0}] & [\dot{\mathbf{\Delta}}] & [\mathbf{0}] & \left[\dot{\mathbf{\Delta}}_2^*\right] & \cdots & [\mathbf{0}] \\ \cdots & \cdots & \cdots & \cdots & \cdots & \cdots \\ [\mathbf{0}] & [\dot{\mathbf{\Delta}}] & [\mathbf{0}] & [\mathbf{0}] & \cdots & \left[\dot{\mathbf{\Delta}}_n^*\right] \end{bmatrix}. \tag{6.111}$$

The matrix differential equation of the motion of the rigid solid is given by the equality (6.15) in which we replace n by $3n$ and p by $2n$.

6.7 The Rigid Solid with a Fixed Point and with Another Point Situated on a Fixed Surface

6.7.1 The Case When the Fixed Surface is Defined by an Implicit Equation

We consider that the point A of the rigid solid, defined in the local reference system by the coordinates x_A, y_A, z_A, coincides to the fixed point having the coordinates X_A, Y_A, Z_A relative to the general fixed reference system. In addition, the point B of the rigid solid, having the local coordinates x_B, y_B, z_B, remains situated on the fixed surface of equation

$$F(X, Y, Z) = 0. \tag{6.112}$$

We use the notations

$$\{\mathbf{r}_A\} = [x_A \quad y_A \quad z_A]^{\mathrm{T}}, [\mathbf{r}_A] = \begin{bmatrix} 0 & -z_A & y_A \\ z_A & 0 & -x_A \\ -y_A & x_A & 0 \end{bmatrix}, \{\mathbf{R}_A\} = [X_A \quad Y_A \quad Z_A]^{\mathrm{T}}, \tag{6.113}$$

$$[\mathbf{\Delta}_A] = [\mathbf{A}][\mathbf{r}_A]^{\mathrm{T}}[\mathbf{Q}], [\mathbf{\Gamma}_A] = [[\mathbf{I}] \ \ [\mathbf{\Delta}_A]], \tag{6.114}$$

$$\{\mathbf{r}_B\} = [x_B \ \ y_B \ \ z_B]^{\mathrm{T}}, [\mathbf{r}_B] = \begin{bmatrix} 0 & -z_B & y_B \\ z_B & 0 & -x_B \\ -y_B & x_B & 0 \end{bmatrix}, \{\mathbf{R}_B\} = [X_B \ \ Y_B \ \ Z_B]^{\mathrm{T}}, \tag{6.115}$$

$$[\mathbf{\Delta}_B] = [\mathbf{A}][\mathbf{r}_B]^{\mathrm{T}}[\mathbf{Q}], [\mathbf{\Gamma}_B] = [[\mathbf{I}] \ \ [\mathbf{\Delta}_B]], \tag{6.116}$$

$$\{\mathbf{F}_p\} = \begin{bmatrix} \frac{\partial F}{\partial X} & \frac{\partial F}{\partial Y} & \frac{\partial F}{\partial Z} \end{bmatrix}^{\mathrm{T}}, [\mathbf{F}_s] = \begin{bmatrix} \frac{\partial^2 F}{\partial X^2} & \frac{\partial^2 F}{\partial X \partial Y} & \frac{\partial^2 F}{\partial X \partial Z} \\ \frac{\partial^2 F}{\partial Y \partial X} & \frac{\partial^2 F}{\partial Y^2} & \frac{\partial^2 F}{\partial Y \partial Z} \\ \frac{\partial^2 F}{\partial Z \partial X} & \frac{\partial^2 F}{\partial Z \partial Y} & \frac{\partial^2 F}{\partial Z^2} \end{bmatrix}. \tag{6.117}$$

Starting from the equalities

$$\{\mathbf{R}_A\} = \{\mathbf{R}_O\} + [\mathbf{A}]\{\mathbf{r}_A\}, \tag{6.118}$$

$$\{\mathbf{R}_B\} = \{\mathbf{R}_O\} + [\mathbf{A}]\{\mathbf{r}_B\}, \tag{6.119}$$

by derivation with respect to time and taking into account the relation

$$\{\mathbf{F}_p\}^{\mathrm{T}}\{\dot{\mathbf{R}}_B\} = 0, \tag{6.120}$$

one obtains the expressions

$$\{\dot{\mathbf{R}}_O\} + [\mathbf{A}][\mathbf{r}_A]^{\mathrm{T}}[\mathbf{Q}]\{\dot{\boldsymbol{\beta}}\} = \{\mathbf{0}\}, \tag{6.121}$$

$$\{\mathbf{F}_p\}^{\mathrm{T}}\left\{\{\dot{\mathbf{R}}_O\} + [\mathbf{A}][\mathbf{r}_B]^{\mathrm{T}}[\mathbf{Q}]\{\dot{\boldsymbol{\beta}}\}\right\} = 0 \tag{6.122}$$

or

$$[\mathbf{\Gamma}_A]\{\dot{\mathbf{q}}\} = \{\mathbf{0}\}, \tag{6.123}$$

$$\{\mathbf{F}_p\}^{\mathrm{T}}[\mathbf{\Gamma}_B]\{\dot{\mathbf{q}}\} = \{\mathbf{0}\}; \tag{6.124}$$

hence, the matrix of constraints reads

$$[\mathbf{B}] = \begin{bmatrix} [\mathbf{\Gamma}_A] \\ \{\mathbf{F}_p\}^{\mathrm{T}}[\mathbf{\Gamma}_B] \end{bmatrix}. \tag{6.125}$$

The derivative of the matrix of constraints with respect to time is

$$[\dot{\mathbf{B}}] = \begin{bmatrix} [\dot{\mathbf{\Gamma}}_A] \\ \{\dot{\mathbf{q}}\}^{\mathrm{T}}[\mathbf{\Gamma}_B][\mathbf{F}_s][\mathbf{\Gamma}_B] + \{\mathbf{F}_p\}^{\mathrm{T}}[\dot{\mathbf{\Gamma}}_B] \end{bmatrix}, \tag{6.126}$$

where

$$[\dot{\mathbf{\Gamma}}_A] = \begin{bmatrix} [\mathbf{0}] & [\dot{\mathbf{A}}][\mathbf{r}_A]^{\mathrm{T}}[\mathbf{Q}] + [\mathbf{A}][\mathbf{r}_A]^{\mathrm{T}}[\dot{\mathbf{Q}}] \end{bmatrix}, \tag{6.127}$$

$$[\dot{\mathbf{\Gamma}}_B] = \begin{bmatrix} [\mathbf{0}] & [\dot{\mathbf{A}}][\mathbf{r}_B]^{\mathrm{T}}[\mathbf{Q}] + [\mathbf{A}][\mathbf{r}_B]^{\mathrm{T}}[\dot{\mathbf{Q}}] \end{bmatrix}. \tag{6.128}$$

If one considers that the fixed point A coincides to the points O and O_0, then one obtains the equalities

$$X_O = Y_O = Z_O = x_A = y_A = z_A = 0, \{\mathbf{q}\} = [0 \;\; 0 \;\; 0 \;\; \psi \;\; \theta \;\; \varphi]^{\mathrm{T}}, \tag{6.129}$$

$$[\mathbf{\Delta}_A] = [\mathbf{0}], [\mathbf{\Gamma}_A] = [[\mathbf{I}] \;\; [\mathbf{0}]], [\dot{\mathbf{\Gamma}}_A] = [[\mathbf{0}] \;\; [\mathbf{0}]] \tag{6.130}$$

and it results

$$[\mathbf{B}] = \begin{bmatrix} [[\mathbf{I}] \;\; [\mathbf{0}]] \\ \{\mathbf{F}_p\}^{\mathrm{T}}[\mathbf{\Gamma}_B] \end{bmatrix}. \tag{6.131}$$

6.7.2 *The Case When the Fixed Surface is Defined by Parametric Equations*

We consider that the parametric equations of the fixed surface are

$$X = X(u,v), \; Y = Y(u,v), \; Z = Z(u,v), \tag{6.132}$$

and we use the notations

$$\{\mathbf{R}_u\} = \begin{bmatrix} \frac{\partial X}{\partial u} & \frac{\partial Y}{\partial u} & \frac{\partial Z}{\partial u} \end{bmatrix}^{\mathrm{T}}, \{\mathbf{R}_v\} = \begin{bmatrix} \frac{\partial X}{\partial v} & \frac{\partial Y}{\partial v} & \frac{\partial Z}{\partial v} \end{bmatrix}^{\mathrm{T}}, [\mathbf{\Delta}_B] = [\{\mathbf{R}_u\} \;\; \{\mathbf{R}_v\}], \tag{6.133}$$

$$\{\mathbf{q}^*\} = [X_O \;\; Y_O \;\; Z_O \;\; \psi \;\; \theta \;\; \varphi \;\; u \;\; v]^{\mathrm{T}}, \tag{6.134}$$

$$[\mathbf{\Gamma}_B^*] = [[\mathbf{I}] \;\; [\mathbf{\Delta}_B] \;\; [\mathbf{\Delta}_B^*]], \tag{6.135}$$

$$[\mathbf{\Gamma}_A^*] = [[\mathbf{I}] \ [\mathbf{\Delta}_A] \ [\mathbf{0}]]. \tag{6.136}$$

Deriving the relations (6.118) and (6.119) with respect to time, one obtains the expressions

$$[\mathbf{\Gamma}_A^*]\{\dot{\mathbf{q}}^*\} = \{\mathbf{0}\}, [\mathbf{\Gamma}_B^*]\{\dot{\mathbf{q}}^*\} = \{\mathbf{0}\}, \tag{6.137}$$

wherefrom one deduces the matrix of constraints

$$[\mathbf{B}] = \begin{bmatrix} [\mathbf{\Gamma}_A^*] \\ [\mathbf{\Gamma}_B^*] \end{bmatrix}. \tag{6.138}$$

To be out to obtain the derivative with respect to time of this matrix, we recall the notations (6.86)–(6.88) and it results

$$[\dot{\mathbf{B}}] = \begin{bmatrix} [\dot{\mathbf{\Gamma}}_A^*] \\ [\dot{\mathbf{\Gamma}}_B^*] \end{bmatrix}, \tag{6.139}$$

where

$$[\dot{\mathbf{\Gamma}}_A^*] = \left[[\mathbf{0}] \ [\dot{\mathbf{\Delta}}_A] \ [\mathbf{0}]\right], \tag{6.140}$$

$$[\dot{\mathbf{\Gamma}}_B^*] = \left[[\mathbf{0}] \ [\dot{\mathbf{\Delta}}_B] \ [\dot{\mathbf{\Delta}}_B^*]\right]. \tag{6.141}$$

6.8 The Rigid Solid with Two Given Points Situated on a Fixed Curve

6.8.1 The Case When the Curve is Defined by Two Implicit Equations

If the implicit equations that define the fixed curve are

$$F_1(X,Y,Z) = 0, F_2(X,Y,Z) = 0, \tag{6.142}$$

while the points A_1, A_2 situated on the fixed curve have the local coordinates x_i, y_i, z_i, $i = 1, 2$, then using the notations from the paragraph 6.6, one successively writes the relations

$$\{\mathbf{R}_i\} = \{\mathbf{R}_O\} + [\mathbf{A}]\{\mathbf{r}_i\}, \ i = 1, 2, \tag{6.143}$$

$$\{\dot{\mathbf{R}}_i\} = \{\dot{\mathbf{R}}_O\} + [\mathbf{A}][\mathbf{r}_i]^{\mathrm{T}}[\mathbf{Q}]\{\dot{\boldsymbol{\beta}}\}, \ i = 1, 2, \tag{6.144}$$

$$\{\mathbf{F}_{ip}\}^{\mathrm{T}}\{\dot{\mathbf{R}}_j\}=0, i=1,2, j=1,2. \tag{6.145}$$

The relations (6.145) may be also written in the form

$$\{\mathbf{F}_{ip}\}^{\mathrm{T}}[\boldsymbol{\Gamma}_j]\{\dot{\mathbf{q}}\}=\{\mathbf{0}\}, \tag{6.146}$$

wherefrom one obtains the matrix of constraints

$$[\mathbf{B}]=\begin{bmatrix}\{\mathbf{F}_{1p}\}^{\mathrm{T}}[\boldsymbol{\Gamma}_1]\\ \{\mathbf{F}_{2p}\}^{\mathrm{T}}[\boldsymbol{\Gamma}_1]\\ \{\mathbf{F}_{1p}\}^{\mathrm{T}}[\boldsymbol{\Gamma}_2]\\ \{\mathbf{F}_{2p}\}^{\mathrm{T}}[\boldsymbol{\Gamma}_2]\end{bmatrix}, \tag{6.147}$$

which has four rows and six columns.

6.8.2 *The Case When the Curve is Defined by Parametric Equations*

We consider that the parametric equations of the fixed curve are

$$X=X(\xi), Y=Y(\xi), Z=Z(\xi). \tag{6.148}$$

Using the notations

$$\{\mathbf{R}_i\}=[X(\xi_i)\quad Y(\xi_i)\quad Z(\xi_i)]^{\mathrm{T}}, i=1,2, \tag{6.149}$$

$$X_p=\frac{\partial X}{\partial \xi}, Y_p=\frac{\partial Y}{\partial \xi}, Z_p=\frac{\partial Z}{\partial \xi}, \tag{6.150}$$

$$X_{ip}=X_p(\xi_i), Y_{ip}=Y_p(\xi_i), Z_{ip}=Z_p(\xi_i), i=1,2, \tag{6.151}$$

$$\{\mathbf{R}_{ip}\}=[X_{ip}\quad Y_{ip}\quad Z_{ip}]^{\mathrm{T}}, \tag{6.152}$$

$$\{\mathbf{q}^*\}=[X_O\quad Y_O\quad Z_O\quad \psi\quad \theta\quad \varphi\quad \xi_1\quad \xi_2]^{\mathrm{T}}, \tag{6.153}$$

$$[\boldsymbol{\Gamma}_1^*]=[[\mathbf{I}]\quad [\boldsymbol{\Delta}_1]\quad -\{\mathbf{R}_{1p}\}\quad \{\mathbf{0}\}], \tag{6.154}$$

$$[\boldsymbol{\Gamma}_2^*]=[[\mathbf{I}]\quad [\boldsymbol{\Delta}_2]\quad \{\mathbf{0}\}\quad -\{\mathbf{R}_{1p}\}], \tag{6.155}$$

from the relations (6.144) we obtain the expressions

$$[\boldsymbol{\Gamma}_1^*]\{\dot{\mathbf{q}}^*\}=\{\mathbf{0}\}, [\boldsymbol{\Gamma}_2^*]\{\dot{\mathbf{q}}^*\}=\{\mathbf{0}\}; \tag{6.156}$$

hence, the matrix of constraints reads

$$[\mathbf{B}] = \begin{bmatrix} [\mathbf{\Gamma}_1^*] \\ [\mathbf{\Gamma}_2^*] \end{bmatrix}. \tag{6.157}$$

6.8.3 The Helical Motion of the Rigid Solid

If the fixed curve is a straight line, then the rigid solid has a helical motion.

We choose the general fixed reference system so that the axis O_0Z coincides to the axis A_1A_2, the mobile reference system $Oxyz$ so that the axis OZ coincides to the axis O_0Z, the point O coincides to the point A_1, while the center of weight of the rigid solid is situated in the plan Oxz. Denoting by h the distance A_1A_2, we deduce the equalities

$$X_O = Y_O = 0,\ \psi = \theta = 0,\ \{\mathbf{r}_1\} = [0\ \ 0\ \ 0]^{\mathrm{T}},\ \{\mathbf{r}_2\} = [0\ \ 0\ \ h]^{\mathrm{T}},\ [\mathbf{A}] = [\boldsymbol{\varphi}], [\mathbf{Q}] = [\boldsymbol{\varphi}]^{\mathrm{T}},$$

$$[\mathbf{\Delta}_1] = [\mathbf{0}], [\mathbf{\Delta}_2] = \begin{bmatrix} 0 & h & 0 \\ -h & 0 & 0 \\ 0 & 0 & 0 \end{bmatrix}, \{\mathbf{F}_{1p}\} = [1\ \ 0\ \ 0]^{\mathrm{T}}, \{\mathbf{F}_{2p}\} = [0\ \ 1\ \ 0]^{\mathrm{T}},$$

$$[\mathbf{B}] = \begin{bmatrix} 1 & 0 & 0 & 0 & 0 & 0 \\ 0 & 1 & 0 & 0 & 0 & 0 \\ 1 & 0 & 0 & 0 & h & 0 \\ 0 & 1 & 0 & -h & 0 & 0 \end{bmatrix}, \{\mathbf{q}\} = [0\ \ 0\ \ Z_O\ \ 0\ \ 0\ \ \varphi]^{\mathrm{T}}. \tag{6.158}$$

In these conditions, the expressions (6.6) and (6.7) become

$$\left\{\widetilde{\mathbf{F}}_{\mathbf{s}}\right\} = m\dot{\varphi}^2[\boldsymbol{\varphi}][x_C\ \ 0\ \ 0]^{\mathrm{T}}, \left\{\widetilde{\mathbf{F}}_{\boldsymbol{\beta}}\right\} = -\dot{\varphi}^2[\boldsymbol{\varphi}][J_{yz}\ \ -J_{xz}\ \ 0]^{\mathrm{T}} \tag{6.159}$$

and from the equality (6.1) one deduces the matrix equations

$$\begin{aligned} m\ddot{Z}_O\begin{bmatrix}0\\0\\1\end{bmatrix} + \ddot{\varphi}[\mathbf{S}]^{\mathrm{T}}\begin{bmatrix}0\\0\\1\end{bmatrix} &= [\boldsymbol{\varphi}]^{\mathrm{T}}\{\mathbf{F}_{\mathbf{s}}\} + m\dot{\varphi}^2\begin{bmatrix}x_C\\0\\0\end{bmatrix} + [\boldsymbol{\varphi}]^{\mathrm{T}}\begin{bmatrix}\lambda_1+\lambda_3\\ \lambda_2+\lambda_4\\ 0\end{bmatrix}, \\ \ddot{Z}_O\begin{bmatrix}0\\0\\1\end{bmatrix} + \ddot{\varphi}[\mathbf{J}_O]\begin{bmatrix}0\\0\\1\end{bmatrix} &= [\boldsymbol{\varphi}]^{\mathrm{T}}\{\mathbf{F}_{\boldsymbol{\beta}}\} - \dot{\varphi}^2\begin{bmatrix}J_{yz}\\ -J_{xz}\\ 0\end{bmatrix} + h[\boldsymbol{\varphi}]^{\mathrm{T}}\begin{bmatrix}\lambda_4\\ -\lambda_3\\ 0\end{bmatrix}. \end{aligned} \tag{6.160}$$

The expressions $[\boldsymbol{\varphi}]^{\mathrm{T}}\{\mathbf{F}_{\mathrm{s}}\}$, and $[\boldsymbol{\varphi}]^{\mathrm{T}}\{\mathbf{F}_{\boldsymbol{\beta}}\}$ represent the column matrices of the projections of the torsor of the external forces onto the axes of the reference system $Oxyz$, that is, the matrices $[F_x \;\; F_y \;\; F_z]^{\mathrm{T}}$, and $[M_x \;\; M_y \;\; M_z]^{\mathrm{T}}$, respectively; analogically, the expressions $[\boldsymbol{\varphi}]^{\mathrm{T}}[\lambda_1+\lambda_3 \;\; \lambda_2+\lambda_4 \;\; 0]^{\mathrm{T}}$, and $[\boldsymbol{\varphi}]^{\mathrm{T}}[\lambda_4 h \;\; -\lambda_3 h \;\; 0]^{\mathrm{T}}$ represent the matrices of the components of the reaction, and reaction moment, respectively, that is, the matrices $\left[F_x^* \;\; F_y^* \;\; 0\right]^{\mathrm{T}}$, and $\left[M_x^* \;\; M_y^* \;\; 0\right]^{\mathrm{T}}$, respectively.

In these conditions, from the expressions (6.160) one deduces the system

$$\begin{aligned}-mx_C\dot{\varphi}^2 = F_x + F_x^*, mx_C\ddot{\varphi} = F_y + F_y^*, m\ddot{Z}_O = F_z, -J_{xz}\ddot{\varphi} + J_{yz}\dot{\varphi}^2 = M_x + M_x^*,\\ -mx_C\ddot{Z}_O - J_{yz}\ddot{\varphi} - J_{xz}\dot{\varphi}^2 = M_y + M_y^*, J_z\ddot{\varphi} = M_z.\end{aligned} \tag{6.161}$$

Further Reading

Amirouche FML (1992). Computational Methods in Multibody Dynamics. Englewood Cliffs: Prentice Hall.

Arnold VI (1997). Mathematical Methods of Classical Mechanics (Graduate Texts in Mathematics, Vol. 60). 2nd ed. New York: Springer.

Ball RS (1998). A Treatise on the Theory of Screws. Cambridge: Cambridge University Press.

Bloch AM, Baillieul J, Crouch P, Marsden J (2007). Nonholonomic Mechanics and Control (Interdisciplinary Applied Mathematics). Berlin: Springer.

Chaudhary H, Saha SH (2008). Dynamics and Balancing of Multibody Systems (Lecture Notes in Applied and Computational Mechanics). Berlin: Springer.

Cortes J (2002). Geometric, Control and Numerical Aspects of Nonholonomic Systems. Berlin: Springer.

Coutinho MG (2001). Dynamic Simulations of Multibody Systems. New York: Springer.

de Jalón JG, Bayo E (2011). Kinematic and Dynamic Simulation of Multibody Systems: The Real-Time Challenge. Berlin: Springer Verlag.

Den Hartog JP (1961). Mechanics. New York: Dover Publications Inc.

Douglas Gregory R (2006). Classical Mechanics. Cambridge: Cambridge University Press.

Edelen DG (2010). Lagrangian Mechanics of Nonconservative Nonholonomic Systems. Leyden: Kluwer.

Eich-Soellner E, Führer C (2013). Numerical Methods in Multibody Dynamics. Wiesbaden: Springer.

Fasano A, Marmi S, Pelloni B (2006). Analytical Mechanics: An Introduction. Oxford: Oxford University Press.

Fearstone R (2007). Rigid body Dynamics Algorithms. New York: Springer.

Gattringer H, Gerstmayr J (2013). Multibody Systems Dynamics, Robotics and Control. Wien: Springer.

Ginsberg JH (1998). Advanced Engineering Dynamics. 2nd ed. Cambridge: Cambridge University Press.

Greenwood DT (2006). Advanced Dynamics. Cambridge: Cambridge University Press.

Greiner W (2009). Classical Mechanics: Systems of Particles and Hamiltonian Dynamics. 2nd ed. Heidelberg: Springer.

Hahn H (2002). Rigid Body Dynamics of Mechanisms 1: Theoretical Basis. Berlin: Springer.

Hahn H (2010). Rigid Body Dynamics of Mechanisms 2: Applications. Berlin: Springer.

Hand LN, Finch JD (1998). Analytical Mechanics. Cambridge: Cambridge University Press.

Karnopp DC, Margolis DL, Rosenberg RC (2012). System Dynamics: Modeling, Simulation, and Control of Mechatronic Systems. Hoboken: John Wiley & Sons.
Kibble TWB, Berkshire FH (2004). Classical Mechanics. 5th ed. London: Imperial College Press.
Lagrange JL (1996). Analytical Mechanics. Dordrecht: Kluwer Academic Publishers,
Lurie AI (2002). Analytical Mechanics. Berlin: Springer.
Meriam JL, Kraige LG (2012). Engineering Mechanics: Dynamics. Hoboken: John Wiley & Sons.
Neimark IuI, Fufaev NA (1972). Dynamics of Nonholonomic Systems (Translations of Mathematical Monographs, V. 33). Providence, Rhode Island: American Mathematical Society.
Nikravesh PE (2007). Planar Multibody Dynamics: Formulation, Programming and Applications. Boca Raton: CRC Press.
O'Reilly OM (2008). Intermediate Dynamics for Engineers: A unified Treatment of Newton-Euler and Lagrangian Mechanics. Cambridge: Cambridge University Press
Pandrea N, Stănescu ND (2002). Mecanica. Bucureăti: Editura Didactică şi Pedagogică (in Romanian).
Papastavridis JG (2002). Analytical Mechanics: A Comprehensive Treatise on the Dynamics of Constrained Systems; For Engineers, Physicists, and Mathematicians. Oxford: Oxford University Press.
Pfeiffer F, Glocker C (1996). Multibody Dynamics with Unilateral Contacts. New York: John Wiley & Sons, Inc.
Roberson RE, Schwertassek R (1988). Dynamics of Multibody Systems. Berlin: Springer Verlag.
Shabana AA (2008). Computational Dynamics. New York: John Wiley & Sons.
Shabana AA (2013). Dynamics of Multibody Systems. Cambridge: Cambridge University Press.
Soltakhanov ShKh, Yushkov MPP Zegzhda SA (2009). Mechanics of non-nolonomic systems: A New Class of control systems (Foundations of Engineering Mechanics). Berlin: Springer.
Spong MW, Hutchinson S, Vidyasagar M (2005). Robot Modeling and Control. New York: John Wiley & Sons.
Stănescu ND (2013). Mecanica sistemelor. Bucureşti: Editura Didactică şi Pedagogică (in Romanian).
Stănescu ND, Munteanu L, Chiroiu V, Pandrea N (2007). Sisteme dinamice. Teorie şi aplicaţii vol. I. Bucureşti: Editura Academiei Române (in Romanian).
Stănescu ND, Munteanu L, Chiroiu V, Pandrea N (2011). Sisteme dinamice. Teorie şi aplicaţii vol. II. Bucureşti: Editura Academiei Române (in Romanian).
Teodorescu PP (2009). Mechanical Systems, Classical Models: Volume 3: Analytical Mechanics (Mathematical and Analytical Techniques with Applications to Engineering). Dordrecht: Springer.
Teodorescu PP, Stănescu ND, Pandrea N (2013). Numerical Analysis with Applications in Mechanics and Engineering. Hoboken: Wiley.
Udwadia FE, Kalaba RE (2007). Analytical Dynamics: A New Approach. Cambridge: Cambridge University Press.
Wittenburg J (2007). Dynamics of Multibody Systems. Berlin: Springer.

7

The Motion of the Rigid Solid with Constraints on Given Proper Curves

In this chapter we discuss the motion of the rigid solid with constraints on given proper curves; the chapter is dedicated to the study of the rigid body at which given curves support on given curves or surfaces.

7.1 General Aspects: Classification

There exist motions of the rigid solid as: the rolling of a disk with and without sliding on a surface (Fig. 7.1), the motion of the journal in an axial bearing with clearance (Fig. 7.2) or the motion of a body (Fig. 7.3) by tripod supporting (tripod coupling), at which the contact between the bodies is realized at certain points situated on curves (straight lines) of the rigid solid.

These motions may be called as motions of the rigid solid with constraints (linkages) on given proper curves.

In a general analysis one may distinguish the following types of contacts:

a. mobile curve supported at a fixed point;
b. mobile curve supported (with and without sliding) on a fixed curve;
c. mobile curve supported (with and without sliding) on a fixed surface.

The mathematical aspects discussed in paragraph 6.1 remaining valid for these motions too, one will pass directly to the determination of the matrices of constraints.

Dynamics of the Rigid Solid with General Constraints by a Multibody Approach, First Edition.
Nicolae Pandrea and Nicolae-Doru Stănescu.

Companion website: www.wiley.com/go/pandrea

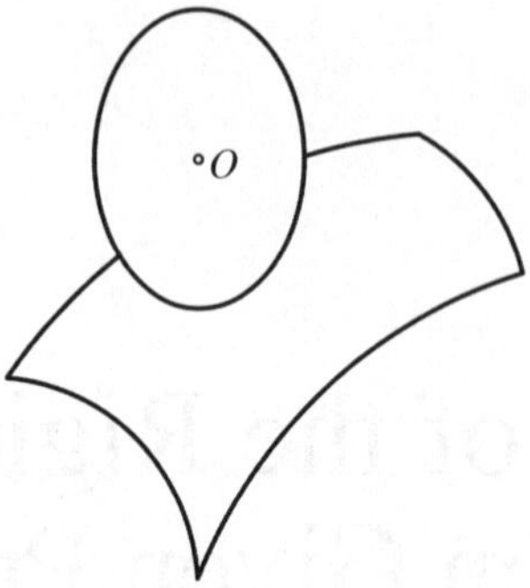

Figure 7.1 Rolling of a disk on a surface.

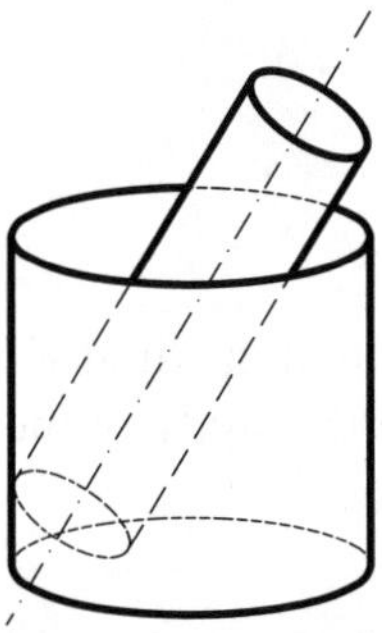

Figure 7.2 The motion of a journal in a bearing with clearance.

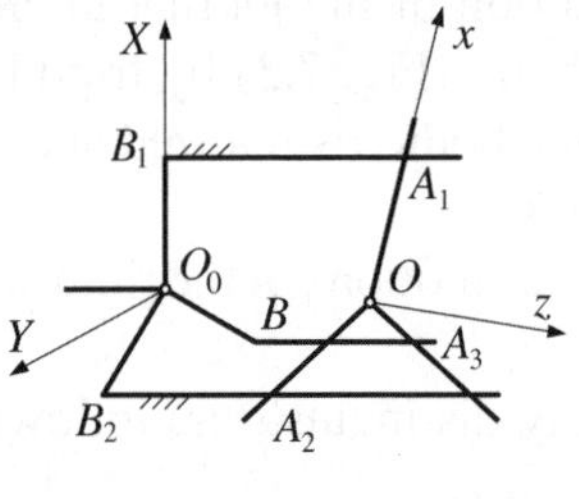

Figure 7.3 The tripod coupling.

7.2 The Rigid Solid Supported at Fixed Points on Given Proper Curves

7.2.1 Notations

One considers that the fixed points A_i have the coordinates X_i, Y_i, Z_i in the general reference system, and the curves are given parametrically in the local system $Oxyz$ by the equations

$$x_i = x_i(\xi_i),\ y_i = y_i(\xi_i),\ z_i = z_i(\xi_i). \tag{7.1}$$

In the calculations will be performed, one uses the general notations previously defined and, in addition, the specific ones for the analyzed cases:

- x_{ip}, y_{ip}, z_{ip} – the first order derivatives of the functions x_i, y_i, z_i with respect to the parameter ξ_i;
- x_{is}, y_{is}, z_{is} – the second order derivatives of the functions x_i, y_i, z_i with respect to the parameter ξ_i;
- $\{\mathbf{r}_i\}$, $\{\mathbf{r}_{ip}\}$, $\{\mathbf{r}_{is}\}$ – the column matrices

$$\{\mathbf{r}_i\} = [x_i \ \ y_i \ \ z_i]^{\mathrm{T}}, \{\mathbf{r}_{ip}\} = [x_{ip} \ \ y_{ip} \ \ z_{ip}]^{\mathrm{T}}, \{\mathbf{r}_{is}\} = [x_{is} \ \ y_{is} \ \ z_{is}]^{\mathrm{T}}; \tag{7.2}$$

- $[\mathbf{r}_i]$, $[\mathbf{r}_{ip}]$ – the skew matrices

$$[\mathbf{r}_i] = \begin{bmatrix} 0 & -z_i & y_i \\ z_i & 0 & -x_i \\ -y_i & x_i & 0 \end{bmatrix}, [\mathbf{r}_{ip}] = \begin{bmatrix} 0 & -z_{ip} & y_{ip} \\ z_{ip} & 0 & -x_{ip} \\ -y_{ip} & x_{ip} & 0 \end{bmatrix}; \tag{7.3}$$

- $[\boldsymbol{\Delta}_i]$, $\{\boldsymbol{\Delta}_i^*\}$ – the matrices

$$[\boldsymbol{\Delta}_i] = [\mathbf{A}][\mathbf{r}_i]^{\mathrm{T}}[\mathbf{Q}], \{\boldsymbol{\Delta}_i^*\} = [\mathbf{A}]\{\mathbf{r}_{ip}\}; \tag{7.4}$$

- $\{\mathbf{R}_i\}$ – the column matrix

$$\{\mathbf{R}_i\} = [X_i \ \ Y_i \ \ Z_i]^{\mathrm{T}}; \tag{7.5}$$

- $\{\mathbf{q}^*\}$ – the column matrix

$$\{\mathbf{q}^*\} = [X_O \ \ Y_O \ \ Z_O \ \ \psi \ \ \theta \ \ \varphi \ \ \xi_1 \ \ \ldots]^{\mathrm{T}}. \tag{7.6}$$

7.2.2 The Matrix of Constraints

Starting from the expression

$$\{\mathbf{R}_i\} = \{\mathbf{R}_O\} + [\mathbf{A}]\{\mathbf{r}_i\}, \tag{7.7}$$

by derivation with respect to time one obtains

$$\{\dot{\mathbf{R}}_O\} + [\dot{\mathbf{A}}]\{\mathbf{r}_i\} + \dot{\xi}_i[\mathbf{A}]\{\mathbf{r}_{ip}\} = \{\mathbf{0}\} \tag{7.8}$$

or

$$\{\dot{\mathbf{R}}_O\} + [\mathbf{A}][\mathbf{r}_i]^{\mathrm{T}}[\mathbf{Q}]\{\boldsymbol{\beta}\} + \dot{\xi}_i[\mathbf{A}]\{\mathbf{r}_{ip}\} = \{\mathbf{0}\}; \tag{7.9}$$

using the notations (7.4) and (7.5), it results

$$\left[[\mathbf{I}]\ \ [\boldsymbol{\Delta}_i]\ \ \{\mathbf{0}\}\ \ \ldots\ \ \{\boldsymbol{\Delta}_i^*\}\ \ \ldots\right]\{\dot{\mathbf{q}}^*\} = \{\mathbf{0}\}. \tag{7.10}$$

The relation (7.10) contains three scalar equations and at least seven parameters; if one eliminates the parameter ξ_i, then one obtains two equations with at least six parameters (X_O, Y_O, Z_O, ψ, θ, φ).

In conclusion, maximum three contacts of this kind may exist and, consequently, the matrix of constraints may be

$$[\mathbf{B}] = \left[[\mathbf{I}]\ \ [\boldsymbol{\Delta}_1]\ \ \{\boldsymbol{\Delta}_1^*\}\right], \tag{7.11}$$

$$[\mathbf{B}] = \begin{bmatrix} [\mathbf{I}] & [\boldsymbol{\Delta}_1] & \{\boldsymbol{\Delta}_1^*\} & \{\mathbf{0}\} \\ [\mathbf{I}] & [\boldsymbol{\Delta}_2] & \{\mathbf{0}\} & \{\boldsymbol{\Delta}_2^*\} \end{bmatrix}, \tag{7.12}$$

$$[\mathbf{B}] = \begin{bmatrix} [\mathbf{I}] & [\boldsymbol{\Delta}_1] & \{\boldsymbol{\Delta}_1^*\} & \{\mathbf{0}\} & \{\mathbf{0}\} \\ [\mathbf{I}] & [\boldsymbol{\Delta}_2] & \{\mathbf{0}\} & \{\boldsymbol{\Delta}_2^*\} & \{\mathbf{0}\} \\ [\mathbf{I}] & [\boldsymbol{\Delta}_3] & \{\mathbf{0}\} & \{\mathbf{0}\} & \{\boldsymbol{\Delta}_3^*\} \end{bmatrix}, \tag{7.13}$$

for one, two or three supports of this kind, respectively.

To obtain the derivative of the matrix of constraints, one has to keep into account the relations

$$[\dot{\boldsymbol{\Delta}}_i] = [\dot{\mathbf{A}}][\mathbf{r}_i]^{\mathrm{T}}[\mathbf{Q}] + \dot{\xi}_i[\mathbf{A}][\mathbf{r}_{ip}][\mathbf{Q}] + [\mathbf{A}][\mathbf{r}_i][\dot{\mathbf{Q}}], \tag{7.14}$$

$$\left[\dot{\boldsymbol{\Delta}}_i^*\right] = [\dot{\mathbf{A}}]\{\mathbf{r}_{ip}\} + \dot{\xi}_i[\mathbf{A}]\{\mathbf{r}_{is}\}. \tag{7.15}$$

7.3 The Rigid Solid at Which Given Proper Curves Support with Sliding on Fixed Curves

7.3.1 Notations

One considers the parametric equations, in the general reference system O_0XYZ, for the fixed curves

$$X_i = X_i(\eta_i),\ Y_i = Y_i(\eta_i),\ Z_i = Z_i(\eta_i), \tag{7.16}$$

and the parametric equations, in the local reference system $Oxyz$, for the curves jointed to the rigid solid,

$$x_i = x_i(\xi_i),\ y_i = y_i(\xi_i),\ z_i = z_i(\xi_i). \tag{7.17}$$

We maintain for the mobile curves the notations (7.2) – (7.4), while for the fixed curves one uses the analogical notations:

- X_{ip}, Y_{ip}, Z_{ip} – the first order derivatives of the functions X_i, Y_i, Z_i with respect to the parameter η_i;
- X_{is}, Y_{is}, Z_{is} – the second order derivatives;
- $\{\mathbf{R}_i\}$, $\{\mathbf{R}_{ip}\}$, $\{\mathbf{R}_{is}\}$ – the column matrices

$$\{\mathbf{R}_i\} = [X_i \ \ Y_i \ \ Z_i]^{\mathrm{T}}, \{\mathbf{R}_{ip}\} = [X_{ip} \ \ Y_{ip} \ \ Z_{ip}]^{\mathrm{T}}, \{\mathbf{R}_{is}\} = [X_{is} \ \ Y_{is} \ \ Z_{is}]^{\mathrm{T}}; \tag{7.18}$$

- $\{\mathbf{q}^*\}$ – the column matrix

$$\{\mathbf{q}^*\} = [X_O \ \ Y_O \ \ Z_O \ \ \psi \ \ \theta \ \ \varphi \ \ \xi_1 \ \ \eta_1 \ \ \ldots]^{\mathrm{T}}. \tag{7.19}$$

7.3.2 The Simple Contact between the Curves

Starting from the expression

$$\{\mathbf{R}_i\} = \{\mathbf{R}_0\} + [\mathbf{A}]\{\mathbf{r}_i\}, \tag{7.20}$$

by derivation with respect to time one obtains

$$\dot{\eta}\{\mathbf{R}_{ip}\} = \{\dot{\mathbf{R}}_O\} + [\boldsymbol{\Delta}_i]\{\dot{\boldsymbol{\beta}}\} + \dot{\xi}\{\boldsymbol{\Delta}_i^*\}. \tag{7.21}$$

From the three scalar equations that result from the relation (7.21), if one eliminates the parameters ξ_i, η_i, then one obtains only one equation in the coordinates X_O, Y_O, Z_O, ψ, θ, φ and, consequently, such a support introduces only one constraint in the case of the contact with sliding; hence, only five supports of this kind are possible.

In conclusion, the matrix of constraints reads

$$[\mathbf{B}] = \left[[\mathbf{I}] \ \ [\boldsymbol{\Delta}_1] \ \ \{\boldsymbol{\Delta}_1^*\} \ \ -\{\mathbf{R}_{ip}\} \right] \tag{7.22}$$

for a support, and

$$[\mathbf{B}]=\begin{bmatrix}[\mathbf{I}] & [\boldsymbol{\Delta}_1] & \{\boldsymbol{\Delta}_1^*\} & -\{\mathbf{R}_{1p}\} & \{\mathbf{0}\} & \{\mathbf{0}\} & \ldots & \{\mathbf{0}\} & \{\mathbf{0}\}\\ [\mathbf{I}] & [\boldsymbol{\Delta}_2] & \{\mathbf{0}\} & \{\mathbf{0}\} & \{\boldsymbol{\Delta}_2^*\} & -\{\mathbf{R}_{2p}\} & \ldots & \{\mathbf{0}\} & \{\mathbf{0}\}\\ \ldots & \ldots & \ldots & \ldots & \ldots & \ldots & \ldots & \ldots & \ldots\\ [\mathbf{I}] & [\boldsymbol{\Delta}_n] & \{\mathbf{0}\} & \{\mathbf{0}\} & \{\mathbf{0}\} & \{\mathbf{0}\} & \ldots & \{\boldsymbol{\Delta}_n^*\} & -\{\mathbf{R}_{np}\}\end{bmatrix} \tag{7.23}$$

for $n \le 5$ supports, respectively.

In the case of one support, the matrix of constraints reads

$$[\mathbf{B}]=\left[[\mathbf{I}]\ \ [\boldsymbol{\Delta}]\ \ \{\boldsymbol{\Delta}^*\}\ \ -\{\mathbf{R}_p\}\right], \tag{7.24}$$

and the equations of motion are

$$[\mathbf{M}]\{\ddot{\mathbf{q}}\}=\{\mathbf{F}\}+\{\widetilde{\mathbf{F}}\}+\begin{bmatrix}[\mathbf{I}]\\ [\boldsymbol{\Delta}]^{\mathrm{T}}\end{bmatrix}\{\boldsymbol{\lambda}\}, \tag{7.25}$$

$$\{\boldsymbol{\Delta}^*\}^{\mathrm{T}}\{\boldsymbol{\lambda}\}=0,\{\mathbf{R}_p\}^{\mathrm{T}}\{\boldsymbol{\lambda}\}=0, \tag{7.26}$$

$$[[\mathbf{I}]\ \ [\boldsymbol{\Delta}]]\{\ddot{\mathbf{q}}\}+[\{\boldsymbol{\Delta}^*\}\ \ -\{\mathbf{R}_p\}]\begin{bmatrix}\ddot{\xi}\\ \ddot{\eta}\end{bmatrix}=-\left[[\mathbf{0}]\ \ [\dot{\boldsymbol{\Delta}}]\right]\{\dot{\mathbf{q}}\}-\left[\{\dot{\boldsymbol{\Delta}}^*\}\ \ -\{\dot{\mathbf{R}}_p\}\right]\begin{bmatrix}\dot{\xi}\\ \dot{\eta}\end{bmatrix}. \tag{7.27}$$

The equalities (7.26) show that the Lagrange multipliers λ_1, λ_2, λ_3 are, in fact, the director parameters of the common normal to the two curves, at the contact point.

The system formed by the relations (7.25) and (7.26) contains 11 equations for the 11 unknowns X_O, Y_O, Z_O, ψ, θ, φ, ξ, η, λ_1, λ_2, λ_3.

If one uses the notation

$$[\widetilde{\mathbf{M}}]=\begin{bmatrix}[\mathbf{M}] & [\mathbf{0}_{62}]\\ [\mathbf{0}_{26}] & [\mathbf{0}_{22}]\end{bmatrix}, \tag{7.28}$$

then the equations (7.24)–(7.27) form the matrix equation

$$\begin{bmatrix}[\widetilde{\mathbf{M}}] & -[\mathbf{B}]^{\mathrm{T}}\\ [\mathbf{B}] & [\mathbf{0}_{33}]\end{bmatrix}\begin{bmatrix}\{\ddot{\mathbf{q}}^*\}\\ \{\boldsymbol{\lambda}\}\end{bmatrix}=\begin{bmatrix}\{\mathbf{F}\}+\{\widetilde{\mathbf{F}}\}\\ \{\mathbf{0}_{21}\}\\ -[\dot{\mathbf{B}}]\{\dot{\mathbf{q}}^*\}\end{bmatrix}, \tag{7.29}$$

where

$$\{\mathbf{q}^*\}=[X_O\ \ Y_O\ \ Z_O\ \ \psi\ \ \theta\ \ \varphi\ \ \xi\ \ \eta]^{\mathrm{T}}. \tag{7.30}$$

In the case of $n \le 5$ supports, using the notation

$$[\widetilde{\mathbf{M}}] = \begin{bmatrix} [\mathbf{M}] & [\mathbf{0}_{6,2n}] \\ [\mathbf{0}_{2n,6}] & [\mathbf{0}_{2n,2n}] \end{bmatrix}, \tag{7.31}$$

one obtains the matrix equation

$$\begin{bmatrix} [\widetilde{\mathbf{M}}] & -[\mathbf{B}]^{\mathrm{T}} \\ [\mathbf{B}] & [\mathbf{0}_{3n,3n}] \end{bmatrix} \begin{bmatrix} \{\ddot{\mathbf{q}}^*\} \\ \{\boldsymbol{\lambda}\} \end{bmatrix} = \begin{bmatrix} \{\mathbf{F}\} + \{\widetilde{\mathbf{F}}\} \\ \{\mathbf{0}_{2n,1}\} \\ -[\dot{\mathbf{B}}]\{\dot{\mathbf{q}}^*\} \end{bmatrix}, \tag{7.32}$$

where

$$\{\mathbf{q}^*\} = [X_O \;\; Y_O \;\; Z_O \;\; \psi \;\; \theta \;\; \varphi \;\; \xi_1 \;\; \eta_1 \;\; \ldots \;\; \xi_n \;\; \eta_n]^{\mathrm{T}}, \tag{7.33}$$

$$\{\boldsymbol{\lambda}\} = [\lambda_1 \;\; \lambda_2 \;\; \ldots \;\; \lambda_{3n}]^{\mathrm{T}}. \tag{7.34}$$

For the calculation of the derivative matrix $[\dot{\mathbf{B}}]$ one has to keep into account the equalities

$$[\dot{\boldsymbol{\Delta}}_i] = [\dot{\mathbf{A}}][\mathbf{r}_i]^{\mathrm{T}}[\mathbf{Q}] + \dot{\xi}_i[\mathbf{A}][\mathbf{r}_{ip}][\mathbf{Q}] + [\mathbf{A}][\mathbf{r}_i][\dot{\mathbf{Q}}], \tag{7.35}$$

$$\{\dot{\boldsymbol{\Delta}}_i^*\} = [\dot{\mathbf{A}}]\{\mathbf{r}_{ip}\} + \dot{\xi}_i[\mathbf{A}]\{\mathbf{r}_{is}\}, \tag{7.36}$$

$$\{\mathbf{R}_{ip}\} = \dot{\eta}_i\{\mathbf{R}_{is}\}. \tag{7.37}$$

Example 7.3.1 Determine the matrix of constraints for the tripod coupling in Fig. 7.3 knowing that the fixed axes (the axes of the tulip) are parallel and arranged at 120^0, while the arms of the tripod are situated in the same plan, perpendicular to the axis of the axle and arranged at 120^0 too.

Solution: Denoting by R the distance $O_0B_1 = O_0B_2 = O_0B_3 = R$, by δ_i the angles 0, $\frac{2\pi}{3}$, $\frac{4\pi}{3}$, by λ_i the distances B_iA_i, by μ_i the distances OA_i and keeping into account the representations in Fig. 7.3, one obtains the equations of contact

$$\begin{bmatrix} R\cos\delta_i \\ R\sin\delta_i \\ \lambda_i \end{bmatrix} = \begin{bmatrix} X_O \\ Y_O \\ Z_O \end{bmatrix} + [\mathbf{A}] \begin{bmatrix} \mu_i\cos\delta_i \\ \mu_i\sin\delta_i \\ 0 \end{bmatrix}, \; i = 1, 2, 3. \tag{7.38}$$

Further on, with the notation

$$[\mathbf{A}]=\begin{bmatrix}\alpha_1 & \alpha_2 & \alpha_3\\ \beta_1 & \beta_2 & \beta_3\\ \gamma_1 & \gamma_2 & \gamma_3\end{bmatrix}, \tag{7.39}$$

from the relation (7.38) one deduces the scalar equalities

$$R\cos\delta_i=X_O+\mu_i(\alpha_1\cos\delta_i+\alpha_2\sin\delta_i),\ R\sin\delta_i=Y_O+\mu_i(\beta_1\cos\delta_i+\beta_2\sin\delta_i),$$
$$\lambda_i=Z_O+\mu_i(\gamma_1\cos\delta_i+\gamma_2\sin\delta_i). \tag{7.40}$$

From the first two equalities (7.40), eliminating the parameter μ_i, one obtains the equations

$$\begin{aligned}&2X_O(\beta_1\cos\delta_i+\beta_2\sin\delta_i)-2Y_O(\alpha_1\cos\delta_i+\alpha_2\sin\delta_i)\\&-R[\beta_1-\alpha_2+(\beta_1+\alpha_2)\cos 2\delta_i+(\beta_2-\alpha_1)\sin 2\delta_i]=0\end{aligned} \tag{7.41}$$

which, by summation and keeping into account the relations

$$\sum_{i=1}^{3}\cos\delta_i=\sum_{i=1}^{3}\sin\delta_i=\sum_{i=1}^{3}\cos 2\delta_i=\sum_{i=1}^{3}\sin 2\delta_i=0, \tag{7.42}$$

lead to the equality

$$\beta_1-\alpha_2=0. \tag{7.43}$$

Making $\delta_i=\delta_2$, $\delta_i=\delta_3$ in the relations (7.41) and then summing and subtracting, one gets the expressions

$$X_O\alpha_2-Y_O\alpha_1-R\alpha_2=0,\ 2X_O\beta_2-2Y_O\alpha_2+R(\beta_2-\alpha_1)=0. \tag{7.44}$$

If the angles ψ, θ, φ are the Euler angles, then from expression (7.43) one deduces $\psi=-\varphi$, and the relations (7.44) become

$$2X_O\cos\theta-R(1-\cos\theta)(\cos 3\varphi\cos\varphi+\cos\theta\sin 3\varphi\sin\varphi)=0,$$
$$2Y_O\cos\theta-R(1-\cos\theta)(-\cos 3\varphi\cos\varphi+\cos\theta\sin 3\varphi\sin\varphi)=0. \tag{7.45}$$

One obtains the matrix of constraints for the parameters X_O, Y_O, Z_O, θ, φ,

$$[\mathbf{B}]=\begin{bmatrix}2\cos\theta & 0 & 0 & B_{14} & B_{15}\\ 0 & 2\cos\theta & 0 & B_{24} & B_{25}\end{bmatrix}, \tag{7.46}$$

where

$$
\begin{aligned}
B_{14} &= -2X_O \sin\theta - R(\cos 3\varphi \cos\varphi + \cos\theta \sin 3\varphi \sin\varphi)\sin\theta \\
&\quad + R(1-\cos\theta)\sin\theta \sin 3\varphi \sin\varphi, \\
B_{15} &= R(1-\cos\theta) \\
&\quad \cdot (3\sin 3\varphi \cos\varphi + \cos 3\varphi \sin\varphi + 3\cos\theta \cos 3\varphi \sin\varphi + \cos\theta \sin 3\varphi \cos\varphi), \\
B_{24} &= -Y_O \sin\theta - R(-\cos 3\varphi \cos\varphi + \cos\theta \sin 3\varphi \sin\varphi)\sin\theta \\
&\quad + R(1-\cos\theta)\sin\theta \sin 3\varphi \sin\varphi, \\
B_{25} &= -R(1-\cos\theta) \\
&\quad \cdot (3\sin 3\varphi \sin\varphi + \cos 3\varphi \sin\varphi - 3\cos\theta \sin 3\varphi \sin\varphi - \cos\theta \sin 3\varphi \cos\varphi).
\end{aligned}
\tag{7.47}
$$

7.3.3 *The Tangency Contact between Spatial Curves*

We consider the case in which the mobile curve of parametric equations (in the local reference system)

$$x = x(\xi), y = y(\xi), z = z(\xi) \tag{7.48}$$

remains tangent to the fixed curve of equations (in the reference system O_0XYZ)

$$X = X(\eta), Y = Y(\eta), Z = Z(\eta). \tag{7.49}$$

Starting from the contact relation

$$\{\mathbf{R}\} = \{\mathbf{R}_O\} + [\mathbf{A}]\{\mathbf{r}\}, \tag{7.50}$$

by derivation with respect to time one obtains the expression

$$\dot{\eta}\{\mathbf{R}_p\} = \{\dot{\mathbf{R}}_O\} + [\mathbf{A}][\mathbf{r}]^{\mathrm{T}}[\mathbf{Q}]\{\dot{\boldsymbol{\beta}}\} + \dot{\xi}[\mathbf{A}]\{\mathbf{r}_p\}. \tag{7.51}$$

The condition of tangency leads to the relation

$$\{\mathbf{R}_p\} = \zeta[\mathbf{A}]\{\mathbf{r}_p\}, \tag{7.52}$$

in which ζ is a factor of proportionality.

By derivation of the relation (7.52) with respect to time one obtains

$$\zeta[\mathbf{A}][\mathbf{r}_p]^{\mathrm{T}}[\mathbf{Q}]\{\dot{\boldsymbol{\beta}}\} + \zeta\dot{\xi}[\mathbf{A}][\mathbf{r}_s] - \dot{\eta}\{\mathbf{R}_s\} + \dot{\zeta}[\mathbf{A}]\{\mathbf{r}_p\} = \{\mathbf{0}\}; \tag{7.53}$$

from the relations (7.51)–(7.53) one deduces the matrix of constraints

$$[\mathbf{B}]=\begin{bmatrix}[\mathbf{I}] & [\mathbf{A}][\mathbf{r}]^{\mathrm{T}}[\mathbf{Q}] & [\mathbf{A}]\{\mathbf{r}_p\} & -\{\mathbf{R}_p\} & \{\mathbf{0}\}\\ [\mathbf{0}] & \zeta[\mathbf{A}][\mathbf{r}_p]^{\mathrm{T}}[\mathbf{Q}] & \zeta[\mathbf{A}][\mathbf{r}_s] & -\zeta\{\mathbf{R}_s\} & [\mathbf{A}]\{\mathbf{r}_p\}\end{bmatrix}, \tag{7.54}$$

where the column matrix of the coordinates is given by

$$\{\mathbf{q}^*\}=[X_O\ Y_O\ Z_O\ \psi\ \theta\ \varphi\ \xi\ \eta\ \zeta]^{\mathrm{T}}. \tag{7.55}$$

Using the notation

$$[\widetilde{\mathbf{M}}]=\begin{bmatrix}[\mathbf{M}] & [\mathbf{0}_{63}]\\ [\mathbf{0}_{36}] & [\mathbf{0}_{33}]\end{bmatrix}, \tag{7.56}$$

one obtains the matrix equation

$$\begin{bmatrix}[\widetilde{\mathbf{M}}] & -[\mathbf{B}]^{\mathrm{T}}\\ [\mathbf{B}] & [\mathbf{0}_{66}]\end{bmatrix}\begin{bmatrix}\{\ddot{\mathbf{q}}^*\}\\ \{\boldsymbol{\lambda}\}\end{bmatrix}=\begin{bmatrix}\{\mathbf{F}\}+\{\widetilde{\mathbf{F}}\}\\ \{\mathbf{0}_{61}\}\\ -[\dot{\mathbf{B}}]\{\dot{\mathbf{q}}^*\}\end{bmatrix}, \tag{7.57}$$

where

$$\{\boldsymbol{\lambda}\}=[\lambda_1\ \lambda_2\ \ldots\ \lambda_6]^{\mathrm{T}}. \tag{7.58}$$

7.3.4 Contact with Sliding between Planar Curves (Rolling with Sliding on the Plan)

In the case of the planar curves the rotational matrix reads

$$[\mathbf{A}]=[\boldsymbol{\varphi}]=\begin{bmatrix}\cos\varphi & -\sin\varphi\\ \sin\varphi & \cos\varphi\end{bmatrix}. \tag{7.59}$$

The derivative matrix with respect to the angle φ is given by

$$[\boldsymbol{\varphi}_p]=\begin{bmatrix}-\sin\varphi & -\cos\varphi\\ \cos\varphi & -\sin\varphi\end{bmatrix}=[\mathbf{U}][\boldsymbol{\varphi}]=[\boldsymbol{\varphi}][\mathbf{U}], \tag{7.60}$$

where the matrix $[\mathbf{U}]$ is given by the expression

$$[\mathbf{U}]=\begin{bmatrix}0 & -1\\ 1 & 0\end{bmatrix}. \tag{7.61}$$

and it has the property

$$[\mathbf{U}]^2 = \begin{bmatrix} -1 & 0 \\ 0 & -1 \end{bmatrix} = -[\mathbf{I}]. \tag{7.62}$$

Starting from the contact condition

$$\begin{bmatrix} X \\ Y \end{bmatrix} = \begin{bmatrix} X_O \\ Y_O \end{bmatrix} + [\boldsymbol{\varphi}]\begin{bmatrix} x \\ y \end{bmatrix}, \tag{7.63}$$

by derivation with respect to time one obtains

$$\dot{\eta}\begin{bmatrix} X_p \\ Y_p \end{bmatrix} = \begin{bmatrix} \dot{X}_O \\ \dot{Y}_O \end{bmatrix} + \dot{\varphi}[\mathbf{U}][\boldsymbol{\varphi}]\begin{bmatrix} x \\ y \end{bmatrix} + \dot{\xi}[\boldsymbol{\varphi}]\begin{bmatrix} x_p \\ y_p \end{bmatrix}. \tag{7.64}$$

The tangency condition reads

$$\begin{bmatrix} X_p \\ Y_p \end{bmatrix} = \zeta[\boldsymbol{\varphi}]\begin{bmatrix} x_p \\ y_p \end{bmatrix}, \tag{7.65}$$

where ζ is a factor of proportionality.

From the relation (7.65) it results

$$[X_p \;\; Y_p][\mathbf{U}][\boldsymbol{\varphi}]\begin{bmatrix} x_p \\ y_p \end{bmatrix} = 0 \tag{7.66}$$

or

$$\{\mathbf{R}_p\}^{\mathrm{T}}[\mathbf{U}][\boldsymbol{\varphi}]\{\mathbf{r}_p\} = 0, \tag{7.67}$$

and by derivation with respect to time one obtains

$$-\dot{\varphi}\{\mathbf{R}_p\}^{\mathrm{T}}[\boldsymbol{\varphi}]\{\mathbf{r}_p\} + \dot{\xi}\{\mathbf{R}_p\}^{\mathrm{T}}[\mathbf{U}][\boldsymbol{\varphi}]\{\mathbf{r}_s\} + \dot{\eta}\{\mathbf{R}_s\}^{\mathrm{T}}[\mathbf{U}][\boldsymbol{\varphi}]\{\mathbf{r}_p\} = 0; \tag{7.68}$$

from the relations (7.64) and (7.68) one deduces the matrix of constraints

$$[\mathbf{B}] = \begin{bmatrix} [\mathbf{I}] & [\mathbf{U}][\boldsymbol{\varphi}]\{\mathbf{r}_p\} & [\boldsymbol{\varphi}]\{\mathbf{r}_p\} & -\{\mathbf{R}_p\} \\ [0 \;\; 0] & -\{\mathbf{R}_p\}^{\mathrm{T}}[\boldsymbol{\varphi}]\{\mathbf{r}_p\} & \{\mathbf{R}_p\}^{\mathrm{T}}[\mathbf{U}][\boldsymbol{\varphi}]\{\mathbf{r}_s\} & \{\mathbf{R}_s\}^{\mathrm{T}}[\mathbf{U}][\boldsymbol{\varphi}]\{\mathbf{r}_p\} \end{bmatrix}, \tag{7.69}$$

the matrix of coordinates being

$$\{\mathbf{q}^*\} = [X_O \;\; Y_O \;\; \varphi \;\; \xi \;\; \eta]^{\mathrm{T}}. \tag{7.70}$$

With the aid of the notations

$$[\mathbf{M}]=\begin{bmatrix} m & 0 & 0 \\ 0 & m & 0 \\ 0 & 0 & J \end{bmatrix},\ [\widetilde{\mathbf{M}}]=\begin{bmatrix} m & 0 & 0 & 0 & 0 \\ 0 & m & 0 & 0 & 0 \\ 0 & 0 & J & 0 & 0 \\ 0 & 0 & 0 & 0 & 0 \\ 0 & 0 & 0 & 0 & 0 \end{bmatrix}, \tag{7.71}$$

one obtains the matrix equation of motion

$$\begin{bmatrix} [\widetilde{\mathbf{M}}] & -[\mathbf{B}]^{\mathrm{T}} \\ [\mathbf{B}] & [\mathbf{0}_{33}] \end{bmatrix}\begin{bmatrix} \{\ddot{\mathbf{q}}^*\} \\ \{\boldsymbol{\lambda}\} \end{bmatrix}=\begin{bmatrix} \{\mathbf{F}\} \\ \{\mathbf{0}_{21}\} \\ -[\dot{\mathbf{B}}]\{\dot{\mathbf{q}}^*\} \end{bmatrix}. \tag{7.72}$$

Example 7.3.2 Write the matrix equation of motion for an elliptic disk of mass m and semi-axes a, b (Fig. 7.4) which rolls without sliding inside a circle of radius R, the disk being acted only by its own weight.

Solution: The parametric equations of the ellipse and circle being

$$x=a\cos\xi,\ y=b\sin\xi,\ X=R\cos\eta,\ Y=R\sin\eta, \tag{7.73}$$

one obtains the matrix of constraints

$$[\mathbf{B}]=\begin{bmatrix} 1 & 0 & -as_1c_2-bc_1s_2 & -ac_1s_2-bs_1c_2 & Rc_3 \\ 0 & 1 & ac_1c_2-bs_1s_2 & -as_1s_2+bc_1c_2 & -Rs_3 \\ 0 & 0 & -as_2s_4+bc_2c_4 & ac_2c_4-bs_2s_4 & as_2s_4-bc_2c_4 \end{bmatrix}, \tag{7.74}$$

where s_1, c_1, s_2, c_2, s_3, c_3, s_4, and c_4 mark $\sin\varphi$, $\cos\varphi$, $\sin\xi$, $\cos\xi$, $\sin\eta$, $\cos\eta$, $\sin(\varphi-\eta)$, and $\cos(\varphi-\eta)$, respectively; then one deduces the matrix equation of motion (7.72), where

$$\{\mathbf{F}\}=[0\ \ mg\ \ 0]^{\mathrm{T}}. \tag{7.75}$$

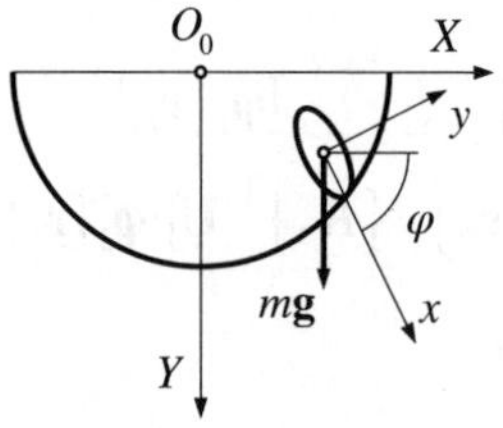

Figure 7.4 Motion of an elliptic disk inside a circle in Example 7.3.2.

7.4 Rolling without Sliding of a Curve on a Fixed Curve

7.4.1 The General Case for Spatial Curves

Maintaining the notations (7.48) and (7.49), one arrives at the contact condition (7.51) which, due to the condition of rolling without sliding,

$$\dot{\eta}\{\mathbf{R}_p\} = \dot{\xi}[\mathbf{A}]\{\mathbf{r}_p\}, \tag{7.76}$$

becomes

$$\{\dot{\mathbf{R}}_O\} + [\mathbf{A}][\mathbf{r}]^{\mathrm{T}}[\mathbf{Q}]\{\dot{\boldsymbol{\beta}}\} = \{\mathbf{0}\}. \tag{7.77}$$

For the variation in the same sense of the parameters ξ, η, the equality (7.76) leads to the relation

$$\dot{\xi}\sqrt{x_p^2 + y_p^2 + z_p^2} - \dot{\eta}\sqrt{X_p^2 + Y_p^2 + Z_p^2} = 0. \tag{7.78}$$

The equality (7.76) represents the tangency condition

$$\{\mathbf{R}_p\} = \zeta[\mathbf{A}]\{\mathbf{r}_p\}, \tag{7.79}$$

at which the condition (7.78) is added.

Deriving the equality (7.79) with respect to time, one deduces the expression

$$\zeta[\mathbf{A}]\{\mathbf{r}_p\}[\mathbf{Q}]\{\dot{\boldsymbol{\beta}}\} + \dot{\xi}\zeta[\mathbf{A}]\{\mathbf{r}_s\} - \dot{\eta}\{\mathbf{R}_s\} + \dot{\zeta}[\mathbf{A}]\{\mathbf{r}_p\} = \{\mathbf{0}\}, \tag{7.80}$$

and from the relations (7.77), (7.78), and (7.80) one obtains the matrix of constraints

$$[\mathbf{B}] = \begin{bmatrix} [\mathbf{I}] & [\mathbf{A}][\mathbf{r}]^{\mathrm{T}}[\mathbf{Q}] & \{\mathbf{0}\} & \{\mathbf{0}\} & \{\mathbf{0}\} \\ [\mathbf{0}] & \zeta[\mathbf{A}][\mathbf{r}_p]^{\mathrm{T}}[\mathbf{Q}] & \zeta[\mathbf{A}]\{\mathbf{r}_s\} & -\{\mathbf{R}_s\} & [\mathbf{A}]\{\mathbf{r}_p\} \\ [\mathbf{0}] & [\mathbf{0}] & \sqrt{x_p^2 + y_p^2 + z_p^2} & -\sqrt{X_p^2 + Y_p^2 + Z_p^2} & 0 \end{bmatrix}, \tag{7.81}$$

where the column matrix of coordinates is given by

$$\{\mathbf{q}^*\} = [X_O \;\; Y_O \;\; Z_O \;\; \psi \;\; \theta \;\; \varphi \;\; \xi \;\; \eta \;\; \zeta]^{\mathrm{T}}. \tag{7.82}$$

Because there exist nine coordinates and seven equations of constraints, it result that the motion has two degrees of freedom.

7.4.2 The Rolling Without Sliding of a Curve on a Fixed Curve in the Plan

Maintaining the notations from the previous paragraph, one obtains the equalities

$$\{\dot{\mathbf{R}}_O\} + \dot{\varphi}[\mathbf{U}][\boldsymbol{\varphi}]\{\mathbf{r}\} = \{\mathbf{0}\}, \tag{7.83}$$

$$\dot{\eta}\{\mathbf{R}_p\} = \dot{\xi}\{\mathbf{r}_p\}. \tag{7.84}$$

The condition (7.79) leads to

$$\dot{\xi}\sqrt{x_p^2 + y_p^2} - \dot{\eta}\sqrt{X_p^2 + Y_p^2} = 0 \tag{7.85}$$

and to the equation (7.68) and, consequently, the matrix of constraints reads

$$[\mathbf{B}] = \begin{bmatrix} [\mathbf{I}] & [\mathbf{U}][\boldsymbol{\varphi}]\{\mathbf{r}\} & \{\mathbf{0}\} & \{\mathbf{0}\} \\ [0\ \ 0] & -\{\mathbf{R}_p\}^{\mathrm{T}}[\boldsymbol{\varphi}]\{\mathbf{r}_p\} & \{\mathbf{R}_p\}[\mathbf{U}][\boldsymbol{\varphi}]\{\mathbf{r}_s\} & \{\mathbf{R}_s\}^{\mathrm{T}}[\mathbf{U}][\boldsymbol{\varphi}]\{\mathbf{r}_p\} \\ [0\ \ 0] & 0 & -\sqrt{x_p^2 + y_p^2} & -\sqrt{X_p^2 + Y_p^2} \end{bmatrix}, \tag{7.86}$$

where the column matrix $\{\mathbf{q}*\}$ is given by

$$\{\mathbf{q}^*\} = [X_O\ \ Y_O\ \ \varphi\ \ \xi\ \ \eta]^{\mathrm{T}}. \tag{7.87}$$

Because there are five coordinates and four equations of constraint, it results that the motion has only one degree of freedom.

Using the notation

$$[\widetilde{\mathbf{M}}] = \begin{bmatrix} m & 0 & 0 & 0 & 0 \\ 0 & m & 0 & 0 & 0 \\ 0 & 0 & J & 0 & 0 \\ 0 & 0 & 0 & 0 & 0 \\ 0 & 0 & 0 & 0 & 0 \end{bmatrix}, \tag{7.88}$$

one obtains the equation of motion

$$\begin{bmatrix} [\widetilde{\mathbf{M}}] & -[\mathbf{B}]^{\mathrm{T}} \\ [\mathbf{B}] & [\mathbf{0}_{44}] \end{bmatrix} \begin{bmatrix} \{\ddot{\mathbf{q}}^*\} \\ \{\boldsymbol{\lambda}\} \end{bmatrix} = \begin{bmatrix} \{\mathbf{F}\} \\ \begin{bmatrix} 0 \\ 0 \end{bmatrix} \\ -[\dot{\mathbf{B}}]\{\dot{\mathbf{q}}^*\} \end{bmatrix}, \tag{7.89}$$

where

$$\{\boldsymbol{\lambda}\} = [\lambda_1 \ \ \lambda_2 \ \ \lambda_3 \ \ \lambda_4]^{\mathrm{T}}. \tag{7.90}$$

The scalar equations that correspond to the fourth and fifth lines of the system (7.89) are

$$\lambda_3\{\mathbf{R}_p\}^{\mathrm{T}}[\mathbf{U}][\boldsymbol{\varphi}]\{\mathbf{r}_s\} + \lambda_4\sqrt{x_p^2 + y_p^2} = 0,\ \lambda_3\{\mathbf{R}_s\}^{\mathrm{T}}[\mathbf{U}][\boldsymbol{\varphi}]\{\mathbf{r}_p\} + \lambda_4\sqrt{X_p^2 + Y_p^2} = 0; \tag{7.91}$$

since the determinant of this algebraic system in λ_3, λ_4 is, in general, nonzero, it results

$$\lambda_3 = \lambda_4 = 0. \tag{7.92}$$

With these considerations in mind and using the notations

$$[\widetilde{\mathbf{B}}] = [[\mathbf{I}] \ \ [\mathbf{U}][\boldsymbol{\varphi}]\{\mathbf{r}\}], [\mathbf{B}_1] = \begin{bmatrix} [\mathbf{I}] & [\mathbf{U}][\boldsymbol{\varphi}]\{\mathbf{r}\} \\ [0 \ \ 0] & -\{\mathbf{R}_p\}^{\mathrm{T}}[\boldsymbol{\varphi}]\{\mathbf{r}_p\} \\ [0 \ \ 0] & 0 \end{bmatrix}, \tag{7.93}$$

$$[\mathbf{B}_2] = \begin{bmatrix} 0 & 0 \\ 0 & 0 \\ \{\mathbf{R}_p\}^{\mathrm{T}}[\mathbf{U}][\boldsymbol{\varphi}]\{\mathbf{r}_s\} & \{\mathbf{R}_s\}^{\mathrm{T}}[\mathbf{U}][\boldsymbol{\varphi}]\{\mathbf{r}_p\} \\ \sqrt{x_p^2 + y_p^2} & -\sqrt{X_p^2 + Y_p^2} \end{bmatrix}, \tag{7.94}$$

$$\{\mathbf{q}\} = [X_O \ \ Y_O \ \ \varphi]^{\mathrm{T}}, \{\widetilde{\mathbf{q}}\} = [\xi \ \ \eta]^{\mathrm{T}}, \tag{7.95}$$

the matrix equation (7.89) decomposes into the system

$$[\mathbf{M}]\{\ddot{\mathbf{q}}\} = \{\mathbf{F}\} + [\widetilde{\mathbf{B}}]^{\mathrm{T}}\begin{bmatrix} \lambda_1 \\ \lambda_2 \end{bmatrix}, [\mathbf{B}_1]\{\ddot{\mathbf{q}}\} + [\mathbf{B}_2]\{\ddot{\widetilde{\mathbf{q}}}\} = -[\dot{\mathbf{B}}]\{\dot{\mathbf{q}}^*\}, \tag{7.96}$$

from which it results (if the matrix $[\mathbf{M}]$ is invertible) the matrix equation

$$\left[[\mathbf{B}_2] \ \ [\mathbf{B}_1][\mathbf{M}]^{-1}[\widetilde{\mathbf{B}}]^{\mathrm{T}}\right]\begin{bmatrix} \ddot{\xi} \\ \ddot{\eta} \\ \lambda_1 \\ \lambda_2 \end{bmatrix} = -[\dot{\mathbf{B}}]\{\dot{\mathbf{q}}^*\} - [\mathbf{B}_1][\mathbf{M}]^{-1}\{\mathbf{F}\}. \tag{7.97}$$

Example 7.4.1 Write the equation of motion (7.97) for the case when (Fig. 7.5) an elliptic disk of mass m and semi-axes a, b, acted by a moment M rolls without sliding inside a circle of radius R.

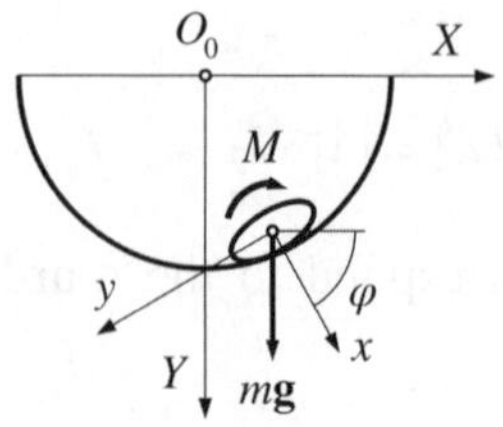

Figure 7.5 Rolling without sliding of an elliptic disk acted by a moment inside a circle in Example 7.4.1.

Solution: The parametric equations of the ellipse and circle are

$$x = a\sin\xi, y = b\cos\xi,\ X = R\sin\eta,\ Y = R\cos\eta; \tag{7.98}$$

with the notations in Fig. 7.5 one successively obtains the equalities

$$\begin{gathered}
\{\mathbf{R}\} = \begin{bmatrix} R\sin\eta \\ R\cos\eta \end{bmatrix}, \{\mathbf{r}\} = \begin{bmatrix} a\sin\xi \\ b\cos\xi \end{bmatrix}, [\boldsymbol{\varphi}] = \begin{bmatrix} \cos\varphi & -\sin\varphi \\ \sin\varphi & \cos\varphi \end{bmatrix}, \\
[\mathbf{U}][\boldsymbol{\varphi}]\{\mathbf{r}\} = \begin{bmatrix} -a\sin\varphi\sin\xi - b\cos\varphi\cos\xi \\ a\cos\varphi\sin\xi - b\sin\varphi\cos\xi \end{bmatrix}, \sqrt{x_p^2 + y_p^2} = \sqrt{a^2\cos^2\xi + b^2\sin^2\xi}, \\
\sqrt{X_p^2 + Y_p^2} = R, -\{\mathbf{R}_p\}^{\mathrm{T}}[\boldsymbol{\varphi}]\{\mathbf{r}_s\} = -R[a\cos\xi\cos(\eta+\varphi) + b\sin\xi\sin(\eta+\varphi)], \\
\{\mathbf{R}_p\}^{\mathrm{T}}[\mathbf{U}][\boldsymbol{\varphi}]\{\mathbf{r}_s\} = -R[a\sin\xi\sin(\eta+\varphi) + b\cos\xi\cos(\eta+\varphi)], \\
\{\mathbf{R}_s\}^{\mathrm{T}}[\mathbf{U}][\boldsymbol{\varphi}]\{\mathbf{r}_p\} = -R[a\cos\xi\cos(\eta+\varphi) + b\sin\xi\sin(\eta+\varphi)],
\end{gathered} \tag{7.99}$$

$$[\widetilde{\mathbf{B}}] = \begin{bmatrix} 1 & 0 & -a\sin\varphi\sin\xi - b\cos\varphi\cos\xi \\ 0 & 1 & a\cos\varphi\sin\xi - b\sin\varphi\sin\xi \end{bmatrix},$$

$$[\mathbf{B}_1] = \begin{bmatrix} 1 & 0 & -a\sin\varphi\sin\xi - b\cos\varphi\cos\xi \\ 0 & 1 & a\cos\varphi\sin\xi - b\sin\varphi\cos\xi \\ 0 & 0 & -R[a\cos\xi\cos(\eta+\varphi) + b\sin\xi\sin(\eta+\varphi)] \\ 0 & 0 & 0 \end{bmatrix},$$

$$[\mathbf{B}_2] = \begin{bmatrix} 0 & 0 \\ 0 & 0 \\ \dfrac{R[as\xi s(\eta+\varphi) + bc\xi c(\eta+\varphi)]}{\sqrt{a^2c^2\xi + b^2s^2\xi}} & -R[ac\xi c(\eta+\varphi) + bs\xi s(\eta+\varphi)] \\ & -R \end{bmatrix},$$

$$\begin{gathered}
[\mathbf{B}] = [[\mathbf{B}_1]\ \ [\mathbf{B}_2]], \\
\{\mathbf{q}^*\} = [X_O\ \ Y_O\ \ Z_O\ \ \xi\ \ \eta]^{\mathrm{T}},
\end{gathered} \tag{7.100}$$

$$[\mathbf{M}] = \begin{bmatrix} m & 0 & 0 \\ 0 & m & 0 \\ 0 & 0 & J \end{bmatrix}, \ [\mathbf{M}]^{-1} = \begin{bmatrix} \frac{1}{m} & 0 & 0 \\ 0 & \frac{1}{m} & 0 \\ 0 & 0 & \frac{1}{J} \end{bmatrix}, \ \{\mathbf{F}\} = \begin{bmatrix} mg \\ 0 \\ M \end{bmatrix}, \tag{7.101}$$

while the matrix equation of motion is the equation (7.97).

For the particular case in which the ellipse becomes a circle, $a = b = r$, from the matrix equation

$$[\mathbf{B}_1]\{\dot{\mathbf{q}}\} + [\mathbf{B}_2]\{\widetilde{\mathbf{q}}\} = \{\mathbf{0}\} \tag{7.102}$$

one obtains the scalar equations

$$\dot{X}_O - r\dot{\varphi}\cos(\xi - \varphi) = 0, \ \dot{Y}_O + r\dot{\varphi}\sin(\xi - \varphi) = 0,$$

$$Rr(\dot{\xi} - \dot{\varphi} - \dot{\eta})\cos(\eta + \varphi - \xi) = 0, \ r\dot{\xi} - R\dot{\eta} = 0; \tag{7.103}$$

considering the initial conditions $\varphi = 0$, $\xi = \varphi = \eta = 0$, $\dot{\xi} = \dot{\varphi} = \dot{\eta} = 0$, result the obvious geometric relations

$$X_O = (R - r)\sin\eta, \ Y_O = (R - r)\cos\eta, \ \xi = \varphi + \eta. \tag{7.104}$$

7.5 The Motion of the Rigid Solid at Which the Curves Jointed to It Support with Sliding on Fixed Surfaces

7.5.1 *The Case of a Single Curve*

One considers that equations in the local reference system $Oxyz$ are parametrically given

$$x = x(\xi), \ y = y(\xi), \ z = z(\xi), \tag{7.105}$$

while the implicit equation of the surface with respect to the fixed reference system O_0XYZ is

$$F(X, Y, Z) = 0. \tag{7.106}$$

With the notations

$$\{\mathbf{R}\}=[X\ Y\ Z]^{\mathrm{T}},\{\mathbf{R}_O\}=[X_O\ Y_O\ Z_O]^{\mathrm{T}},\{\mathbf{r}\}=[x\ y\ z]^{\mathrm{T}},\{\mathbf{r}_p\}=[x_p\ y_p\ z_p]^{\mathrm{T}},$$

$$[\mathbf{r}]=\begin{bmatrix}0 & -z & y\\ z & 0 & -x\\ -y & x & 0\end{bmatrix},[\mathbf{r}_p]=\begin{bmatrix}0 & -z_p & y_p\\ z_p & 0 & -x_p\\ -y_p & x_p & 0\end{bmatrix},[\mathbf{r}_s]=\begin{bmatrix}0 & -z_s & y_s\\ z_s & 0 & -x_s\\ -y_s & x_s & 0\end{bmatrix},$$

$$\{\mathbf{F}_p\}=[F_{,X}\ F_{,Y}\ F_{,Z}]^{\mathrm{T}}=\begin{bmatrix}\frac{\partial F}{\partial X} & \frac{\partial F}{\partial Y} & \frac{\partial F}{\partial Z}\end{bmatrix}^{\mathrm{T}},[\mathbf{F}_s]=\begin{bmatrix}F_{,X^2} & F_{,XY} & F_{,XZ}\\ F_{,XY} & F_{,Y^2} & F_{,YZ}\\ F_{,XZ} & F_{,YZ} & F_{,Z^2}\end{bmatrix} \tag{7.107}$$

the contact and tangency conditions read

$$\{\mathbf{R}\}=\{\mathbf{R}_O\}+[\mathbf{A}]\{\mathbf{r}\}, \tag{7.108}$$

$$\{\mathbf{F}_p\}^{\mathrm{T}}[\mathbf{A}]\{\mathbf{r}\}=0; \tag{7.109}$$

deriving these expressions with respect to time, one gets the relations

$$\{\dot{\mathbf{R}}\}=\{\dot{\mathbf{R}}_O\}+[\mathbf{A}][\mathbf{r}]^{\mathrm{T}}[\mathbf{Q}]\{\dot{\boldsymbol{\beta}}\}+\dot{\xi}[\mathbf{A}]\{\mathbf{r}_p\}=\{\mathbf{0}\}, \tag{7.110}$$

$$\{\mathbf{r}_p\}^{\mathrm{T}}[\mathbf{A}]^{\mathrm{T}}\{\dot{\mathbf{F}}_p\}+\{\mathbf{F}_p\}[\mathbf{A}][\mathbf{r}_p]^{\mathrm{T}}[\mathbf{Q}]\{\dot{\boldsymbol{\beta}}\}+\dot{\xi}\{\mathbf{F}_p\}^{\mathrm{T}}[\mathbf{A}]\{\mathbf{r}_s\}=0. \tag{7.111}$$

Keeping into account the relations

$$\{\mathbf{F}_p\}^{\mathrm{T}}\{\dot{\mathbf{R}}\}=0,\{\dot{\mathbf{F}}_p\}=[\mathbf{F}_s]\{\dot{\mathbf{R}}\} \tag{7.112}$$

and the notations

$$[\mathbf{B}_{\boldsymbol{\beta}}]=\{\mathbf{r}_p\}^{\mathrm{T}}[\mathbf{A}]^{\mathrm{T}}[\mathbf{F}_s][\mathbf{A}][\mathbf{r}]^{\mathrm{T}}[\mathbf{Q}]+\{\mathbf{F}_p\}^{\mathrm{T}}[\mathbf{A}][\mathbf{r}_p]^{\mathrm{T}}[\mathbf{Q}], \tag{7.113}$$

$$B_{\xi}=\{\mathbf{F}_p\}^{\mathrm{T}}[\mathbf{A}]^{\mathrm{T}}[\mathbf{F}_s][\mathbf{A}]\{\mathbf{r}_p\}[\mathbf{Q}]+\{\mathbf{F}_p\}^{\mathrm{T}}[\mathbf{A}]\{\mathbf{r}_s\}, \tag{7.114}$$

from the equalities (7.110) and (7.111) one obtains the matrix of constraints

$$[\mathbf{B}]=\begin{bmatrix}\{\mathbf{F}_p\}^{\mathrm{T}} & \{\mathbf{F}_p\}^{\mathrm{T}}[\mathbf{A}][\mathbf{r}]^{\mathrm{T}}[\mathbf{Q}] & 0\\ \{\mathbf{r}_p\}^{\mathrm{T}}[\mathbf{A}]^{\mathrm{T}}[\mathbf{F}_s] & [\mathbf{B}_{\boldsymbol{\beta}}] & B_{\xi}\end{bmatrix}; \tag{7.115}$$

this matrix has two lines and seven columns and, consequently, in this case the rigid solid has five degrees of freedom.

For the calculation of the matrix $[\dot{\mathbf{B}}]$ one uses the known notations and, in addition, the equalities

$$[\mathbf{F}_{s1}] = \begin{bmatrix} F_{,X^3} & F_{,X^2Y} & F_{,X^2Z} \\ F_{,X^2Y} & F_{,XY^2} & F_{,XYZ} \\ F_{,X^2Z} & F_{,XYZ} & F_{,XZ^2} \end{bmatrix}, \ [\mathbf{F}_{s2}] = \begin{bmatrix} F_{,X^2Y} & F_{,XY^2} & F_{,XYZ} \\ F_{,XY^2} & F_{,Y^3} & F_{,Y^2Z} \\ F_{,XYZ} & F_{,Y^2Z} & F_{,YZ^2} \end{bmatrix},$$

$$[\mathbf{F}_{s3}] = \begin{bmatrix} F_{,X^2Z} & F_{,XYZ} & F_{,XZ^2} \\ F_{,XYZ} & F_{,Y^2Z} & F_{,YZ^2} \\ F_{,XZ^2} & F_{,YZ^2} & F_{,Z^3} \end{bmatrix} \tag{7.116}$$

$$[\dot{\mathbf{F}}_s] = \left[[\mathbf{F}_{s1}]\{\dot{\mathbf{R}}\} \ \ [\mathbf{F}_{s2}]\{\dot{\mathbf{R}}\} \ \ [\mathbf{F}_{s3}]\{\dot{\mathbf{R}}\} \right]. \tag{7.117}$$

Further on, with the aid of the notations

$$[\widetilde{\mathbf{M}}] = \begin{bmatrix} [\mathbf{M}] & \{\mathbf{0}_{61}\} \\ [\mathbf{0}_{61}] & 0 \end{bmatrix}, \tag{7.118}$$

$$\{\mathbf{q}^*\} = [X_O \ \ Y_O \ \ Z_O \ \ \psi \ \ \theta \ \ \varphi \ \ \xi]^{\mathrm{T}}, \tag{7.119}$$

$$\{\boldsymbol{\lambda}\} = [\lambda_1 \ \ \lambda_2]^{\mathrm{T}}, \tag{7.120}$$

one obtains the matrix differential equation

$$\begin{bmatrix} [\widetilde{\mathbf{M}}] & -[\mathbf{B}]^{\mathrm{T}} \\ [\mathbf{B}] & [\mathbf{0}_{22}] \end{bmatrix} \begin{bmatrix} \{\ddot{\mathbf{q}}^*\} \\ \{\boldsymbol{\lambda}\} \end{bmatrix} = \begin{bmatrix} \{\mathbf{F}\} + \{\widetilde{\mathbf{F}}\} \\ 0 \\ -[\dot{\mathbf{B}}]\{\dot{\mathbf{q}}^*\} \end{bmatrix}. \tag{7.121}$$

If the matrix [**M**] is invertible, then with the aid of the notations

$$[\mathbf{B}_1] = \left[\{\mathbf{F}_p\}^{\mathrm{T}} \ \ \{\mathbf{F}_p\}^{\mathrm{T}}[\mathbf{A}][\mathbf{r}]^{\mathrm{T}}[\mathbf{Q}] \right], \ [\mathbf{B}_2] = \left[\{\mathbf{r}_p\}^{\mathrm{T}}[\mathbf{A}]^{\mathrm{T}}[\mathbf{F}_s] \ \ [\mathbf{B}_\beta] \right], \tag{7.122}$$

$$\{\mathbf{q}\} = [X_O \ \ Y_O \ \ Z_O \ \ \psi \ \ \theta \ \ \varphi]^{\mathrm{T}}, \tag{7.123}$$

from the relation (7.121) one obtains the results

$$\lambda_1 = -\left[[\mathbf{B}_1][\mathbf{M}]^{-1}[\mathbf{B}_1]^{\mathrm{T}} \right] \left[[\dot{\mathbf{B}}_1]\{\dot{\mathbf{q}}\} + [\mathbf{B}_1][\mathbf{M}]^{-1}\{\{\mathbf{F}\} + \{\widetilde{\mathbf{F}}\}\} \right], \ \lambda_2 = 0,$$
$$\{\ddot{\mathbf{q}}\} = [\mathbf{M}]^{-1}\{\{\mathbf{F}\} + \{\widetilde{\mathbf{F}}\}\} + \lambda_1[\mathbf{M}]^{-1}[\mathbf{B}_1]^{\mathrm{T}}, \ \ddot{\xi} = -\frac{1}{B_\xi}\left[\dot{B}_\xi \dot{\xi} + [\mathbf{B}_2]\{\ddot{\mathbf{q}}\} + [\dot{\mathbf{B}}_2]\{\dot{\mathbf{q}}\} \right]. \tag{7.124}$$

7.5.2 *The Case of the Supporting with Sliding by Curves on Surface*

It was proved that the supporting between a curve and a surface reduces the number of the degrees of freedom by one unity, wherefrom it results that, theoretically, the rigid solid may be supported with maximum five proper curves on five given fixed curves.

Indexing by $i = 1, 2, \ldots, n, n \le 5$, the mobile curves and the fixed surfaces in contact, respectively, from the relations (7.113) and (7.114) one deduces the expressions

$$[\mathbf{B}_{i\boldsymbol{\beta}}] = \{\mathbf{r}_{ip}\}^{\mathrm{T}}[\mathbf{A}]^{\mathrm{T}}[\mathbf{F}_{is}][\mathbf{A}][\mathbf{r}_i]^{\mathrm{T}}[\mathbf{Q}] + \{\mathbf{F}_{ip}\}^{\mathrm{T}}[\mathbf{A}][\mathbf{r}_{ip}]^{\mathrm{T}}[\mathbf{Q}], \tag{7.125}$$

$$B_{i\xi} = \{\mathbf{r}_{ip}\}^{\mathrm{T}}[\mathbf{A}]^{\mathrm{T}}[\mathbf{F}_{is}][\mathbf{A}]\{\mathbf{r}_{ip}\} + \{\mathbf{F}_{ip}\}^{\mathrm{T}}[\mathbf{A}]\{\mathbf{r}_{is}\}. \tag{7.126}$$

With the notations

$$[\mathbf{B}^{(1)}] = \begin{bmatrix} \{\mathbf{F}_{1p}\}^{\mathrm{T}} & \{\mathbf{F}_{1p}\}^{\mathrm{T}}[\mathbf{A}][\mathbf{r}_1]^{\mathrm{T}}[\mathbf{Q}] \\ \{\mathbf{r}_{1p}\}^{\mathrm{T}}[\mathbf{A}]^{\mathrm{T}}[\mathbf{F}_{1s}] & [\mathbf{B}_{1\boldsymbol{\beta}}] \\ \cdots & \cdots \\ \{\mathbf{F}_{np}\}^{\mathrm{T}} & \{\mathbf{F}_{np}\}^{\mathrm{T}}[\mathbf{A}][\mathbf{r}_n]^{\mathrm{T}}[\mathbf{Q}] \\ \{\mathbf{r}_{np}\}^{\mathrm{T}}[\mathbf{A}]^{\mathrm{T}}[\mathbf{F}_{ns}] & [\mathbf{B}_{n\boldsymbol{\beta}}] \end{bmatrix}, \tag{7.127}$$

$$[\mathbf{B}^{(2)}] = \begin{bmatrix} 0 & 0 & \cdots & 0 \\ B_{1\xi} & 0 & \cdots & 0 \\ \cdots & \cdots & \cdots & \cdots \\ 0 & 0 & 0 & 0 \\ 0 & 0 & 0 & B_{n\xi} \end{bmatrix}, \tag{7.128}$$

$$\{\mathbf{q}^*\} = [X_O \ \ Y_O \ \ Z_O \ \ \psi \ \ \theta \ \ \varphi \ \ \xi_1 \ \ \ldots \ \ \xi_n]^{\mathrm{T}}, \tag{7.129}$$

$$[\widetilde{\mathbf{M}}] = \begin{bmatrix} [\mathbf{M}] & [\mathbf{0}_{6,n}] \\ [\mathbf{0}_{n,6}] & [\mathbf{0}_{n,n}] \end{bmatrix}, \tag{7.130}$$

$$\{\boldsymbol{\lambda}\} = [\lambda_1 \ \ \lambda_2 \ \ \ldots \ \ \lambda_n]^{\mathrm{T}}, \tag{7.131}$$

one firstly obtains the matrix of constraints

$$[\mathbf{B}] = \left[[\mathbf{B}^{(1)}] \ \ [\mathbf{B}^{(2)}] \right] \tag{7.132}$$

and then the matrix differential equation of motion

$$\begin{bmatrix} [\widetilde{\mathbf{M}}] & -[\mathbf{B}]^{\mathrm{T}} \\ [\mathbf{B}] & [\mathbf{0}_{2n,2n}] \end{bmatrix} \begin{bmatrix} \{\ddot{\mathbf{q}}^*\} \\ \{\boldsymbol{\lambda}\} \end{bmatrix} = \begin{bmatrix} \{\mathbf{F}\} + \{\widetilde{\mathbf{F}}\} \\ \{\mathbf{0}_{n,1}\} \\ -[\dot{\mathbf{B}}]\{\dot{\mathbf{q}}^*\} \end{bmatrix}. \tag{7.133}$$

Example 7.5.1 Determine the matrix of constraints for a cylinder of radius r and high $2h$ supported with the end circles (Fig. 7.6) on the plans $Z=0$ and $Y=0$, respectively.

Solution: The coordinates of the contact points A_1, A_2 in the local reference systems are $x_1 = r\cos\xi_1$, $y_1 = r\sin\xi_1$, $z_1 = -h$, $x_2 = r\cos\xi_2$, $y_2 = r\sin\xi_2$, $z_2 = h$.

The tangency condition at the point A_1 reads

$$[0 \;\; 0 \;\; 1][\mathbf{A}] \begin{bmatrix} -r\sin\xi_1 \\ r\cos\xi_1 \\ 0 \end{bmatrix} = 0; \tag{7.134}$$

if ψ, θ, φ are the Euler angles, then from the equality (7.134) one obtains the relation $\cos(\varphi + \xi_1) = 0$, wherefrom it results $\xi_1 = -\dfrac{\pi}{2} - \varphi$.

From the third line of the contact condition at the point A_1

$$\begin{bmatrix} X_1 \\ Y_1 \\ 0 \end{bmatrix} = \begin{bmatrix} X_O \\ Y_O \\ Z_O \end{bmatrix} [\mathbf{A}] \begin{bmatrix} r\cos\xi_1 \\ r\sin\xi_1 \\ -h \end{bmatrix} \tag{7.135}$$

one obtains the equation

$$Z_O - r\sin\theta - h\cos\theta = 0. \tag{7.136}$$

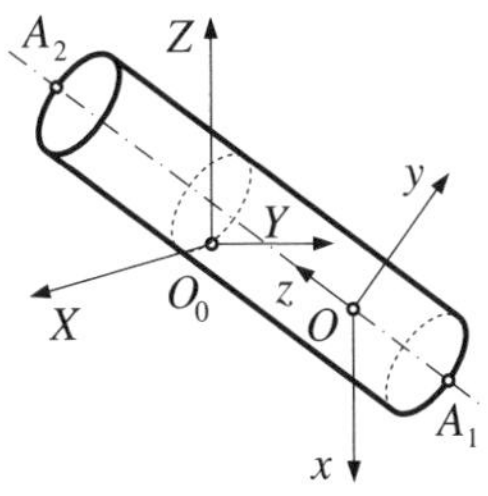

Figure 7.6 A cylinder supported with the end circles on two plans in Example 7.5.1.

The tangency condition at the point A_2

$$[0\ \ 1\ \ 0][\mathbf{A}]\begin{bmatrix} -r\sin\xi_2 \\ r\cos\xi_2 \\ 0 \end{bmatrix}=0 \tag{7.137}$$

leads to the equality

$$-\sin\psi\sin(\varphi+\xi_2)+\cos\psi\cos\theta\cos(\varphi+\xi_2)=0, \tag{7.138}$$

while from the second line of the contact condition at the point A_2

$$\begin{bmatrix} X_2 \\ 0 \\ Z_2 \end{bmatrix}=\begin{bmatrix} X_O \\ Y_O \\ Z_O \end{bmatrix}[\mathbf{A}]\begin{bmatrix} r\cos\xi_2 \\ r\sin\xi_2 \\ h \end{bmatrix} \tag{7.139}$$

one gets the equality

$$r\sin\psi\cos(\varphi+\xi_2)+r\cos\psi\cos\theta\sin(\varphi+\xi_2)-h\cos\psi\sin\theta+Y_O=0. \tag{7.140}$$

Eliminating the parameter ξ_2 from the relations (7.138) and (7.140), one obtains the expression

$$r^2\left(\cos^2\psi\cos^2\theta+\sin^2\psi\right)-\left(Y_O-h\cos\psi\sin\theta\right)^2=0. \tag{7.141}$$

Deriving the equations (7.136) and (7.141) with respect to time and using the notations

$$\begin{aligned}
B_{22}&=-(Y_O-h\cos\psi\sin\theta),\\
B_{24}&=r^2\cos\psi\sin\psi\sin^2\theta-(Y_O-h\cos\psi\sin\theta)h\sin\psi\sin\theta,\\
B_{25}&=-r^2\cos^2\psi\cos\theta\sin\theta+(Y_O-h\cos\psi\cos\theta)h\cos\psi\cos\theta,
\end{aligned} \tag{7.142}$$

one obtains the matrix of constraints

$$[\mathbf{B}]=\begin{bmatrix} 0 & 0 & 1 & 0 & -r\cos\theta+h\sin\theta & 0 \\ 0 & B_{22} & 0 & B_{24} & B_{25} & 0 \end{bmatrix}, \tag{7.143}$$

the matrix of coordinates being

$$\{\mathbf{q}\}=[X_O\ \ Y_O\ \ Z_O\ \ \psi\ \ \theta\ \ \varphi]^{\mathrm{T}}. \tag{7.144}$$

7.6 The Rolling without Sliding of a Disk Bounded by a Spatial Curve on a Fixed Surface

7.6.1 The Matrix Differential Equation of Motion

We consider the variant in which the surface is defined in the reference system O_0XYZ by the implicit equation

$$F(X,Y,Z)=0, \tag{7.145}$$

while the spatial curve is defined in the local reference system $Oxyz$ by the parametric equations

$$x=x(\xi), y=y(\xi), z=z(\xi). \tag{7.146}$$

Starting from the matrix equation of contact

$$\{\mathbf{R}\}=\{\mathbf{R}_O\}+[\mathbf{A}]\{\mathbf{r}\}, \tag{7.147}$$

by derivation with respect to time one obtains

$$\{\dot{\mathbf{R}}\}=\{\dot{\mathbf{R}}_O\}+[\mathbf{A}][\mathbf{r}]^{\mathrm{T}}[\mathbf{Q}]\{\dot{\boldsymbol{\beta}}\}+\dot{\xi}[\mathbf{A}]\{\mathbf{r}_p\}=\{\mathbf{0}\}. \tag{7.148}$$

The condition of rolling without sliding states that the relative velocity at the contact point vanishes and from here it results the matrix equation

$$\{\dot{\mathbf{R}}_O\}+[\mathbf{A}][\mathbf{r}]^{\mathrm{T}}[\mathbf{Q}]\{\dot{\boldsymbol{\beta}}\}=\{\mathbf{0}\}, \tag{7.149}$$

so that from the relation (7.148) it results

$$\{\dot{\mathbf{R}}\}=\dot{\xi}[\mathbf{A}]\{\mathbf{r}_p\}. \tag{7.150}$$

The tangency condition reads

$$\{\mathbf{F}_p\}^{\mathrm{T}}[\mathbf{A}]\{\mathbf{r}_p\}=0; \tag{7.151}$$

keeping into account the equalities

$$\{\dot{\mathbf{F}}_p\}=[\mathbf{F}_s]\{\dot{\mathbf{R}}\}=\dot{\xi}[\mathbf{F}_s][\mathbf{A}]\{\mathbf{r}_p\} \tag{7.152}$$

and deriving the relation (7.151) with respect to time, we get

$$\{\mathbf{F}_p\}[\mathbf{A}]\left[\mathbf{r}_p\right]^{\mathrm{T}}[\mathbf{Q}]\{\dot{\boldsymbol{\beta}}\}+\dot{\xi}\left[\{\mathbf{F}_p\}^{\mathrm{T}}[\mathbf{A}]\{\mathbf{r}_s\}+\{\mathbf{r}_p\}^{\mathrm{T}}[\mathbf{A}][\mathbf{F}_s][\mathbf{A}]\{\mathbf{r}_p\}\right]=0. \tag{7.153}$$

Using the notations

$$[\mathbf{B}_1]=[\mathbf{A}][\mathbf{r}]^{\mathrm{T}}[\mathbf{Q}],[\mathbf{B}_2]=\{\mathbf{F}_p\}^{\mathrm{T}}[\mathbf{A}][\mathbf{r}_p]^{\mathrm{T}}[\mathbf{Q}], \tag{7.154}$$

$$B_3=\{\mathbf{F}_p\}^{\mathrm{T}}[\mathbf{A}]\{\mathbf{r}_s\}+\{\mathbf{r}_p\}^{\mathrm{T}}[\mathbf{A}]^{\mathrm{T}}[\mathbf{F}_s][\mathbf{A}]\{\mathbf{r}_p\}, \tag{7.155}$$

$$\{\mathbf{q}^*\}=[X_O\ Y_O\ Z_O\ \psi\ \theta\ \varphi\ \xi]^{\mathrm{T}}, \tag{7.156}$$

from the relations (7.149) and (7.153) one obtains the equation

$$[\mathbf{B}]\{\dot{\mathbf{q}}^*\}=\{\mathbf{0}\}, \tag{7.157}$$

where

$$[\mathbf{B}]=\begin{bmatrix}[\mathbf{I}] & [\mathbf{B}_1] & \{\mathbf{0}\}\\ [\mathbf{0}_{13}] & [\mathbf{B}_2] & B_3\end{bmatrix} \tag{7.158}$$

is the matrix of constraints.

Further on, with the known notations

$$[\mathbf{M}]=\begin{bmatrix}[\mathbf{m}] & [\mathbf{A}][\mathbf{S}]^{\mathrm{T}}[\mathbf{Q}]\\ [\mathbf{Q}]^{\mathrm{T}}[\mathbf{S}][\mathbf{A}]^{\mathrm{T}} & [\mathbf{Q}]^{\mathrm{T}}[\mathbf{J}_O][\mathbf{Q}]\end{bmatrix}, \tag{7.159}$$

$$\{\tilde{\mathbf{F}}_{\mathbf{s}}\}=-\left[[\mathbf{A}][\mathbf{S}]^{\mathrm{T}}[\dot{\mathbf{Q}}]+[\dot{\mathbf{A}}][\mathbf{S}]^{\mathrm{T}}[\mathbf{Q}]\right]\{\boldsymbol{\beta}\}, \tag{7.160}$$

$$\{\tilde{\mathbf{F}}_{\boldsymbol{\beta}}\}=-\left[[\mathbf{Q}]^{\mathrm{T}}[\mathbf{J}_O][\dot{\mathbf{Q}}]+[\mathbf{Q}]^{\mathrm{T}}[\boldsymbol{\omega}][\mathbf{J}_O][\mathbf{Q}]\right]\{\dot{\boldsymbol{\beta}}\}, \tag{7.161}$$

$$\{\tilde{\mathbf{F}}\}=\begin{bmatrix}\{\tilde{\mathbf{F}}_{\mathbf{s}}\}\\ \{\tilde{\mathbf{F}}_{\boldsymbol{\beta}}\}\end{bmatrix} \tag{7.162}$$

and using the extended matrix of inertia

$$[\tilde{\mathbf{M}}]=\begin{bmatrix}[\mathbf{M}] & \{\mathbf{0}_{61}\}\\ [\mathbf{0}_{16}] & 0\end{bmatrix}, \tag{7.163}$$

one obtains the matrix equation of motion

$$\begin{bmatrix} [\tilde{\mathbf{M}}] & -[\mathbf{B}]^{\mathrm{T}} \\ [\mathbf{B}] & [\mathbf{0}_{44}] \end{bmatrix} \begin{bmatrix} \{\ddot{\mathbf{q}}^*\} \\ \{\boldsymbol{\lambda}\} \end{bmatrix} = \begin{bmatrix} \{\mathbf{F}\} + \{\tilde{\mathbf{F}}\} \\ 0 \\ -[\dot{\mathbf{B}}]\{\dot{\mathbf{q}}^*\} \end{bmatrix}, \tag{7.164}$$

where

$$\{\boldsymbol{\lambda}\} = [\lambda_1 \ \ \lambda_2 \ \ \lambda_3 \ \ \lambda_4]^{\mathrm{T}}, \tag{7.165}$$

while $\{\mathbf{F}\}$ is the matrix of the given generalized forces.

7.6.2 *The Forces at the Contact Point*

From the seventh line of the equation (7.164) one deduces the equality $\lambda_4 = 0$ and, consequently, the matrix $\{\boldsymbol{\lambda}\}$ writes in the form

$$\{\boldsymbol{\lambda}\} = [\lambda_1 \ \ \lambda_2 \ \ \lambda_3]^{\mathrm{T}}. \tag{7.166}$$

From the first three lines of the equation (7.164) one deduces that λ_1, λ_2, λ_3 are the reactions that act onto the directions of the axes O_0X, O_0Y, O_0Z.

The unit vectors $\boldsymbol{\tau}$, and $\mathbf{n}$ of the tangent to the curve, and normal to the surface, respectively, have the components

$$\{\boldsymbol{\tau}\} = \begin{bmatrix} \tau_X \\ \tau_Y \\ \tau_Z \end{bmatrix} = \frac{[\mathbf{A}]\{\mathbf{r}_p\}}{\sqrt{x_p^2 + y_p^2 + z_p^2}}, \tag{7.167}$$

$$\{\mathbf{n}\} = \begin{bmatrix} n_X \\ n_Y \\ n_Z \end{bmatrix} = \frac{\{\mathbf{F}_p\}}{\sqrt{F_X'^2 + F_Y'^2 + F_Z'^2}}, \tag{7.168}$$

while the unit vector of the perpendicular to tangent, situated in the tangent plan at surface has the components

$$\{\mathbf{b}\} = \begin{bmatrix} b_X \\ b_Y \\ b_Z \end{bmatrix} = \begin{bmatrix} \tau_Y n_Z - \tau_Z n_Y \\ \tau_Z n_X - \tau_X n_Z \\ \tau_X n_Y - \tau_Y n_X \end{bmatrix}. \tag{7.169}$$

One obtains the forces

$$F_\tau = \{\boldsymbol{\lambda}\}^{\mathrm{T}}\{\boldsymbol{\tau}\}, F_n = \{\boldsymbol{\lambda}\}^{\mathrm{T}}\{\mathbf{n}\}, F_b = \{\boldsymbol{\lambda}\}^{\mathrm{T}}\{\mathbf{b}\}. \quad (7.170)$$

7.7 The Rolling without Sliding of a Planar Circle Disk on a Horizontal Plan

We consider the disk of radius r_0 and mass m (Fig. 7.7) acted only by its own weight and rolling without sliding on the plan O_0XY.

We choose the local reference system $Oxyz$ so that the point O is the center of the disk, the axes Ox, Oy are situated in the plan of the circle, while the rotational angles are the Euler angles ψ, θ, φ.

In these conditions the supporting surface has the equation

$$Z = 0, \quad (7.171)$$

and the parametric equations of the circle may be written in the form

$$x = r_0 \cos \xi, y = -r_0 \sin \xi, z = 0. \quad (7.172)$$

We successively write the relations

$$[\boldsymbol{\psi}] = \begin{bmatrix} \cos\psi & -\sin\psi & 0 \\ \sin\psi & \cos\psi & 0 \\ 0 & 0 & 1 \end{bmatrix}, [\boldsymbol{\theta}] = \begin{bmatrix} 1 & 0 & 0 \\ 0 & \cos\theta & -\sin\theta \\ 0 & \sin\theta & \cos\theta \end{bmatrix}, [\boldsymbol{\varphi}] = \begin{bmatrix} \cos\varphi & -\sin\varphi & 0 \\ \sin\varphi & \cos\varphi & 0 \\ 0 & 0 & 1 \end{bmatrix},$$

$$[\mathbf{A}] = [\boldsymbol{\psi}][\boldsymbol{\theta}][\boldsymbol{\varphi}], [\mathbf{U}_\psi] = [\mathbf{U}_\varphi] = \begin{bmatrix} 0 & -1 & 0 \\ 1 & 0 & 0 \\ 0 & 0 & 0 \end{bmatrix}, [\mathbf{U}_\theta] = \begin{bmatrix} 0 & 0 & 0 \\ 0 & 0 & -1 \\ 0 & 1 & 0 \end{bmatrix}, [\mathbf{A}_\psi] = [\mathbf{U}_\psi][\mathbf{A}],$$

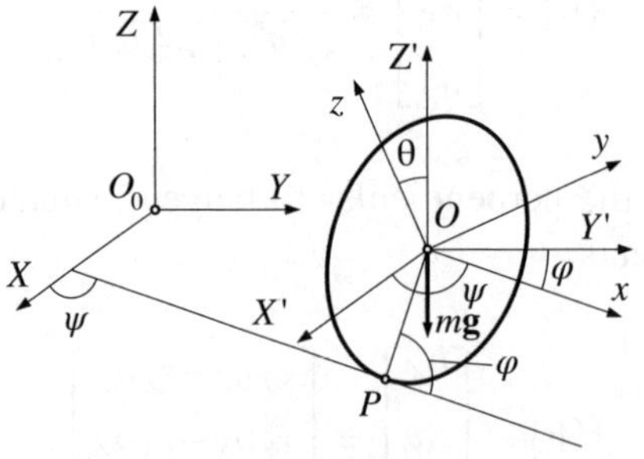

Figure 7.7 The circular circle disk rolling without sliding of a horizontal plan.

$$[\mathbf{A}_\theta] = [\mathbf{A}][\boldsymbol{\varphi}]^{\mathrm{T}}[\mathbf{U}_\varphi][\boldsymbol{\varphi}],\ [\mathbf{A}_\varphi] = [\mathbf{A}][\mathbf{U}_\varphi],\ [\dot{\mathbf{A}}] = \dot{\psi}[\mathbf{A}_\psi] + \dot{\theta}[\mathbf{A}_\theta] + \dot{\varphi}[\mathbf{A}_\varphi],$$

$$[\mathbf{Q}] = \begin{bmatrix} \sin\varphi\sin\theta & \cos\varphi & 0 \\ \cos\varphi\sin\theta & -\sin\varphi & 0 \\ \cos\theta & 0 & 1 \end{bmatrix},\ [\mathbf{Q}_\theta] = \begin{bmatrix} \sin\varphi\cos\theta & 0 & 0 \\ \cos\varphi\cos\theta & 0 & 0 \\ -\sin\theta & 0 & 0 \end{bmatrix},$$

$$[\mathbf{Q}_\varphi] = \begin{bmatrix} \cos\varphi\sin\theta & -\sin\varphi & 0 \\ -\sin\varphi\sin\theta & -\cos\varphi & 0 \\ 0 & 0 & 0 \end{bmatrix},\ [\dot{\mathbf{Q}}] = \dot{\theta}[\mathbf{Q}_\theta] + \dot{\varphi}[\mathbf{Q}_\varphi],\ [\boldsymbol{\omega}] = [\mathbf{A}]^{\mathrm{T}}[\dot{\mathbf{A}}],$$

$$\{\mathbf{r}\} = \begin{bmatrix} r_0\cos\xi \\ -r_0\sin\xi \\ 0 \end{bmatrix},\ \{\mathbf{r}_p\} = \begin{bmatrix} -r_0\sin\xi \\ -r_0\cos\xi \\ 0 \end{bmatrix},\ \{\mathbf{r}_s\} = -\{\mathbf{r}\},\ \{\mathbf{r}_t\} = -\{\mathbf{r}_p\},$$

$$[\mathbf{r}] = r_0\begin{bmatrix} 0 & 0 & -\sin\xi \\ 0 & 0 & -\cos\xi \\ \sin\xi & \cos\xi & 0 \end{bmatrix},\ [\mathbf{r}_p] = r_0\begin{bmatrix} 0 & 0 & -\cos\xi \\ 0 & 0 & \sin\xi \\ \cos\xi & -\sin\xi & 0 \end{bmatrix},\ [\mathbf{r}_s] = -[\mathbf{r}],\ [\mathbf{r}_t] = -[\mathbf{r}_p],$$

$$\{\mathbf{F}_p\} = [0\ \ 0\ \ 1],\ [\mathbf{F}_s] = [\mathbf{0}]. \tag{7.173}$$

With these notations, from the tangency condition (7.151) one obtains the equality

$$\xi = \frac{\pi}{2} + \varphi. \tag{7.174}$$

The first relation (7.154) becomes

$$[\mathbf{B}_1] = [\mathbf{A}][\mathbf{r}]^{\mathrm{T}}[\mathbf{Q}] = r_0\begin{bmatrix} \cos\psi\cos\theta & -\sin\psi\sin\theta & \cos\psi \\ \sin\psi\cos\theta & \cos\psi\sin\theta & \sin\psi \\ 0 & -\cos\theta & 0 \end{bmatrix}, \tag{7.175}$$

while the matrix of constraints reads

$$[\mathbf{B}] = \begin{bmatrix} 1 & 0 & 0 & r_0\cos\psi\cos\theta & -r_0\sin\psi\sin\theta & r_0\cos\psi \\ 0 & 1 & 0 & r_0\sin\psi\cos\theta & r_0\cos\psi\sin\theta & r_0\sin\psi \\ 0 & 0 & 1 & 0 & -r_0\cos\theta & 0 \end{bmatrix}. \tag{7.176}$$

The inertial constants are m, $J_x = J_y = \dfrac{mr_0^2}{4}$, $J_z = \dfrac{mr_0^2}{2}$, and the point O is also the center of weight; from the relation (7.159) we get

$$[\mathbf{S}] = [\mathbf{0}],\ [\mathbf{m}] = \begin{bmatrix} m & 0 & 0 \\ 0 & m & 0 \\ 0 & 0 & m \end{bmatrix},\ [\mathbf{J}] = \begin{bmatrix} J_x & 0 & 0 \\ 0 & J_y & 0 \\ 0 & 0 & J_z \end{bmatrix},\ [\mathbf{M}] = \begin{bmatrix} [\mathbf{m}] & [\mathbf{0}] \\ [\mathbf{0}] & [\mathbf{Q}]^{\mathrm{T}}[\mathbf{J}][\mathbf{Q}] \end{bmatrix}. \tag{7.177}$$

Since the disk is acted only by its own weight, it results

$$\{\mathbf{F}\} = [0\ \ 0\ \ -mg\ \ 0\ \ 0\ \ 0]^{\mathrm{T}}. \tag{7.178}$$

From the relation (7.176) one calculates the derivatives

$$[\mathbf{B}_{\psi}] = \begin{bmatrix} 0 & 0 & 0 & -r_0 \sin\psi\cos\theta & -r_0\cos\psi\sin\theta & -r_0\sin\psi \\ 0 & 0 & 0 & r_0\cos\psi\cos\theta & -r_0\sin\psi\sin\theta & r_0\cos\psi \\ 0 & 0 & 0 & 0 & 0 & 0 \end{bmatrix},$$

$$[\mathbf{B}_{\theta}] = \begin{bmatrix} 0 & 0 & 0 & -r_0\cos\psi\sin\theta & -r_0\sin\psi\cos\theta & 0 \\ 0 & 0 & 0 & -r_0\sin\psi\sin\theta & r_0\cos\psi\cos\theta & 0 \\ 0 & 0 & 0 & 0 & r_0\sin\theta & 0 \end{bmatrix}, [\dot{\mathbf{B}}] = \dot{\psi}[\mathbf{B}_{\psi}] + \dot{\theta}[\mathbf{B}_{\theta}]; \tag{7.179}$$

from the relation (7.160) it results $\{\widetilde{\mathbf{F}}_{\mathbf{s}}\} = \{\mathbf{0}\}$, while from the relations (7.161) and (7.162) one deduces

$$\{\widetilde{\mathbf{F}}\} = \begin{bmatrix} \{\mathbf{0}\} \\ \{\widetilde{\mathbf{F}}_{\beta}\} \end{bmatrix}. \tag{7.180}$$

Finally, we get the matrix differential equation of motion

$$\begin{bmatrix} [\widetilde{\mathbf{M}}] & -[\mathbf{B}]^{\mathrm{T}} \\ [\mathbf{B}] & [\mathbf{0}] \end{bmatrix} \begin{bmatrix} \{\ddot{\mathbf{q}}^*\} \\ \{\boldsymbol{\lambda}\} \end{bmatrix} = \begin{bmatrix} \{\mathbf{F}\} + \{\widetilde{\mathbf{F}}\} \\ -[\dot{\mathbf{B}}]\{\dot{\mathbf{q}}^*\} \end{bmatrix}, \tag{7.181}$$

where

$$\{\mathbf{q}\} = [X_O\ \ Y_O\ \ Z_O\ \ \psi\ \ \theta\ \ \varphi]^{\mathrm{T}}, \{\boldsymbol{\lambda}\} = [\lambda_1\ \ \lambda_2\ \ \lambda_3]^{\mathrm{T}}. \tag{7.182}$$

For the numerical case described by $m = 8\,\mathrm{kg}$, $r_0 = 1\,\mathrm{m}$, $J_x = J_y = 2\,\mathrm{kgm}^2$, $J_z = 4\,\mathrm{kgm}^2$ and the initial conditions $t = 0\,\mathrm{s}$, $\psi = \frac{\pi}{2}\,\mathrm{rad}$, $\theta = \frac{2\pi}{5}\,\mathrm{rad}$, $\varphi = 0\,\mathrm{rad}$, $\dot{\psi} = \dot{\theta} = 0\,\mathrm{rad/s}$, $\dot{\varphi} = -2\,\mathrm{rad/s}$, $X_O = -r_0\cos\frac{2\pi}{5}\,\mathrm{m}$, $Y_O = 0\,\mathrm{m}$, $Z_O = -r_0\sin\frac{2\pi}{5}\,\mathrm{m}$, $\dot{X}_O = 0\,\mathrm{m/s}$, $\dot{Y}_O = 2\,\mathrm{m/s}$, $\dot{Z}_O = 0\,\mathrm{m/s}$ one obtains the diagrams captured in the following figures (Fig. 7.8 to Fig. 7.18).

The reader can easily observe a certain quasi-periodicity of the motion as a result of the periodicity of different parameters that characterize the motion as a function of time, and the quasi-incommensurable values of their periods. As a consequence of this quasi-periodicity one deduces that both the trajectory of the point O and the trajectory of the contact point are bounded. In addition, the forces at the contact point have also a quasi-periodic characteristic. Moreover, the weight of the disk is equal to 78.452 N, while the forces at the contact point rich values of $\approx 120\,\mathrm{N}$ (1.5 time greater) and their values are both positive and negative.

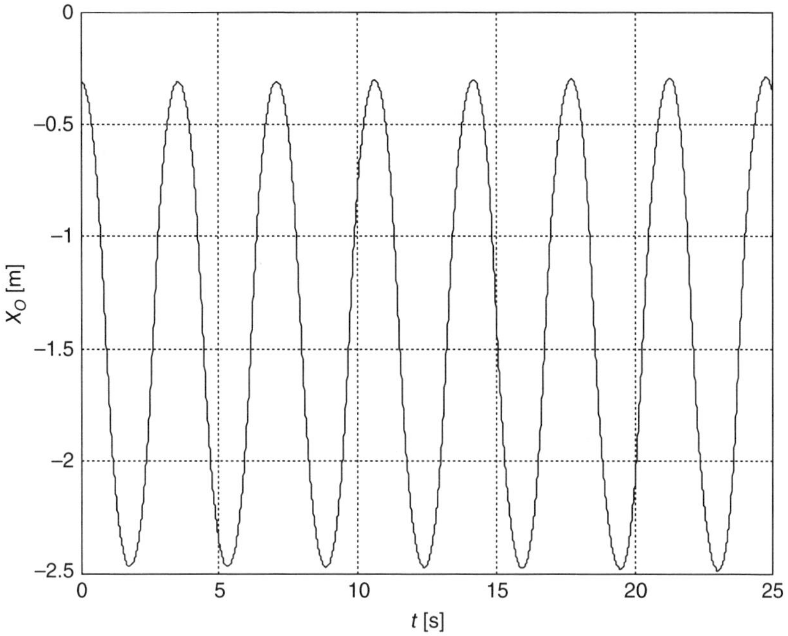

Figure 7.8 Time history $X_O = X_O(t)$ for $0 \leq t \leq 25$ s.

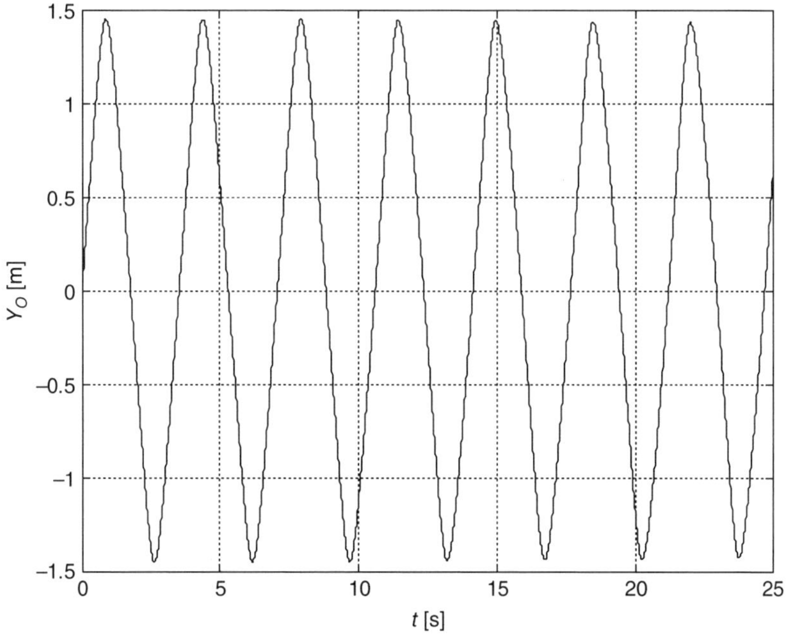

Figure 7.9 Time history $Y_O = Y_O(t)$ for $0 \leq t \leq 25$ s.

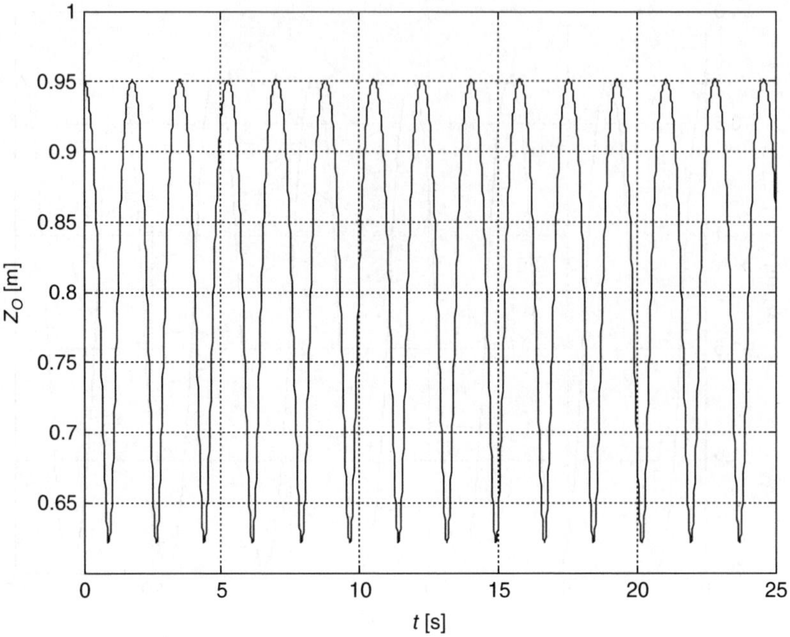

Figure 7.10 Time history $Z_O = Z_O(t)$ for $0 \le t \le 25$ s.

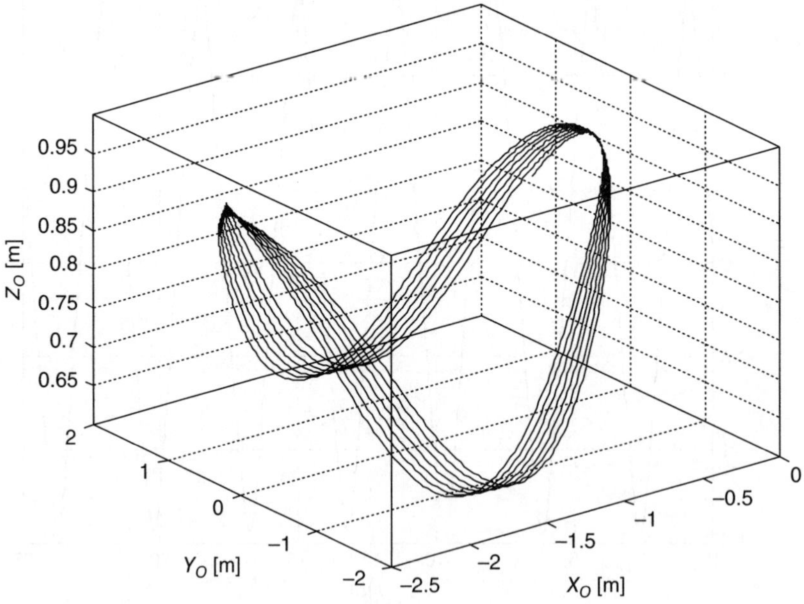

Figure 7.11 Trajectory of the point O (center of disk) in the first 25 seconds.

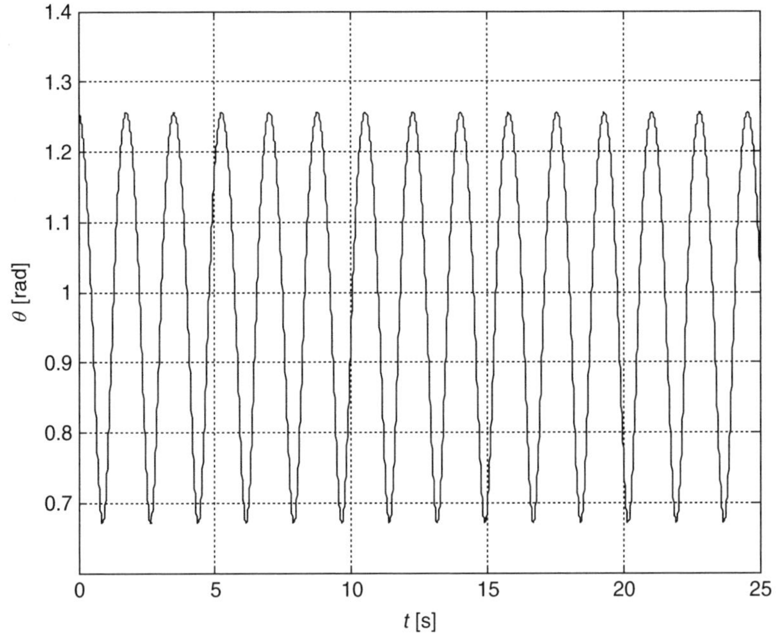

Figure 7.12 Time history $\theta = \theta(t)$ for $0 \le t \le 25$ s.

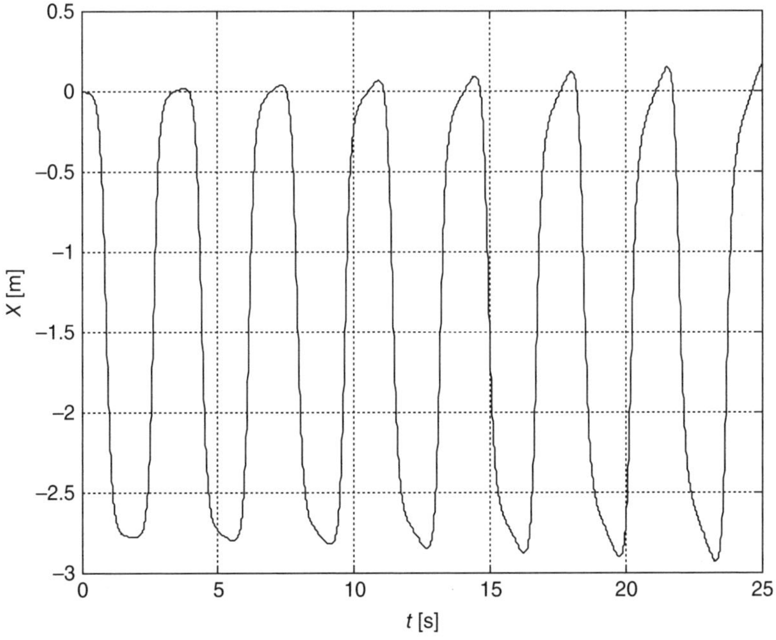

Figure 7.13 Time history $X = X(t)$ for $0 \le t \le 25$ s.

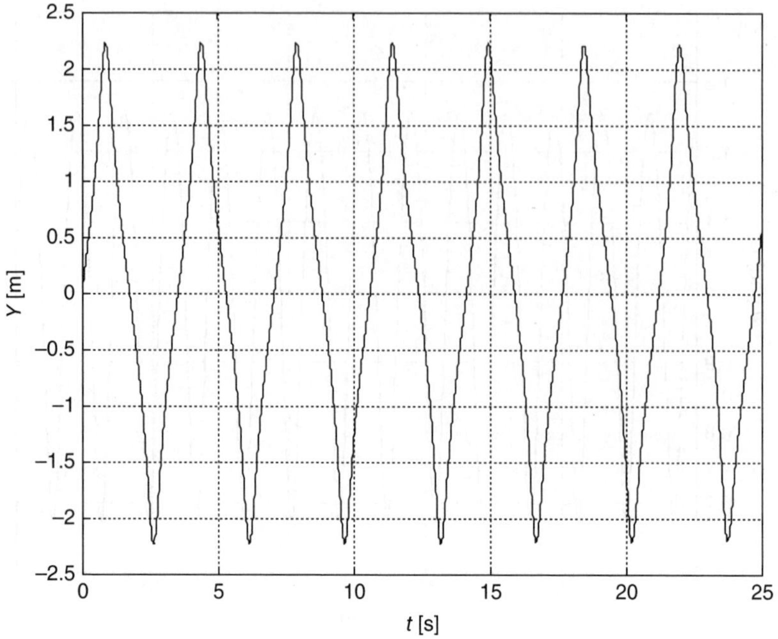

Figure 7.14 Time history $Y = Y(t)$ for $0 \le t \le 25$ s.

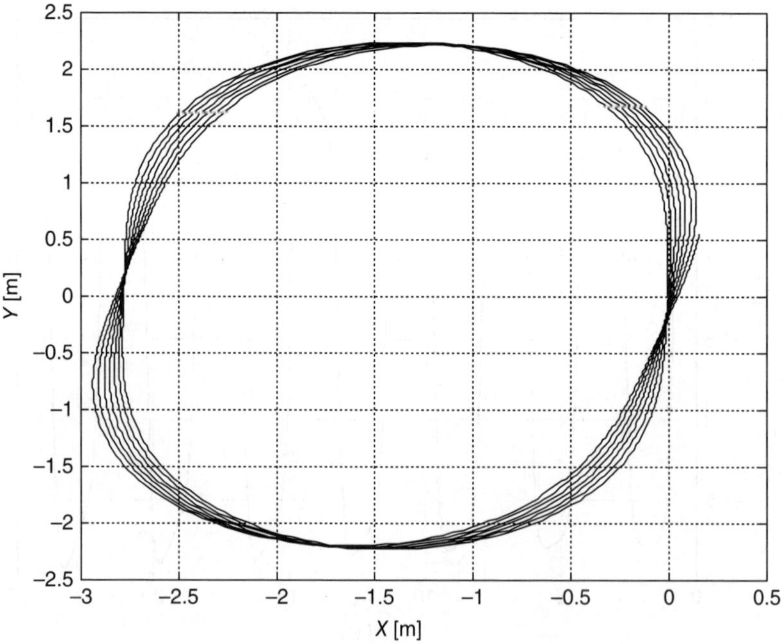

Figure 7.15 Trajectory of the contact point in the first 25 seconds.

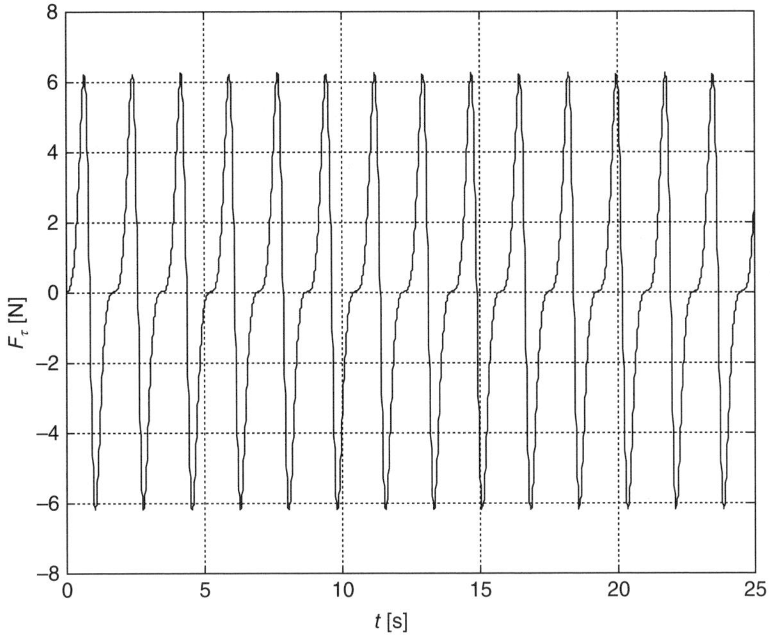

Figure 7.16 Time history $F_\tau = F_\tau(t)$ for $0 \le t \le 25$ s.

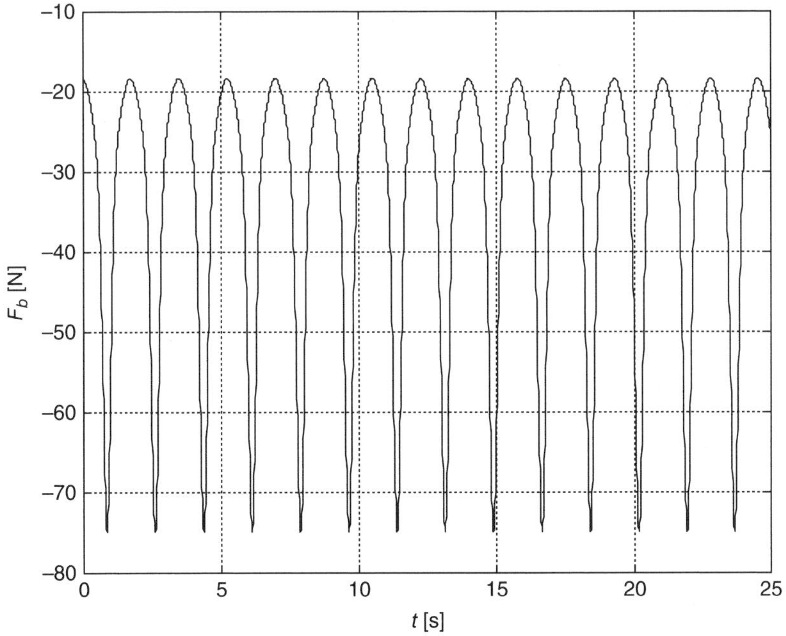

Figure 7.17 Time history $F_b = F_b(t)$ for $0 \le t \le 25$ s.

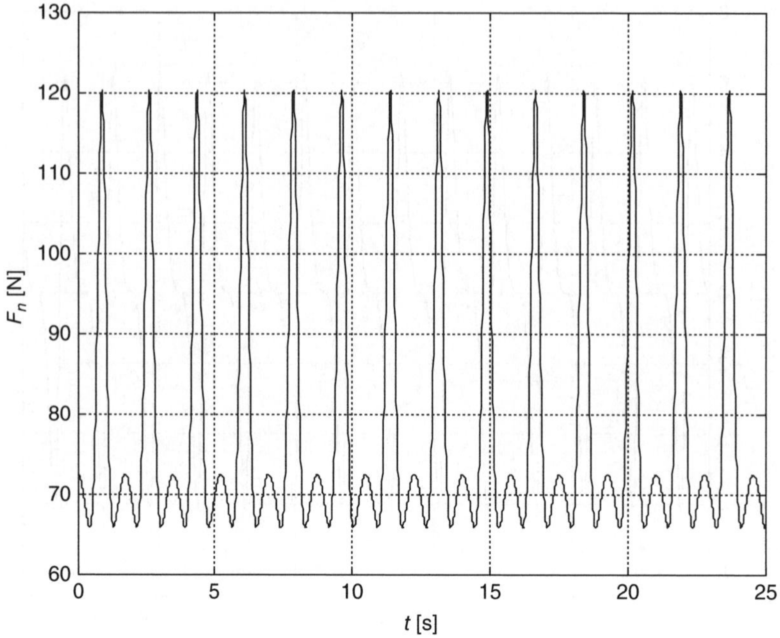

Figure 7.18 Time history $F_n = F_n(t)$ for $0 \le t \le 25$ s.

7.8 The Rolling without Sliding of a Planar Elliptic Disk on a Horizontal Plan

We consider the elliptic disk of mass m and semi-axes a, b, which rolls without sliding on the horizontal plan O_0XY.

One chooses the mobile reference system Oxy so that the point O coincides to the center of ellipse, while the axes Ox, Oy coincide to the axes of symmetry of the ellipse.

In these conditions, the support surface has the equation

$$Z = 0, \tag{7.183}$$

while the parametric equations of the ellipse are

$$x = a\cos\xi,\ y = -b\sin\xi,\ z = 0. \tag{7.184}$$

We recall the relations (7.173) with the modifications

$$\{\mathbf{r}\} = \begin{bmatrix} a\cos\xi \\ -b\sin\xi \\ 0 \end{bmatrix}, \{\mathbf{r}_p\} = \begin{bmatrix} -a\sin\xi \\ -b\cos\xi \\ 0 \end{bmatrix}, \{\mathbf{r}_s\} = -\{\mathbf{r}\}, \{\mathbf{r}_t\} = -\{\mathbf{r}_p\},$$

$$[\mathbf{r}] = \begin{bmatrix} 0 & 0 & -b\sin\xi \\ 0 & 0 & -a\cos\xi \\ b\sin\xi & a\cos\xi & 0 \end{bmatrix}, [\mathbf{r}_p] = \begin{bmatrix} 0 & 0 & -b\cos\xi \\ 0 & 0 & a\sin\xi \\ b\cos\xi & -a\sin\xi & 0 \end{bmatrix}, [\mathbf{r}_s] = -[\mathbf{r}],$$

$$[\mathbf{r}_t] = -[\mathbf{r}_p]. \tag{7.185}$$

From the tangency condition (7.151) one deduces the equality

$$a\sin\varphi\sin\xi + b\cos\varphi\cos\xi = 0 \tag{7.186}$$

which, by derivation with respect to time, becomes

$$(a\cos\varphi\sin\xi - b\sin\varphi\cos\xi)\dot{\varphi} + (a\sin\varphi\cos\xi - b\cos\varphi\sin\xi)\dot{\xi} = 0. \tag{7.187}$$

With the notations

$$\begin{aligned} &[\mathbf{B}_1] = [\mathbf{A}][\mathbf{r}]^{\mathrm{T}}[\mathbf{Q}], [\mathbf{B}_2] = [0 \;\; 0 \;\; a\cos\varphi\sin\xi - b\sin\varphi\cos\xi], \\ &B_3 = a\sin\varphi\cos\xi - b\cos\varphi\cos\xi, \end{aligned} \tag{7.188}$$

the matrix of constraints reads

$$[\mathbf{B}] = \begin{bmatrix} [\mathbf{I}] & [\mathbf{B}_1] & \{\mathbf{0}\} \\ [\mathbf{0}] & [\mathbf{B}_2] & B_3 \end{bmatrix}; \tag{7.189}$$

keeping into account the equalities

$$\begin{aligned} &[\dot{\mathbf{B}}_1] = [\dot{\mathbf{A}}][\mathbf{r}]^{\mathrm{T}}[\mathbf{Q}] + [\mathbf{A}][\mathbf{r}]^{\mathrm{T}}[\dot{\mathbf{Q}}] + \dot{\xi}[\mathbf{A}][\mathbf{r}_p]^{\mathrm{T}}[\mathbf{Q}], \\ &[\dot{\mathbf{B}}_2] = [0 \;\; 0 \;\; \dot{\xi}(a\cos\varphi\cos\xi + b\cos\varphi\sin\xi)], \dot{B}_3 = \dot{\varphi}(a\cos\varphi\cos\xi + b\cos\varphi\sin\xi), \end{aligned} \tag{7.190}$$

one deduces the expression of the derivative of the matrix of constraints

$$[\dot{\mathbf{B}}] = \begin{bmatrix} [\mathbf{0}] & [\dot{\mathbf{B}}_1] & \{\mathbf{0}\} \\ [\mathbf{0}] & [\dot{\mathbf{B}}_2] & \dot{B}_3 \end{bmatrix}. \tag{7.191}$$

Further on, using the notations (7.177), (7.178), and (7.180), where

$$J_x = \frac{mb^2}{4}, J_y = \frac{ma^2}{4}, J_z = \frac{m(a^2 + b^2)}{4}, \tag{7.192}$$

we write the matrix $[\widetilde{\mathbf{M}}]$,

$$\left[\widetilde{\mathbf{M}}\right] = \begin{bmatrix} [\mathbf{M}] & \{\mathbf{0}_{61}\} \\ [\mathbf{0}_{16}] & 0 \end{bmatrix} \tag{7.193}$$

and the differential equation of motion becomes

$$\begin{bmatrix} \left[\widetilde{\mathbf{M}}\right] & -[\mathbf{B}]^{\mathrm{T}} \\ [\mathbf{B}] & [\mathbf{0}_{44}] \end{bmatrix} \begin{bmatrix} \{\ddot{\mathbf{q}}^*\} \\ \{\boldsymbol{\lambda}\} \end{bmatrix} = \begin{bmatrix} \{\mathbf{F}\} + \left\{\widetilde{\mathbf{F}}\right\} \\ 0 \\ -\left[\dot{\mathbf{B}}\right]\{\dot{\mathbf{q}}^*\} \end{bmatrix}, \tag{7.194}$$

with

$$\{\mathbf{q}^*\} = [X_O\ Y_O\ Z_O\ \psi\ \theta\ \varphi\ \xi]^{\mathrm{T}},\ \{\boldsymbol{\lambda}\} = [\lambda_1\ \lambda_2\ \lambda_3\ \lambda_4]^{\mathrm{T}}. \tag{7.195}$$

In the numerical case characterized by $m = 8\,\mathrm{kg}$, $a = 1.2\,\mathrm{m}$, $b = 1\,\mathrm{m}$, $J_x = 2\,\mathrm{kgm}^2$, $J_y = 2.88\,\mathrm{kgm}^2$, $J_z = 4.88\,\mathrm{kgm}^2$ and the initial conditions $t = 0\,\mathrm{s}$, $\psi = \frac{\pi}{2}\,\mathrm{rad}$, $\theta = \frac{2\pi}{5}\,\mathrm{rad}$, $\varphi = 0\,\mathrm{rad}$, $\xi = \frac{\pi}{2}\,\mathrm{rad}$, $\dot{\psi} = \dot{\theta} = 0\,\mathrm{rad/s}$, $\dot{\varphi} = -2\,\mathrm{rad/s}$, $\dot{\xi} = -2.4\,\mathrm{rad/s}$, $X_O = -b\cos\frac{2\pi}{5}\,\mathrm{m}$, $Y_O = 0\,\mathrm{m}$, $Z_O = b\sin\frac{2\pi}{5}\,\mathrm{m}$, $\dot{X}_O = 0\,\mathrm{m/s}$, $\dot{Y}_O = 2\,\mathrm{m/s}$, $\dot{Z}_O = 0\,\mathrm{m/s}$ one obtains the results captured in the following diagrams (Fig. 7.19 to Fig. 7.28).

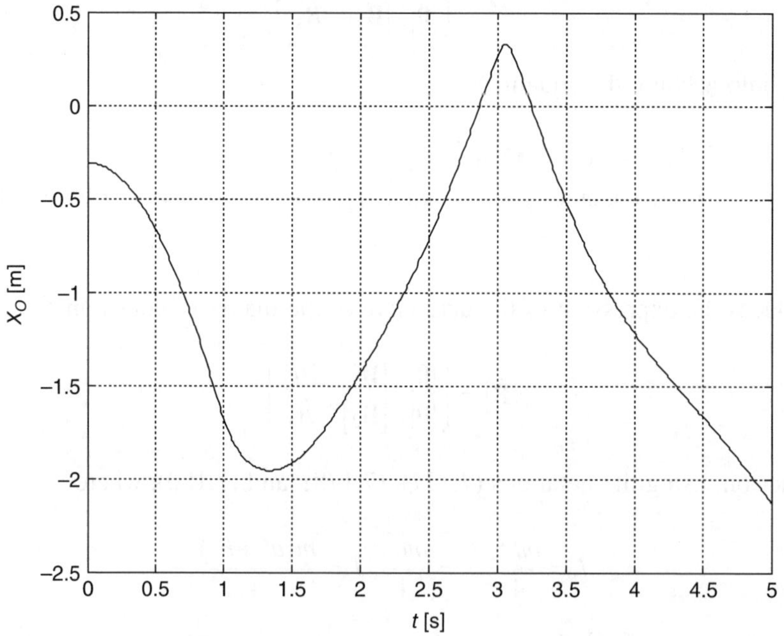

Figure 7.19 Time history $X_O = X_O(t)$ for $0 \le t \le 5\,\mathrm{s}$.

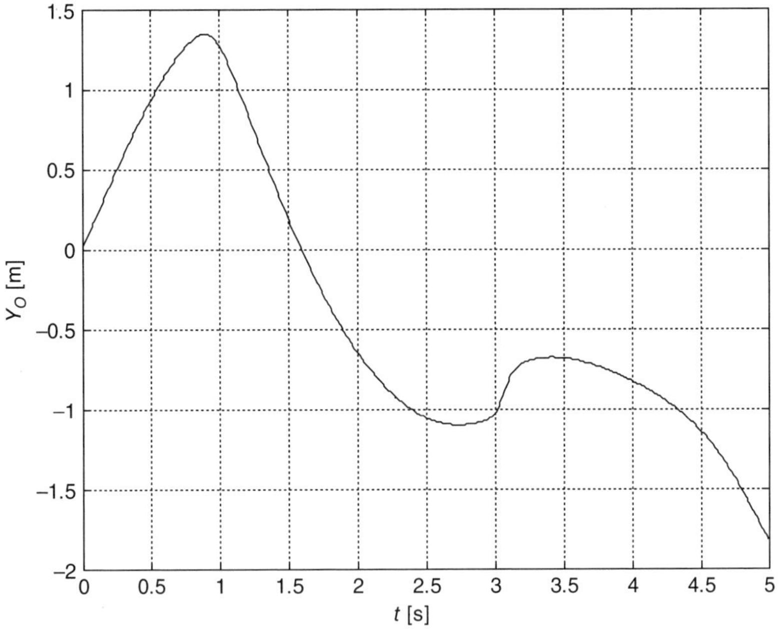

Figure 7.20 Time history $Y_O = Y_O(t)$ for $0 \le t \le 5$ s.

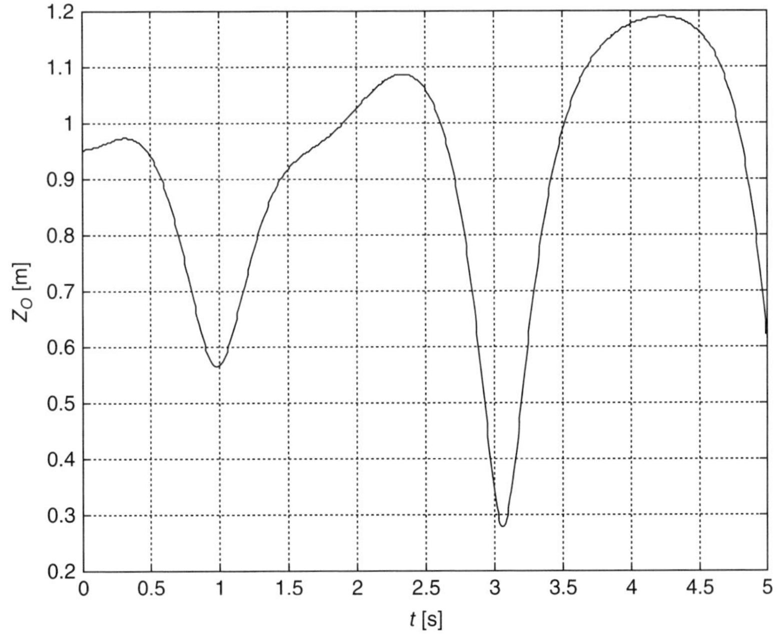

Figure 7.21 Time history $Z_O = Z_O(t)$ for $0 \le t \le 5$ s.

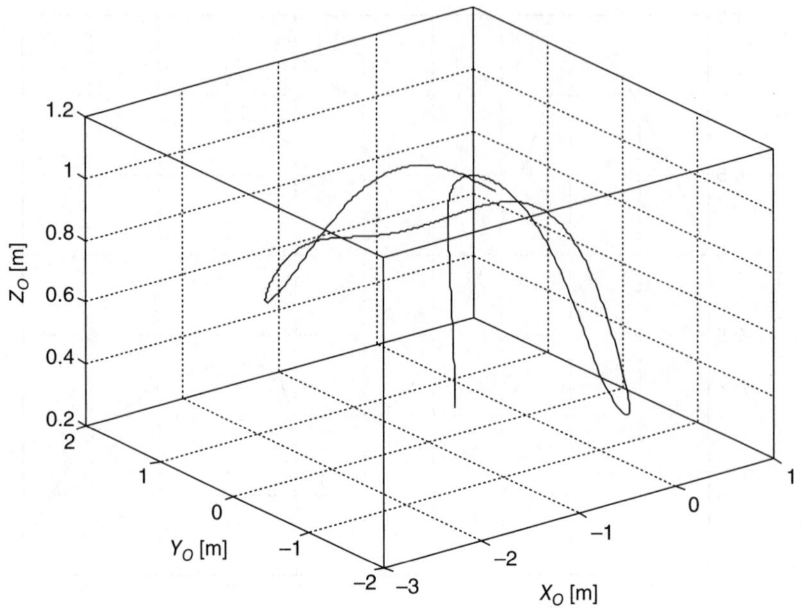

Figure 7.22 Trajectory of the point O (center of disk) in the first 5 seconds.

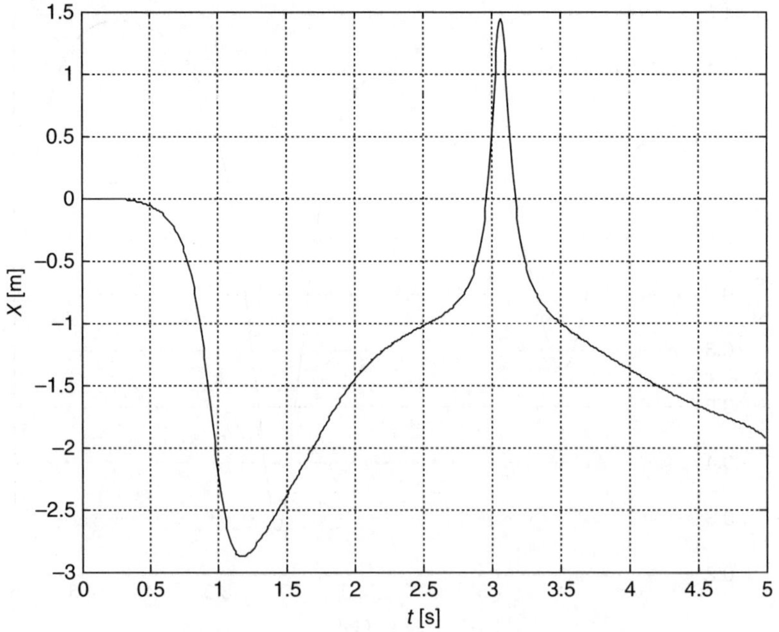

Figure 7.23 Time history $X = X(t)$ for $0 \le t \le 5$ s.

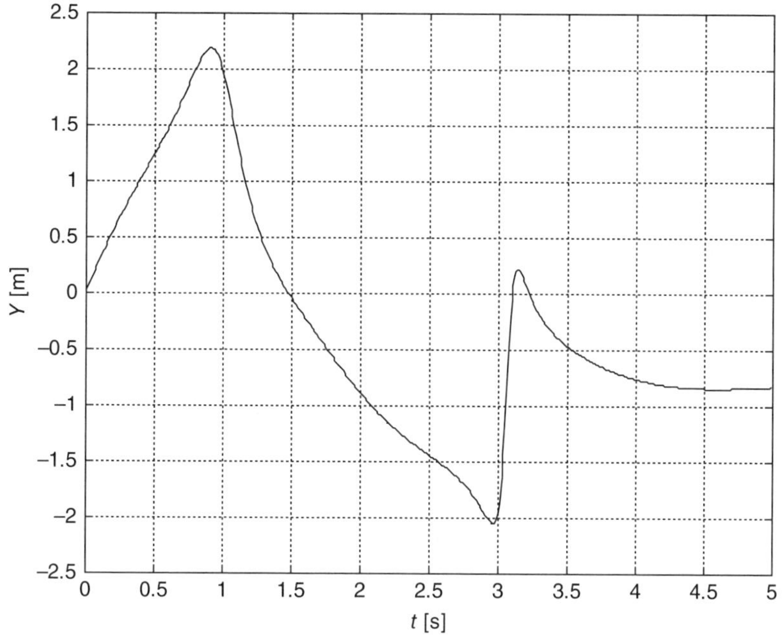

Figure 7.24 Time history $Y = Y(t)$ for $0 \leq t \leq 5$ s.

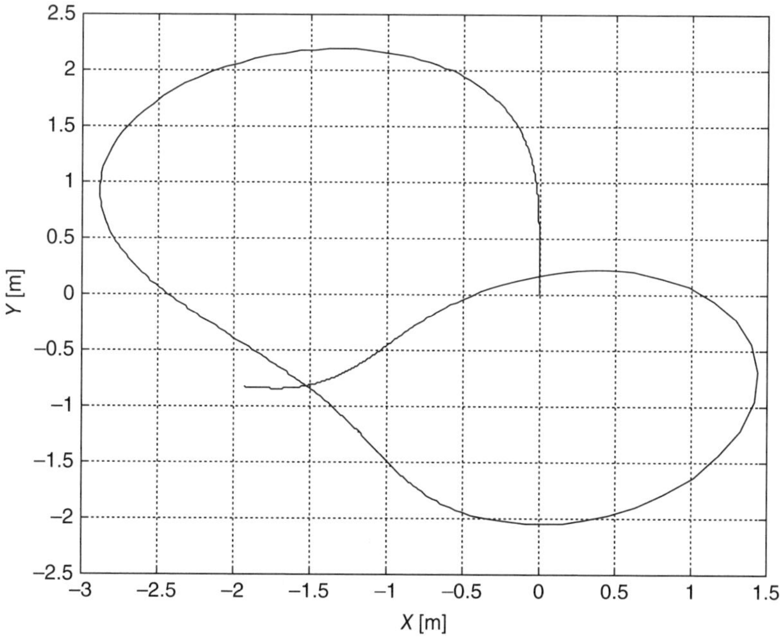

Figure 7.25 Trajectory of the contact point in the first 5 seconds.

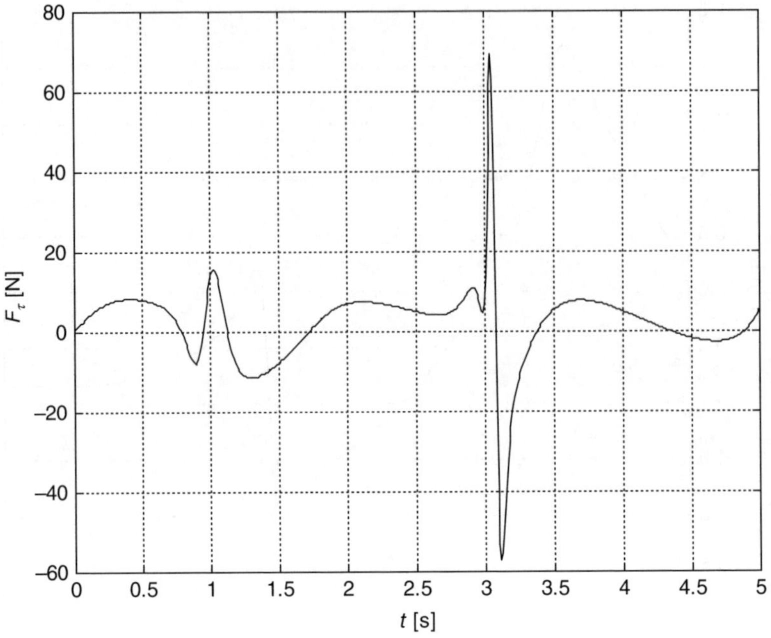

Figure 7.26 Time history $F_\tau = F_\tau(t)$ for $0 \le t \le 5$ s.

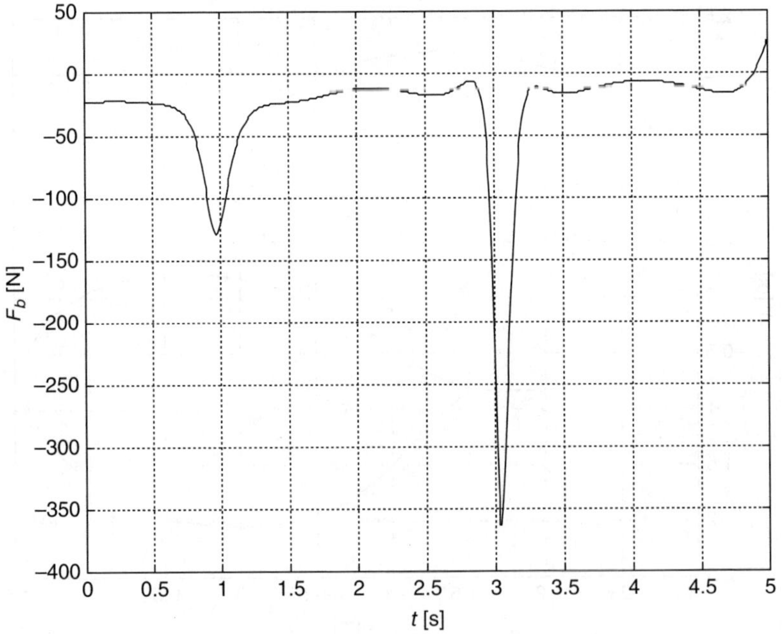

Figure 7.27 Time history $F_b = F_b(t)$ for $0 \le t \le 5$ s.

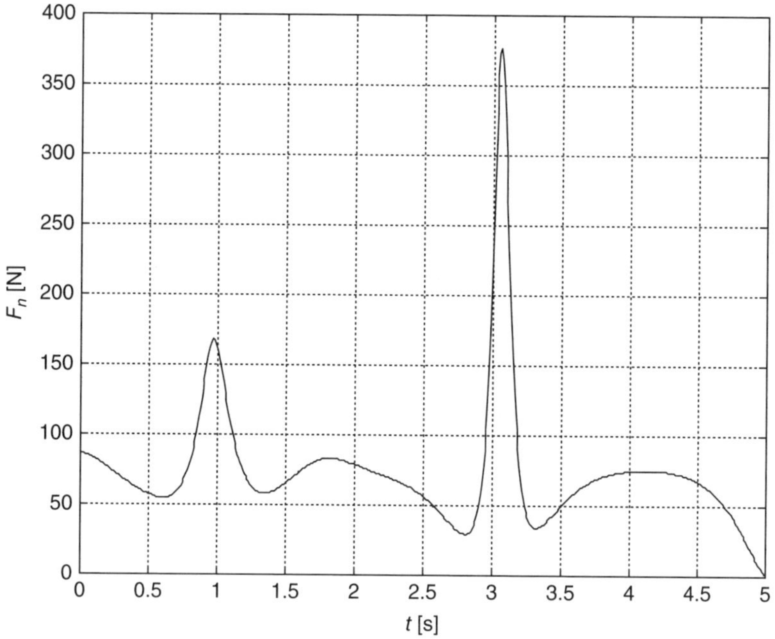

Figure 7.28 Time history $F_n = F_n(t)$ for $0 \le t \le 5$ s.

The quasi-periodic character of the motion does not appear anymore, and the forces have values up to 400 N, that is, 5 times greater than the own weight of the disk.

Observation 7.8.1. Obviously, for $a = b$ one obtains the case discussed in the paragraph 7.7, the circle disk on the horizontal plan.

7.9 The Rolling without Sliding of a Hyperboidic Curve on a Horizontal Plan

7.9.1 Hyperboidic Curves

The hyperboidic curve is the curve of intersection between the hyperbolic paraboloid

$$z = -2\zeta xy \tag{7.196}$$

and the cylinder of equation

$$x^2 + y^2 - r_0^2 = 0. \tag{7.197}$$

The parametric equations of this curve are

$$x = r_0 \cos \xi, \; y = -r_0 \sin \xi, \; z = \zeta r_0 \cos 2\xi, \tag{7.198}$$

and its graphical representation is given in Fig. 7.29.

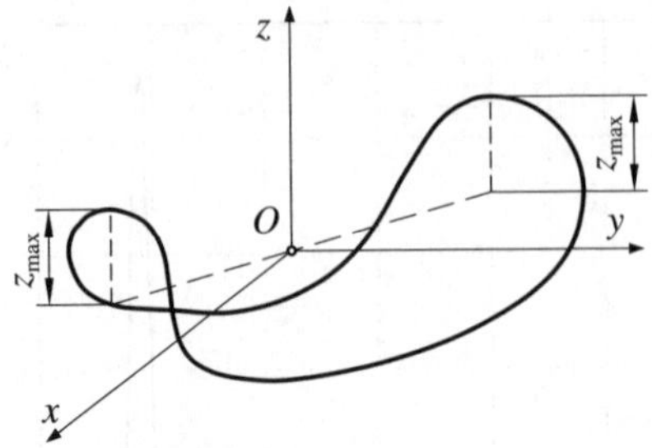

Figure 7.29 The hyperboidic curve.

In the first and the third quadrant the curve is situated under the *Oxy* plan, while in the second and the fourth quadrant it is situated over the *Oxy* plan (the quadrants are defined relative to the *Oxy* plan); the points of extreme high ($z_{max} = \zeta r_0$, $z_{min} = -\zeta r_0$) are situated in the bisector plans.

One may also consider the variant of a conic shell (with the apex at the point O), which is bounded by the hyperboidic curve, case in which the equation of the conical surface reads

$$z = \frac{2\zeta xy}{\sqrt{x^2 + y^2}}. \tag{7.199}$$

Since the Oz-axis is one of symmetry, the center of weight of the curve is situated on this axis; by calculation it results that the center of weight coincides to the point O.

In the case when the parameter ζ is small, the length L of the curve may be approximated by the expression

$$L \approx 2\pi r_0 \left(1 + \zeta^2\right), \tag{7.200}$$

while the moments of inertia read

$$J_x = J_y \approx \frac{m r_0^2}{2}\left(1 + \zeta^2\right),\ J_z = m r_0^2. \tag{7.201}$$

7.9.2 The Matrix Differential Equation of Motion

We recall the equations (7.173) with the modifications

$$\{\mathbf{r}\} = r_0 \begin{bmatrix} \cos\xi \\ -\sin\xi \\ \zeta \sin 2\xi \end{bmatrix},\ \{\mathbf{r}_p\} = r_0 \begin{bmatrix} -\sin\xi \\ -\cos\xi \\ 2\zeta\cos 2\xi \end{bmatrix},\ \{\mathbf{r}_s\} = r_0 \begin{bmatrix} -\cos\xi \\ \sin\xi \\ -4\zeta \sin 2\xi \end{bmatrix},\ \{\mathbf{r}_t\} = r_0 \begin{bmatrix} \sin\xi \\ \cos\xi \\ -8\zeta\cos 2\xi \end{bmatrix},$$

$$[\mathbf{r}] = r_0 \begin{bmatrix} 0 & -\zeta\sin 2\xi & -\sin\xi \\ \zeta\sin 2\xi & 0 & -\cos\xi \\ \sin\xi & \cos\xi & 0 \end{bmatrix},\ [\mathbf{r}_p] = r_0 \begin{bmatrix} 0 & -2\zeta\cos 2\xi & -\cos\xi \\ 2\zeta\cos 2\xi & 0 & \sin\xi \\ \cos\xi & -\sin\xi & 0 \end{bmatrix},$$

$$[\mathbf{r}_s]=r_0\begin{bmatrix}0 & 4\zeta\sin 2\xi & \sin\xi\\ -4\zeta\sin 2\xi & 0 & \cos\xi\\ -\sin\xi & -\cos\xi & 0\end{bmatrix},[\mathbf{r}_t]=r_0\begin{bmatrix}0 & 8\zeta\cos 2\xi & \cos\xi\\ -8\zeta\cos 2\xi & 0 & -\sin\xi\\ -\cos\xi & \sin\xi & 0\end{bmatrix}. \quad (7.202)$$

From the tangency condition (7.151) one deduces the equality

$$-\sin\theta\cos(\xi-\varphi)+2\zeta\cos\theta\cos 2\xi=0, \quad (7.203)$$

which, by derivation with respect to time, becomes

$$\begin{aligned}&[\sin\theta\cos(\xi-\varphi)+2\zeta\sin\theta\cos 2\xi]\dot{\theta}+\dot{\varphi}\sin\theta\sin(\xi-\varphi)\\&+[-\sin\theta\sin(\xi-\varphi)+4\zeta\cos\theta\sin 2\xi]\dot{\xi}=0.\end{aligned} \quad (7.204)$$

With the notations

$$\begin{aligned}&[\mathbf{B}_1]=[\mathbf{A}][\mathbf{r}]^{\mathrm{T}}[\mathbf{Q}],[\mathbf{B}_2]=[0\ \ \sin\theta\cos(\xi-\varphi)+2\zeta\cos\theta\cos 2\xi\ \sin\theta\sin(\xi-\varphi)],\\&B_3=-\sin\theta\sin(\xi-\varphi)+4\zeta\sin\theta\sin 2\xi,\end{aligned} \quad (7.205)$$

the matrix of constraints reads

$$[\mathbf{B}]=\begin{bmatrix}[\mathbf{I}] & [\mathbf{B}_1] & \{\mathbf{0}\}\\ [\mathbf{0}] & [\mathbf{B}_2] & B_3\end{bmatrix}; \quad (7.206)$$

keeping into account the equalities

$$\begin{aligned}&[\dot{\mathbf{B}}_1]=[\dot{\mathbf{A}}][\mathbf{r}]^{\mathrm{T}}[\mathbf{Q}]+[\mathbf{A}][\mathbf{r}]^{\mathrm{T}}[\dot{\mathbf{Q}}]+\dot{\xi}[\mathbf{A}][\mathbf{r}_p]^{\mathrm{T}}[\mathbf{Q}],\\&B_{22}=\dot{\varphi}\cos\theta\sin(\xi-\varphi)-\dot{\xi}[\cos\theta\sin(\xi-\varphi)+4\zeta\sin\theta\sin 2\xi],\\&B_{23}=\dot{\theta}\cos\theta\sin(\xi-\varphi)+(\dot{\xi}-\dot{\varphi})\sin\theta\cos(\xi-\varphi),\ [\dot{\mathbf{B}}_2]=[0\ \ B_{22}\ \ B_{23}],\\&\dot{B}_3=-\dot{\theta}[\cos\theta\sin(\xi-\varphi)+4\zeta\cos\theta\sin 2\xi]+\dot{\varphi}\sin\theta\cos(\xi-\varphi)\\&\quad-\dot{\xi}[\sin\theta\cos(\xi-\varphi)+8\zeta\cos\theta\cos 2\xi],\end{aligned} \quad (7.207)$$

one deduces the derivative of the matrix of constraints

$$[\dot{\mathbf{B}}]=\begin{bmatrix}[\mathbf{0}] & [\dot{\mathbf{B}}_1] & \{\mathbf{0}\}\\ [\mathbf{0}] & [\dot{\mathbf{B}}_2] & \dot{B}_3\end{bmatrix}. \quad (7.208)$$

Further on, using the notations (7.177), (7.178), and (7.180), one writes the matrix $[\widetilde{\mathbf{M}}]$,

$$[\widetilde{\mathbf{M}}] = \begin{bmatrix} [\mathbf{M}] & \{\mathbf{0}_{61}\} \\ [\mathbf{0}_{16}] & 0 \end{bmatrix}, \tag{7.209}$$

and the differential equation of motion becomes

$$\begin{bmatrix} [\widetilde{\mathbf{M}}] & -[\mathbf{B}]^{\mathrm{T}} \\ [\mathbf{B}] & [\mathbf{0}_{44}] \end{bmatrix} \begin{bmatrix} \{\ddot{\mathbf{q}}^*\} \\ \{\boldsymbol{\lambda}\} \end{bmatrix} = \begin{bmatrix} \{\mathbf{F}\} + \{\widetilde{\mathbf{F}}\} \\ 0 \\ -[\dot{\mathbf{B}}]\{\dot{\mathbf{q}}^*\} \end{bmatrix}, \tag{7.210}$$

where

$$\{\mathbf{q}^*\} = [X_O \;\; Y_O \;\; Z_O \;\; \psi \;\; \theta \;\; \varphi \;\; \xi]^{\mathrm{T}}, \{\boldsymbol{\lambda}\} = [\lambda_1 \;\; \lambda_2 \;\; \lambda_3 \;\; \lambda_4]^{\mathrm{T}}. \tag{7.211}$$

To be out to establish the initial conditions, one chooses the contact point, respectively one imposes the values for the initial x, y coordinates, and then the initial values for the Euler angles. From the equation (7.203) and the matrix equation

$$\begin{bmatrix} X \\ Y \\ 0 \end{bmatrix} = \begin{bmatrix} X_O \\ Y_O \\ Z_O \end{bmatrix} + [\mathbf{A}]\{\mathbf{r}\} \tag{7.212}$$

one determines the values of the parameters X_O, Y_O, Z_O, ξ.

For instance, if $X = Y = 0\,\mathrm{m}$, $\psi = \frac{\pi}{2}\,\mathrm{rad}$, $\theta = \frac{2\pi}{5}\,\mathrm{rad}$, $\varphi = 0\,\mathrm{rad}$, then from the equation (7.203) one obtains the solution

$$\xi = \frac{\pi}{2} + \arcsin\frac{4\zeta}{\tan\theta + \sqrt{\tan^2\theta + 32\zeta^2}}, \tag{7.213}$$

while from the equation (7.212) result the values

$$X_O = -r_0(\cos\theta\sin\xi + \zeta\sin\theta\sin 2\xi), Y_O = -r_0\cos\xi,$$
$$Z_O = r_0(\sin\theta\cos\xi + \zeta\cos\theta\sin 2\xi). \tag{7.214}$$

For the initial velocities, if one chooses the values $\dot{\psi}$, $\dot{\theta}$, $\dot{\varphi}$, then from the equation (7.204) one determines $\dot{\xi}$, while from the matrix equation

$$\begin{bmatrix} \dot{X}_O \\ \dot{Y}_O \\ \dot{Z}_O \end{bmatrix} + [\mathbf{A}][\mathbf{r}]^{\mathrm{T}}[\mathbf{Q}] \begin{bmatrix} \dot{\psi} \\ \dot{\theta} \\ \dot{\varphi} \end{bmatrix} = 0 \tag{7.215}$$

one determines the initial velocities $\dot{X}_O$, $\dot{Y}_O$, $\dot{Z}_O$.

For instance, if $\dot{\psi} = \dot{\theta} = 0$ rad/s, $\dot{\varphi} = -2$ rad/s, then from the equation (7.204) it results

$$\dot{\xi} = \dot{\varphi}\frac{\sin\theta\sin\xi}{\sin\theta\sin\xi - 4\zeta\cos\theta\sin 2\xi}, \tag{7.216}$$

while from the equation (7.215) one obtains the values

$$\dot{X}_O = r_0\dot{\varphi}\cos\theta\cos\xi, \dot{Y}_O = -r_0\dot{\varphi}\sin\xi, \dot{Z}_O = -r_0\dot{\varphi}\sin\theta\cos\xi. \tag{7.217}$$

In the numerical case defined by $m = 8$ kg, $r_0 = 1$ m, $\zeta = 0$, $J_x = J_y = 4.04$ kgm^2, $J_z = 8$ kgm^2 and the initial values $t = 0$ s, $\psi = \frac{\pi}{2}$ rad, $\theta = \frac{2\pi}{5}$ rad, $\varphi = 0$ rad, $\xi = 1.635285$ rad, $\dot{\psi} = \dot{\theta} = 0$ rad/s, $\dot{\varphi} = -2$ rad/s, $\dot{\xi} = -1.967049$ rad/s, $X_O = -0.296142$ m, $Y_O = 0.064444$ m, $Z_O = -0.953054$ m, $\dot{X}_O = 0.039829$ m/s, $\dot{Y}_O = 1.995843$ m/s, $\dot{Z}_O = -0.122580$ m/s one obtains the diagrams presented in the Fig. 30 to Fig. 39.

From these diagrams it results that both the trajectory of the point O, and the trajectory of the contact point between the curve and the plan are curves contained in a bounded spatial domain (along the all three axes), and bounded planar domain (along

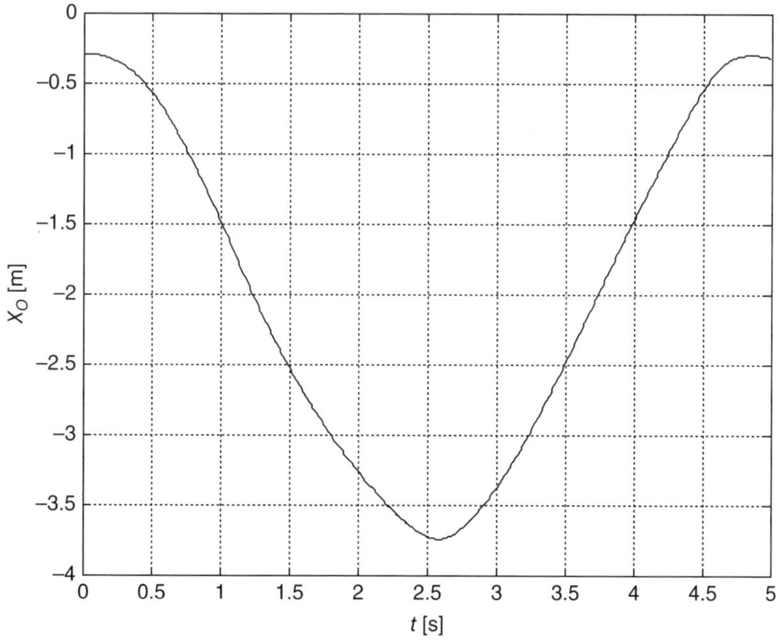

Figure 7.30 Time history $X_O = X_O(t)$ for $0 \le t \le 5$ s.

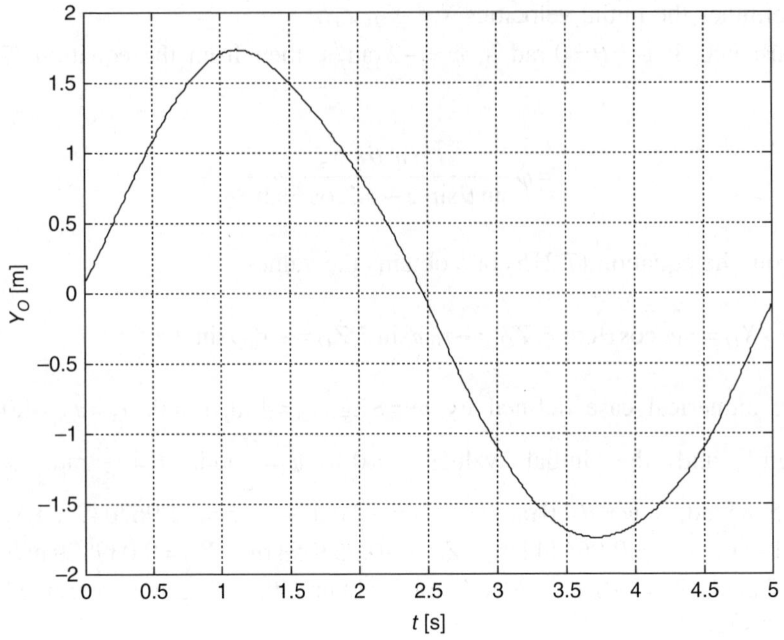

Figure 7.31 Time history $Y_O = Y_O(t)$, for $0 \le t \le 5$ s.

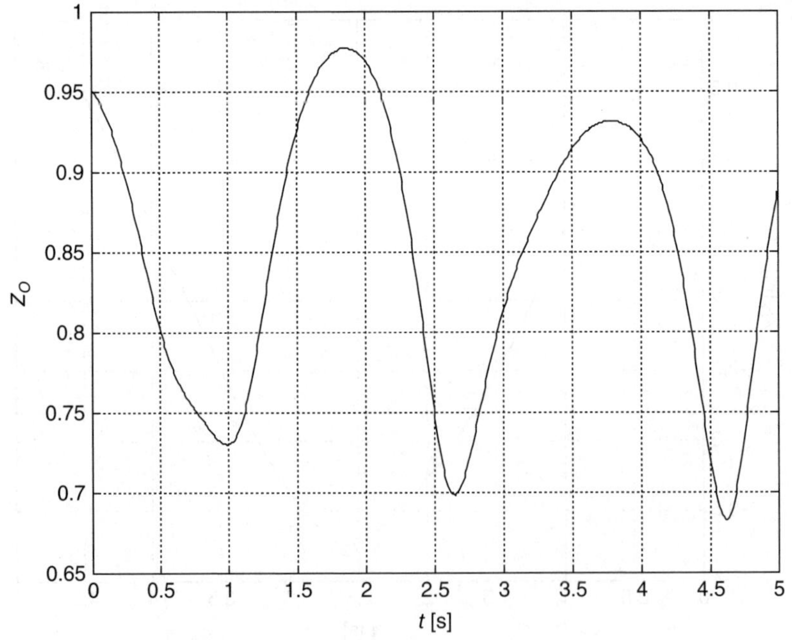

Figure 7.32 Time history $Z_O = Z_O(t)$ for $0 \le t \le 5$ s.

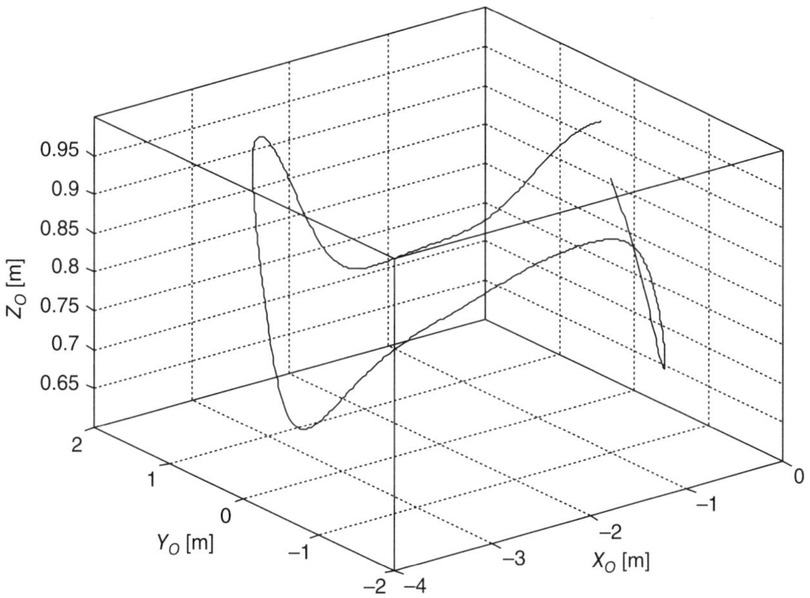

Figure 7.33 Trajectory of the point O (center of curve) in the first 5 seconds.

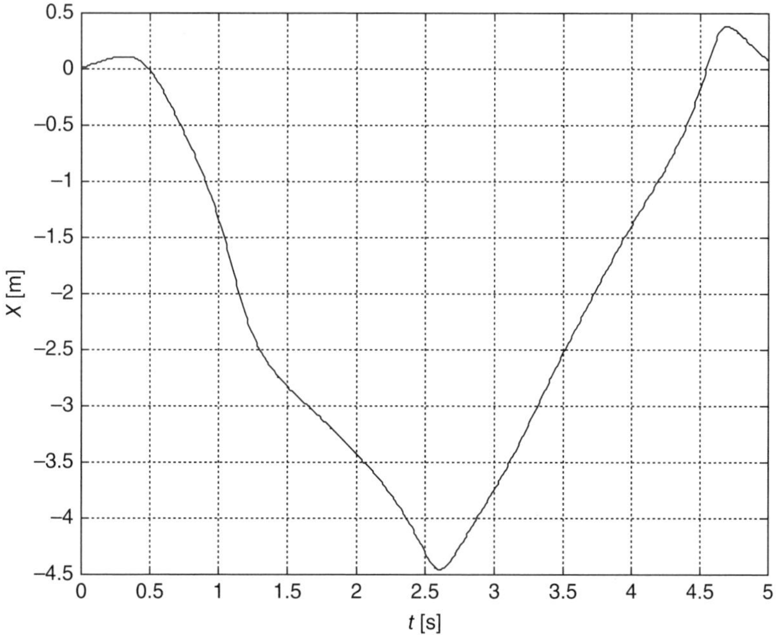

Figure 7.34 Time history $X = X(t)$ for $0 \leq t \leq 5$ s.

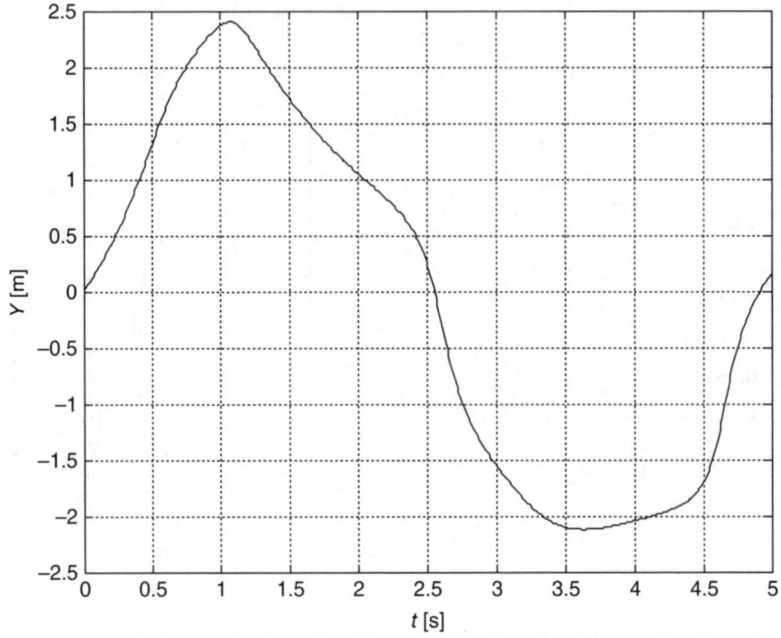

Figure 7.35 Time history $Y = Y(t)$ for $0 \le t \le 5$ s.

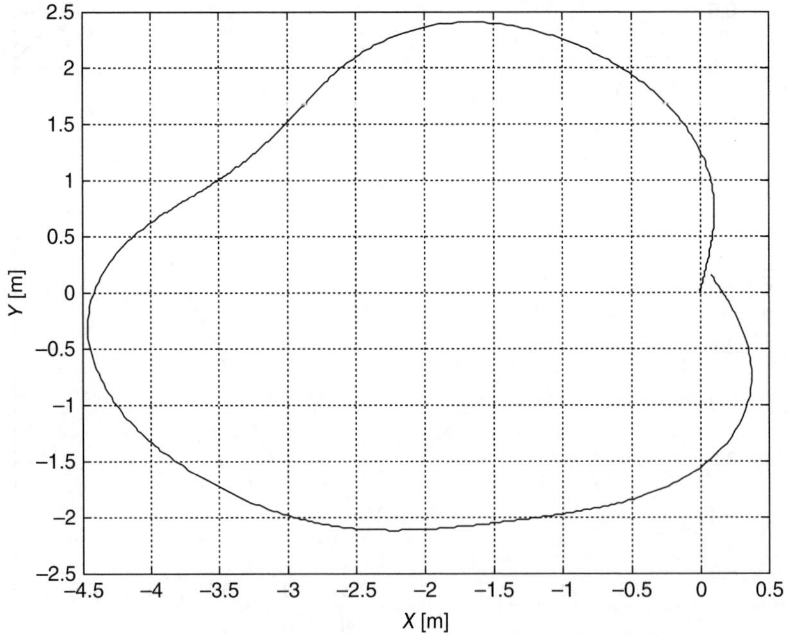

Figure 7.36 Trajectory of the contact point in the first 5 seconds.

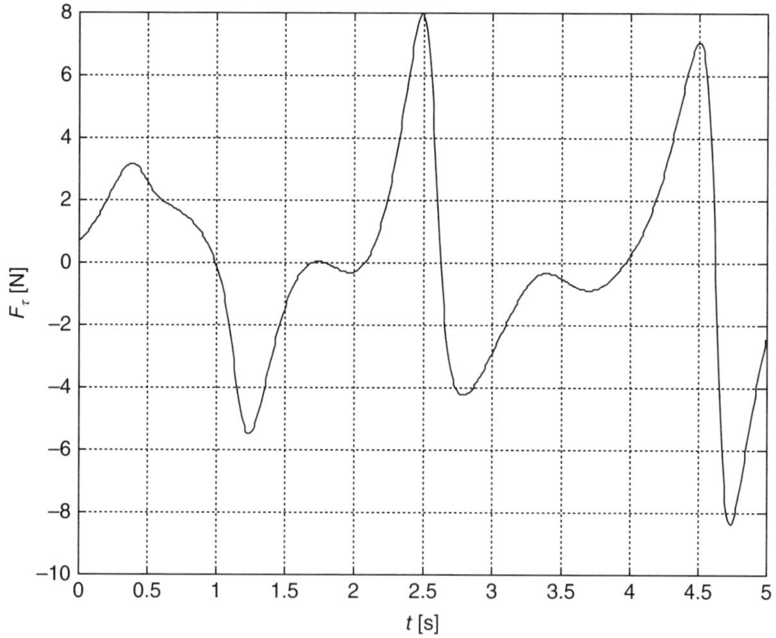

Figure 7.37 Time history $F_\tau = F_\tau(t)$ for $0 \le t \le 5$ s.

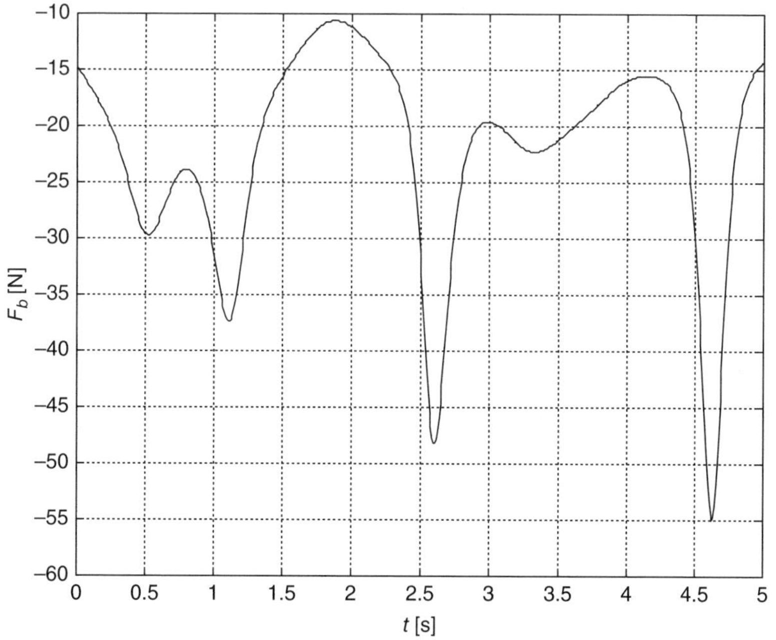

Figure 7.38 Time history $F_b = F_b(t)$ for $0 \le t \le 5$ s.

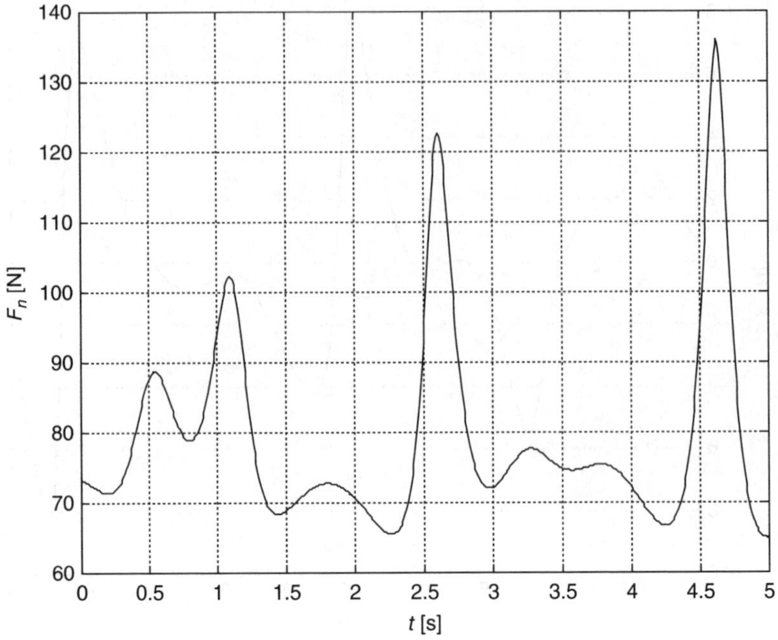

Figure 7.39 Time history $F_n = F_n(t)$ for $0 \leq t \leq 5$ s.

the two axes), respectively. We can state nothing about the existence of a periodic or quasi-periodic character of the time variations of certain characteristic parameters of the motion.

Moreover, the forces of constraints have maximum values of approximate 140 N, that is, 1.8 times greater than the own weight of the curve.

7.10 The Rolling without Sliding of a Planar Circle Disk on a Cylindrical Surface with Horizontal Generatrices

One considers the disk of radius r_0 and mass m (Fig. 7.40) acted only by its own weight and rolling without sliding on the cylindrical surface of equation

$$Z = -\frac{\zeta}{2}Y^2. \tag{7.218}$$

Writing for the circle contour the parametric equations in the reference system $Oxyz$

$$x = r_0 \cos \xi, y = -r_0 \sin \xi, z = 0, \tag{7.219}$$

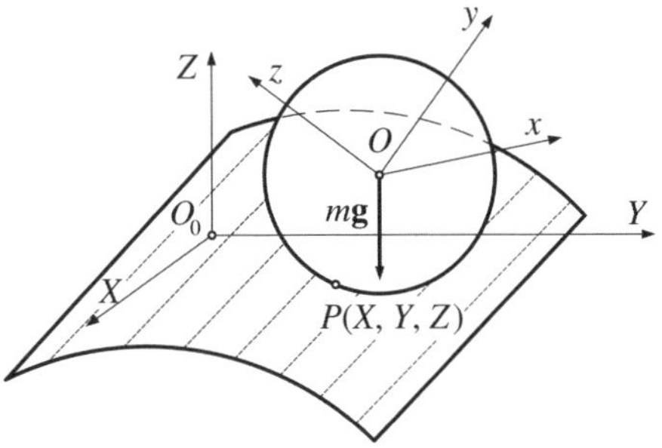

Figure 7.40 The circle disk rolling without sliding on a cylinder with horizontal generatrices.

and using the notations (7.173) with the modifications

$$\{\mathbf{F}_p\} = [0 \ \ \zeta Y \ \ 1]^{\mathrm{T}}, [\mathbf{F}_s] = \begin{bmatrix} 0 & 0 & 0 \\ 0 & \zeta & 0 \\ 0 & 0 & 0 \end{bmatrix}, \tag{7.220}$$

and the tangency condition

$$\{\mathbf{F}_p\}^{\mathrm{T}}[\mathbf{A}]\{\mathbf{r}_p\} = 0, \tag{7.221}$$

one obtains the equation

$$\zeta Y[\sin\theta \sin(\xi - \varphi) + \cos\psi \cos\theta \cos(\xi - \varphi)] + \sin\theta \cos(\xi - \varphi) = 0. \tag{7.222}$$

In the current calculation and also to establish the initial conditions, the parameters X, Y and their derivatives are deduced from the expressions

$$\begin{bmatrix} X \\ Y \\ Z \end{bmatrix} = \begin{bmatrix} X_O \\ Y_O \\ Z_O \end{bmatrix} + [\mathbf{A}]\{\mathbf{r}\}, \begin{bmatrix} \dot{X} \\ \dot{Y} \\ \dot{Z} \end{bmatrix} = \dot{\xi}[\mathbf{A}]\{\mathbf{r}_p\}. \tag{7.223}$$

Further on, one uses the notations (7.154), (7.155), and (7.158), and determines the matrix of constraints.

For the determination of the derivative of the matrix of constraints with respect to time, one firstly calculates the derivatives

$$\begin{aligned}
[\dot{\mathbf{B}}_1] &= [\dot{\mathbf{A}}][\mathbf{r}]^{\mathrm{T}}[\mathbf{Q}] + [\mathbf{A}][\mathbf{r}]^{\mathrm{T}}[\dot{\mathbf{Q}}] + \dot{\xi}[\mathbf{A}][\mathbf{r}_p]^{\mathrm{T}}[\mathbf{Q}], \{\dot{\mathbf{F}}_p\} = \dot{\xi}[\mathbf{F}_s][\mathbf{A}]\{\mathbf{r}_p\}, \\
[\dot{\mathbf{B}}_2] &= \{\dot{\mathbf{F}}_p\}^{\mathrm{T}}[\mathbf{A}][\mathbf{r}_p]^{\mathrm{T}}[\mathbf{Q}] + \{\mathbf{F}_p\}^{\mathrm{T}}[\dot{\mathbf{A}}][\mathbf{r}_p]^{\mathrm{T}}[\mathbf{Q}] + \dot{\xi}\{\mathbf{F}_p\}^{\mathrm{T}}[\mathbf{A}][\mathbf{r}_s]^{\mathrm{T}}[\mathbf{Q}] \\
&\quad + \{\mathbf{F}_p\}^{\mathrm{T}}[\mathbf{A}][\mathbf{r}_p]^{\mathrm{T}}[\dot{\mathbf{Q}}],
\end{aligned}$$

$$\dot{B}_3=\{\dot{\mathbf{F}}_p\}^{\mathrm{T}}[\mathbf{A}]\{\mathbf{r}_s\}+\{\dot{\mathbf{F}}_p\}^{\mathrm{T}}[\dot{\mathbf{A}}]\{\mathbf{r}_s\}+\dot{\xi}\{\mathbf{F}_p\}^{\mathrm{T}}[\mathbf{A}]\{\mathbf{r}_t\}$$
$$+2\dot{\xi}\{\mathbf{r}_s\}^{\mathrm{T}}[\mathbf{A}]^{\mathrm{T}}[\mathbf{F}_s][\mathbf{A}]\{\mathbf{r}_p\}+2\{\mathbf{r}_p\}^{\mathrm{T}}[\dot{\mathbf{A}}]^{\mathrm{T}}[\mathbf{F}_s][\mathbf{A}]\{\mathbf{r}_p\} \qquad (7.224)$$

and it results

$$[\dot{\mathbf{B}}]=\begin{bmatrix}[\mathbf{0}] & [\dot{\mathbf{B}}_1] & \{\mathbf{0}\}\\ [\mathbf{0}] & [\dot{\mathbf{B}}_2] & \dot{B}_3\end{bmatrix}. \qquad (7.225)$$

Further on, one calculates the matrices

$$[\mathbf{M}]=\begin{bmatrix}[\mathbf{m}] & [\mathbf{0}]\\ [\mathbf{0}] & [\mathbf{Q}]^{\mathrm{T}}[\mathbf{J}][\mathbf{Q}]\end{bmatrix},\ [\widetilde{\mathbf{M}}]=\begin{bmatrix}[\mathbf{M}] & \{\mathbf{0}_{61}\}\\ [\mathbf{0}_{16}] & 0\end{bmatrix}, \qquad (7.226)$$

$$\{\boldsymbol{\beta}\}=[\psi\ \ \theta\ \ \varphi]^{\mathrm{T}}, \qquad (7.227)$$

$$\{\widetilde{\mathbf{F}}_{\boldsymbol{\beta}}\}=-\left[[\mathbf{Q}]^{\mathrm{T}}[\mathbf{J}][\dot{\mathbf{Q}}]+[\mathbf{Q}]^{\mathrm{T}}[\boldsymbol{\omega}][\mathbf{J}][\mathbf{Q}]\right]\{\dot{\boldsymbol{\beta}}\},\{\widetilde{\mathbf{F}}\}=\begin{bmatrix}\{\mathbf{0}\}\\ \{\widetilde{\mathbf{F}}_{\boldsymbol{\beta}}\}\end{bmatrix}, \qquad (7.228)$$

$$\{\mathbf{F}\}=[0\ \ 0\ \ -mg\ \ 0\ \ 0\ \ 0]^{\mathrm{T}} \qquad (7.229)$$

and one obtains the matrix equation

$$\begin{bmatrix}[\widetilde{\mathbf{M}}] & -[\mathbf{B}]^{\mathrm{T}}\\ [\mathbf{B}] & [\mathbf{0}_{44}]\end{bmatrix}\begin{bmatrix}\{\ddot{\mathbf{q}}^*\}\\ \{\boldsymbol{\lambda}\}\end{bmatrix}=\begin{bmatrix}\{\mathbf{F}\}+\{\widetilde{\mathbf{F}}\}\\ 0\\ -[\dot{\mathbf{B}}]\{\dot{\mathbf{q}}^*\}\end{bmatrix}, \qquad (7.230)$$

where

$$\{\mathbf{q}^*\}=[X_O\ \ Y_O\ \ Z_O\ \ \psi\ \ \theta\ \ \varphi\ \ \xi]^{\mathrm{T}},\{\boldsymbol{\lambda}\}=[\lambda_1\ \ \lambda_2\ \ \lambda_3\ \ \lambda_4]^{\mathrm{T}}. \qquad (7.231)$$

To determine the trajectory of the contact point one calculates the coordinates X, Y, Z from the first relation (7.223).

One calculates the column matrices of the unit vectors $\boldsymbol{\tau}$, $\mathbf{n}$, $\mathbf{b}$ with the relations (7.167)–(7.169), and the forces at the contact point with the relation (7.170).

To be out to establish the initial conditions one chooses:

- the coordinates X, Y, Z of the initial contact point;
- the values of the Euler angles and their derivatives with respect to time.

In these conditions, from the equations (7.222) and (7.223) one determines the values of the parameters X_O, Y_O, Z_O, ξ; from the matrix equation

$$\begin{bmatrix} \dot{X}_O \\ \dot{Y}_O \\ \dot{Z}_O \end{bmatrix} + [\mathbf{A}][\mathbf{r}]^{\mathrm{T}}[\mathbf{Q}] \begin{bmatrix} \dot{\psi} \\ \dot{\theta} \\ \dot{\varphi} \end{bmatrix} = 0 \tag{7.232}$$

one determines the values $\dot{X}_O$, $\dot{Y}_O$, $\dot{Z}_O$, while from the equation (7.222) derived with respect to time and from the second relation (7.223) one determines $\dot{\xi}$.

In the case when at $t=0\,\mathrm{s}$ correspond the values $X=Y=Z=0\,\mathrm{m}$, $\psi=\frac{\pi}{2}\,\mathrm{rad}$, $\theta=\frac{2\pi}{5}\,\mathrm{rad}$, $\varphi=0\,\mathrm{rad}$, $\dot{\psi}=\dot{\theta}=0\,\mathrm{rad/s}$, $\dot{\varphi}=-2\,\mathrm{rad/s}$, one obtains from the relation (7.222) the value $\xi=\frac{\pi}{2}\,\mathrm{rad}$; from the first relation (7.223) one deduces the values $X_O=-r_0\cos\frac{2\pi}{5}\,\mathrm{m}$, $Y_O=0\,\mathrm{m}$, $Z_O=r_0\sin\frac{2\pi}{5}\,\mathrm{m}$; from the second relation (7.223) one obtains $\dot{Y}=-r_0\dot{\xi}\,\mathrm{m}$; from the expression (7.222) derived with respect to time one deduces $\dot{\xi}=\dot{\varphi}\frac{\sin\theta}{\sin\theta+\zeta r_0}\,\mathrm{rad/s}$, while from the relation (7.232) it results $\dot{X}_O=0\,\mathrm{m/s}$, $\dot{Y}_O=-r_0\dot{\varphi}\,\mathrm{m/s}$, $\dot{Z}_O=0\,\mathrm{m/s}$.

For a numerical application defined by $m=8\,\mathrm{kg}$, $J_x=J_y=2\,\mathrm{kgm}^2$, $J_z=4\,\mathrm{kgm}^2$, $\zeta=0.02$, $r_0=2\,\mathrm{m}$ and the initial values $t=0\,\mathrm{s}$, $X=0\,\mathrm{m}$, $Y=0\,\mathrm{m}$, $Z=0\,\mathrm{m}$, $\psi=\frac{\pi}{2}\,\mathrm{rad}$, $\theta=\frac{2\pi}{5}\,\mathrm{rad}$, $\varphi=0\,\mathrm{rad}$, $\dot{\psi}=\dot{\theta}=0\,\mathrm{rad/s}$, $\dot{\varphi}=-2\,\mathrm{rad/s}$, $\xi=\frac{\pi}{2}\,\mathrm{rad}$, $X_O=-0.618034\,\mathrm{m}$, $Y_O=0\,\mathrm{m}$, $Z_O=1.902113\,\mathrm{m}$, $\dot{\xi}=-1.919278\,\mathrm{rad/s}$, $\dot{X}_O=0\,\mathrm{m/s}$, $\dot{Y}_O=4\,\mathrm{m/s}$, $\dot{Z}_O=0\,\mathrm{m/s}$ one obtains the results represented in the following diagrams (Fig. 7.41 to Fig. 7.49).

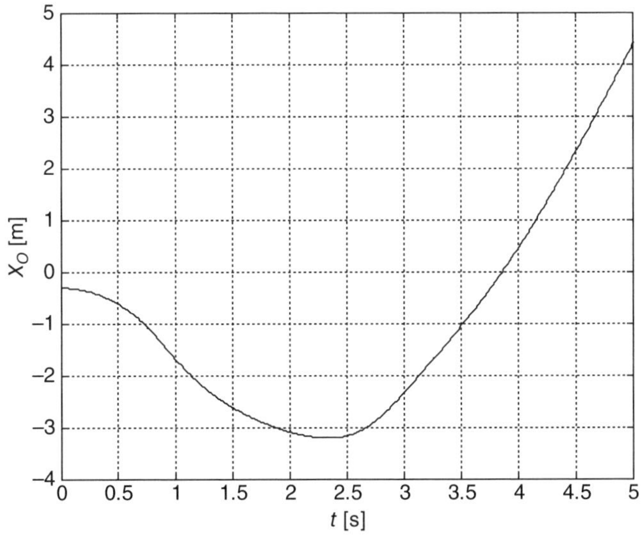

Figure 7.41 Time history $X_O=X_O(t)$ for $0\le t\le 5\,\mathrm{s}$.

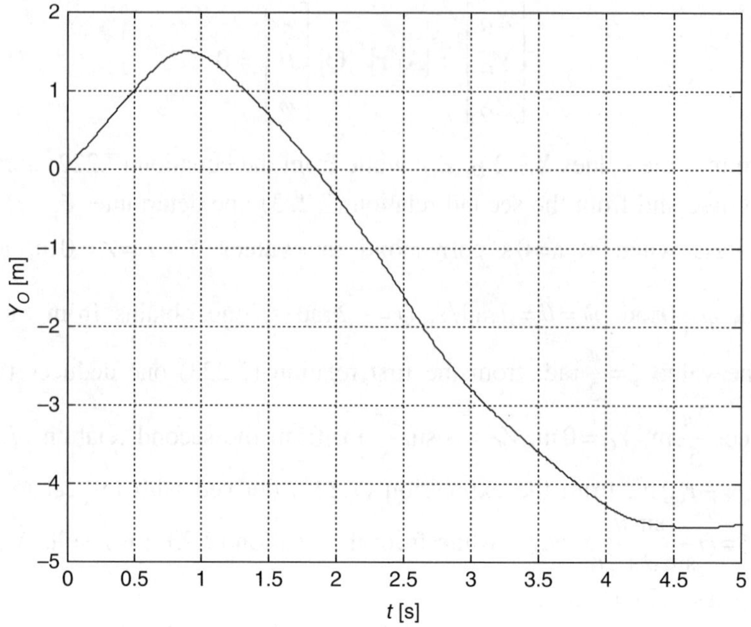

Figure 7.42 Time history $Y_O = Y_O(t)$ for $0 \le t \le 5$ s.

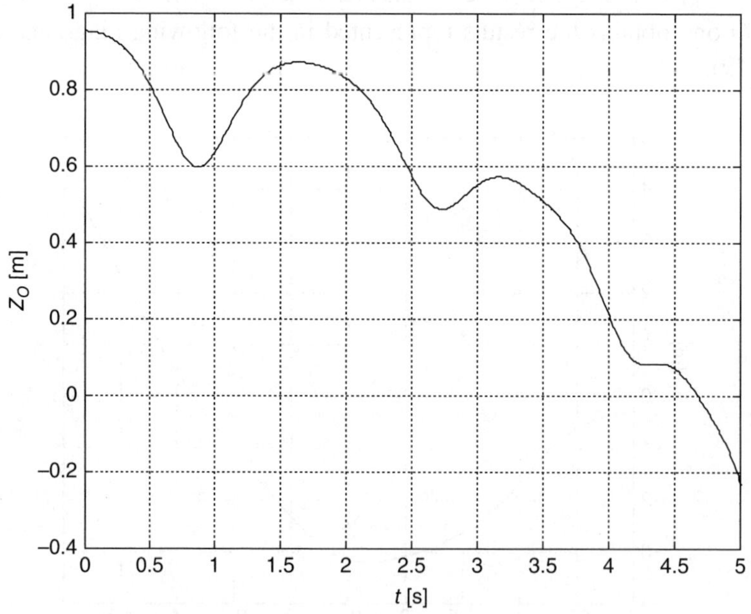

Figure 7.43 Time history $Z_O = Z_O(t)$ for $0 \le t \le 5$ s.

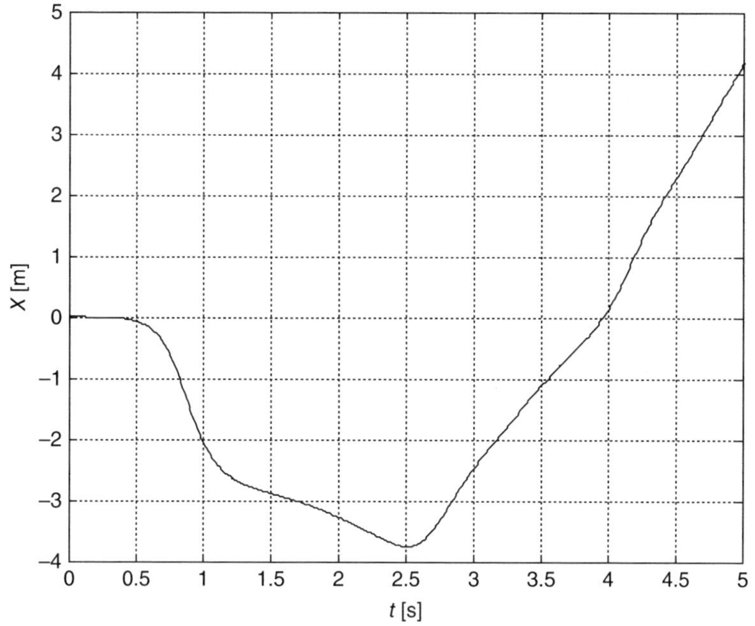

Figure 7.44 Time history $X = X(t)$ for $0 \le t \le 5$ s.

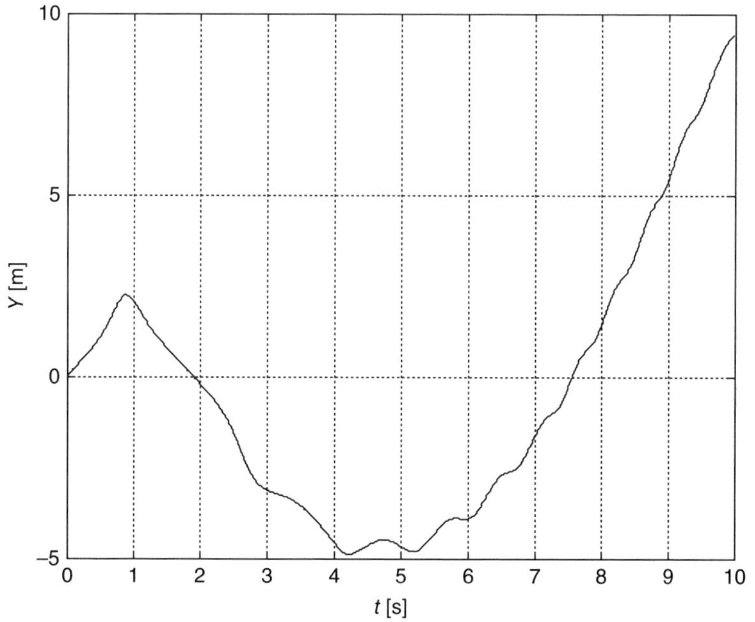

Figure 7.45 Time history $Y = Y(t)$ for $0 \le t \le 5$ s.

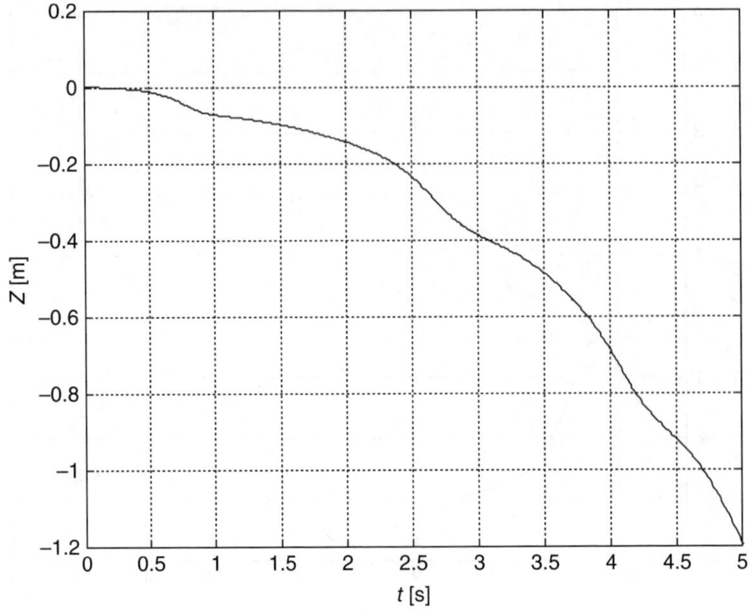

Figure 7.46 Time history $Z = Z(t)$ for $0 \le t \le 5$ s.

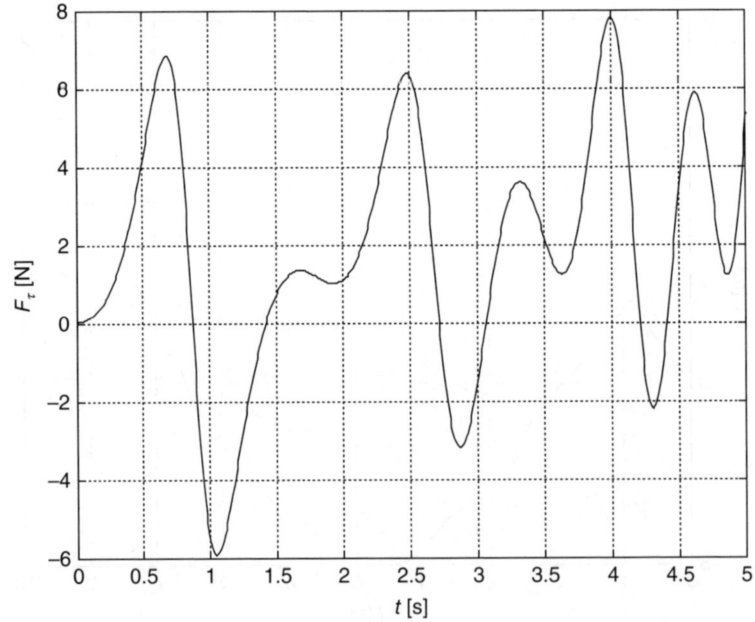

Figure 7.47 Time history $F_\tau = F_\tau(t)$ for $0 \le t \le 5$ s.

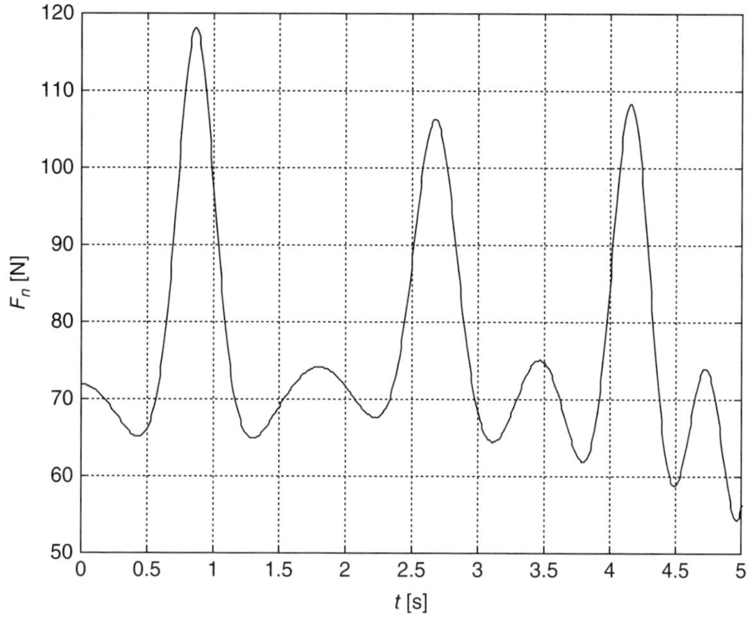

Figure 7.48 Time history $F_n = F_n(t)$ for $0 \le t \le 5$ s.

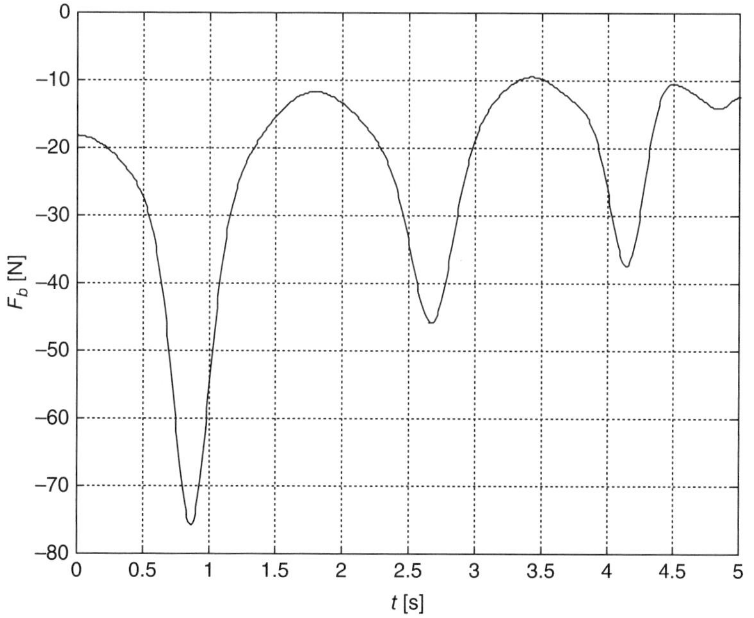

Figure 7.49 Time history $F_b = F_b(t)$ for $0 \le t \le 5$ s.

The diagrams suggest the absence of any periodic or quasi-periodic pattern of the studied motion. One observes that for the point O the coordinate X has an increasing tendency, while the coordinates Y and Z of the same point have a decreasing tendency. For the contact point the coordinates X and Y manifest an increasing tendency, while the coordinate Z has a decreasing tendency.

Concerning the forces that appear at the contact point we may observe an oscillating pattern, with maximums and minimums for which we can not deduce a mathematical expression of periodicity. The maximum force has a value of approximate 120 N, that is, 1.5 times greater than the own weight of the curve.

7.11 The Rolling without Sliding of Two Curves of a Rigid Solid on a Fixed Surface

7.11.1 General Aspects

We consider the case in which a rigid solid, with the local reference frame $Oxyz$ (Fig. 7.50) moves so that that curves (Γ_1), (Γ_2) of parametric equations

$$x=x(\xi_1), y=y(\xi_1), z=z(\xi_1), \tag{7.233}$$

$$x=x(\xi_2), y=y(\xi_2), z=z(\xi_2), \tag{7.234}$$

roll without sliding on the fixed surface (S) of equation

$$F(X,Y,Z)=0. \tag{7.235}$$

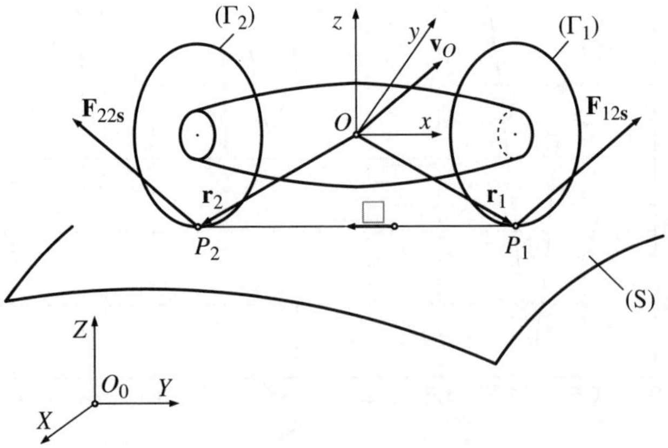

Figure 7.50 The rigid solid with two curves rolling without sliding on a fixed surface.

Since the contact points P_i, $i=1, 2$, have in the local reference system the coordinates $x_i=x(\xi_i)$, $y_i=y(\xi_i)$, $x_i=x(\xi_i)$, $i=1, 2$, and in the reference system O_0XYZ the coordinates X_i, Y_i, Z_i, we may write the matrices

$$\{\mathbf{r}_i\}=\begin{bmatrix}x_i\\ y_i\\ z_i\end{bmatrix},\{\mathbf{R}_i\}=\begin{bmatrix}X_i\\ Y_i\\ Z_i\end{bmatrix},[\mathbf{r}_i]=\begin{bmatrix}0 & -z_i & y_i\\ z_i & 0 & -x_i\\ -y_i & x_i & 0\end{bmatrix}, i=1, 2. \tag{7.236}$$

The contact conditions become

$$\{\mathbf{R}_i\}=\{\mathbf{R}_O\}+[\mathbf{A}]\{\mathbf{r}_i\}, i=1, 2, \tag{7.237}$$

while the tangency conditions are

$$\{\mathbf{F}_{ip}\}^{\mathrm{T}}[\mathbf{A}]\{\mathbf{r}_{ip}\}=0, i=1, 2, \tag{7.238}$$

where

$$\{\mathbf{F}_{ip}\}=[F_{,X_i}\ \ F_{,Y_i}\ \ F_{,Z_i}]^{\mathrm{T}}. \tag{7.239}$$

The rolling without sliding implies that at the contact points P_1, P_2 the relative velocities vanish, that is

$$\{\dot{\mathbf{R}}_O\}+[\mathbf{A}][\mathbf{r}_1]^{\mathrm{T}}[\mathbf{Q}]\{\dot{\boldsymbol{\beta}}\}=\{\mathbf{0}\},\{\dot{\mathbf{R}}_O\}+[\mathbf{A}][\mathbf{r}_2]^{\mathrm{T}}[\mathbf{Q}]\{\dot{\boldsymbol{\beta}}\}=\{\mathbf{0}\}; \tag{7.240}$$

deriving the relations (7.237) with respect to time, one obtains the equalities

$$\{\dot{\mathbf{R}}_i\}=\dot{\xi}_i[\mathbf{A}]\{\mathbf{r}_{ip}\}, i=1, 2. \tag{7.241}$$

From the relations (7.240) by summation and then by multiplication at the left by $[\mathbf{A}]^{\mathrm{T}}$ one deduces the equality

$$\left[[\mathbf{r}_1]^{\mathrm{T}}-[\mathbf{r}_2]^{\mathrm{T}}\right][\mathbf{Q}]\{\dot{\boldsymbol{\beta}}\}=\{\mathbf{0}\}, \tag{7.242}$$

from which it results

$$[\mathbf{Q}]\{\dot{\boldsymbol{\beta}}\}=\zeta\{\{\mathbf{r}_1\}-\{\mathbf{r}_2\}\} \tag{7.243}$$

or

$$\{\boldsymbol{\omega}\}=\zeta\{\{\mathbf{r}_1\}-\{\mathbf{r}_2\}\}, \tag{7.244}$$

where the parameter ζ is a factor of proportionality.

In these conditions, from the first (or the second) equality (7.240) it results

$$\{\dot{\mathbf{R}}_O\} = \zeta[\mathbf{A}][\mathbf{r}_2]^{\mathrm{T}}\{\mathbf{r}_1\}. \tag{7.245}$$

The expressions (7.244) and (7.245) are equivalent to the vector relations

$$\boldsymbol{\omega} = \zeta(\mathbf{r}_1 - \mathbf{r}_2), \tag{7.246}$$

$$\mathbf{v}_O = \boldsymbol{\omega} \times \mathbf{r}_1 = \boldsymbol{\omega} \times \mathbf{r}_2, \tag{7.247}$$

where the velocity of the point O was denoted by $\mathbf{v}_O$.

It is obvious that these relations could be written from the very beginning because the straight line P_1P_2 is the instantaneous rotational axis of the rigid solid since the points P_1, P_2 have the velocities equal to zero.

The forces of constraints (reactions) $\overline{\mathbf{F}}^*_{1is}$, $\overline{\mathbf{F}}^*_{2is}$ (Fig. 7.50) reduce at the point O at the resultant force given by $\mathbf{F}^*_{2s} = \overline{\mathbf{F}}^*_{12s} + \overline{\mathbf{F}}^*_{22s}$ and at the resultant moment of expression $\mathbf{F}^*_{2\beta} = \mathbf{r}_1 \times \overline{\mathbf{F}}^*_{12s} + \mathbf{r}_2 \times \overline{\mathbf{F}}^*_{22s}$.

The fact that the forces $\mathbf{F}^*_{12s}$, $\overline{\mathbf{F}}^*_{22s}$ intersects the straight line P_1P_2 may be written by the relations

$$\mathbf{F}^*_{i2s}(\mathbf{r}_2 \times \mathbf{r}_1) + (\mathbf{r}_i \times \mathbf{F}^*_{i2s})(\mathbf{r}_1 - \mathbf{r}_2) = 0,\ i = 1,\ 2, \tag{7.248}$$

from which results the equality

$$\mathbf{F}^*_{2s}(\mathbf{r}_2 \times \mathbf{r}_1) + \mathbf{F}^*_{2\beta}(\mathbf{r}_1 - \mathbf{r}_2) = 0, \tag{7.249}$$

equivalent to

$$\mathbf{F}^*_{2s}\mathbf{v}_O + \mathbf{F}^*_{2\beta}\boldsymbol{\omega} = 0; \tag{7.250}$$

the last relation states that the work of the forces of constraints is equal to zero.

Considering the notations λ_1, λ_2, λ_3, and λ_4, λ_5, λ_6 for the projections of the vectors $\mathbf{F}^*_{2s}$ and $\mathbf{F}^*_{2\beta}$, respectively, onto the axes of the reference system $Oxyz$, that is, denoting by $\{\boldsymbol{\lambda}\}$ the column matrix

$$\{\boldsymbol{\lambda}\} = [\lambda_1\ \ \lambda_2\ \ \lambda_3\ \ \lambda_4\ \ \lambda_5\ \ \lambda_6]^{\mathrm{T}}, \tag{7.251}$$

from the relations (7.249) and (7.250) one obtains the equality

$$\left[\{\mathbf{r}_1\}^{\mathrm{T}}[\mathbf{r}_2]\ \ \{\mathbf{r}_1\}^{\mathrm{T}} - \{\mathbf{r}_2\}^{\mathrm{T}}\right]\{\boldsymbol{\lambda}\} = 0, \tag{7.252}$$

which shows that maximum five out of the six components λ_i, $i=\overline{1,6}$, are independent.

7.11.2 The Differential Equations of Motion

To be out to establish the differential equations of motion one uses the theorems of momentum and moment of momentum (3.61) and (3.69) which, for the case when the point O is also the center of weight, read as

$$[\mathbf{m}]\{\ddot{\mathbf{R}}_O\} = \{\mathbf{F}_s\} + \{\mathbf{F}_{2s}\}, \tag{7.253}$$

$$[\mathbf{J}]\{\dot{\boldsymbol{\omega}}\} + [\boldsymbol{\omega}][\mathbf{J}]\{\boldsymbol{\omega}\} = \{\mathbf{F}_\beta\} + \{\mathbf{F}_{2\beta}\}, \tag{7.254}$$

where:

- $\{\mathbf{F}_s\}$, $\{\mathbf{F}_{2s}\}$ are the column matrices of the components onto the axis O_0X, O_0Y, O_0Z of the resultant of the given forces and of the forces of constraints, respectively;
- $\{\mathbf{F}_\beta\}$, $\{\mathbf{F}_{2\beta}\}$ are the column matrices of the components onto the axes Ox, Oy, Oz of the resultant moments of the given forces and of the forces of constraints, respectively.

The equations (7.253) and (7.254), with the previous notations for the components of the resultant and resultant moment of the forces of constraints, may be put in the forms

$$[\mathbf{m}]\{\ddot{\mathbf{R}}_O\} = \{\mathbf{F}_s\} + [\mathbf{A}][\lambda_1\ \ \lambda_2\ \ \lambda_3]^{\mathrm{T}}, \tag{7.255}$$

$$[\mathbf{J}]\{\dot{\boldsymbol{\omega}}\} + [\boldsymbol{\omega}][\mathbf{J}]\{\boldsymbol{\omega}\} = \{\mathbf{F}_\beta\} + [\lambda_4\ \ \lambda_5\ \ \lambda_6]^{\mathrm{T}}. \tag{7.256}$$

By derivation of the relations (7.245) and (7.246) with respect to time it results

$$\{\ddot{\mathbf{R}}_O\} = \dot{\zeta}[\mathbf{A}][\mathbf{r}_2]\{\mathbf{r}_1\} + \zeta[\dot{\mathbf{A}}][\mathbf{r}_2]\{\mathbf{r}_1\} + \zeta[\mathbf{A}][\dot{\mathbf{r}}_2]\{\mathbf{r}_1\} + \zeta[\mathbf{A}][\mathbf{r}_2]\{\dot{\mathbf{r}}_1\}, \tag{7.257}$$

$$\{\dot{\boldsymbol{\omega}}\} = \dot{\zeta}\{\{\mathbf{r}_1\}\{\mathbf{r}_2\}\} + \zeta\{\{\dot{\mathbf{r}}_1\} - \{\dot{\mathbf{r}}_2\}\}. \tag{7.258}$$

Using the notations:

$$\{\widetilde{\mathbf{F}}_s\} = -\zeta[\mathbf{m}]\{[\dot{\mathbf{A}}][\mathbf{r}_2]\{\mathbf{r}_1\} + [\mathbf{A}][\dot{\mathbf{r}}_2]\{\mathbf{r}_1\} + [\mathbf{A}][\mathbf{r}_2]\{\dot{\mathbf{r}}_1\}\}, \tag{7.259}$$

$$\{\widetilde{\mathbf{F}}_\beta\} = -\zeta^2[[\mathbf{r}_1] - [\mathbf{r}_2]][\mathbf{J}]\{\{\mathbf{r}_1\} - \{\mathbf{r}_2\}\} - \zeta[\mathbf{J}]\{\{\dot{\mathbf{r}}_1\} - \{\dot{\mathbf{r}}_2\}\}, \tag{7.260}$$

$$\{\widetilde{\mathbf{F}}\} = \begin{bmatrix} \{\widetilde{\mathbf{F}}_s\} \\ \{\widetilde{\mathbf{F}}_\beta\} \end{bmatrix}, \{\mathbf{F}\} = \begin{bmatrix} \{\mathbf{F}_s\} \\ \{\mathbf{F}_\beta\} \end{bmatrix}, \tag{7.261}$$

$$[\mathbf{M}^*] = \begin{bmatrix} [\mathbf{m}][\mathbf{A}][\mathbf{r}_2]\{\mathbf{r}_1\} \\ [\mathbf{J}]\{\{\mathbf{r}_1\} - \{\mathbf{r}_2\}\} \end{bmatrix}, \tag{7.262}$$

$$[\mathbf{B}_1] = -\begin{bmatrix} [\mathbf{A}] & [\mathbf{0}] \\ [\mathbf{0}] & [\mathbf{I}] \end{bmatrix}, [\mathbf{B}_2] = \left[\{\mathbf{r}_1\}^{\mathrm{T}}[\mathbf{r}_2]^{\mathrm{T}} \quad \{\mathbf{r}_1\}^{\mathrm{T}} - \{\mathbf{r}_2\}^{\mathrm{T}}\right], \tag{7.263}$$

from the relations (7.252), (7.255) – (7.263) one obtains the matrix equation

$$\begin{bmatrix} [\mathbf{M}^*] & [\mathbf{B}_1] \\ [\mathbf{0}] & [\mathbf{B}_2] \end{bmatrix} \begin{bmatrix} \dot{\zeta} \\ \{\boldsymbol{\lambda}\} \end{bmatrix} = \begin{bmatrix} \{\mathbf{F}\} + \{\widetilde{\mathbf{F}}\} \\ 0 \end{bmatrix}, \tag{7.264}$$

from which result the unknowns $\{\boldsymbol{\lambda}\}$, $\dot{\zeta}$ and then, by integration, the new value of the parameter ζ; with the aid of this new value, from the expressions (7.243) and (7.245) one obtains, by integration, the new values of the matrices $\{\boldsymbol{\beta}\}$ and $\{\mathbf{R}_O\}$.

For the determination of the parameters ξ_i, $i = 1, 2$, one considers the tangency conditions

$$\{\mathbf{F}_{ip}\}^{\mathrm{T}}[\mathbf{A}]\{\mathbf{r}_{ip}\} = 0, i = 1, 2, \tag{7.265}$$

by derivation of which with respect to time one obtains the relations

$$\dot{\xi}_i = -\frac{\zeta\{\mathbf{F}_{ip}\}^{\mathrm{T}}[\dot{\mathbf{A}}]\{\mathbf{r}_{ip}\}}{\{\mathbf{F}_{ip}\}^{\mathrm{T}}[\mathbf{A}]\{\mathbf{r}_{is}\} + \{\mathbf{r}_{ip}\}^{\mathrm{T}}[\mathbf{A}]^{\mathrm{T}}[\mathbf{F}_{is}][\mathbf{A}]\{\mathbf{r}_{ip}\}}, i = 1, 2, \tag{7.266}$$

wherefrom, by integration, one determines the new values of the parameters ξ_i, $i = 1, 2$.

7.11.3 The Algorithm of Numerical Calculation

At the initial moment $t = 0$ one considers as known the parameters X_O, Y_O, Z_O, ψ, θ, φ, ξ_1, ξ_2, $\dot{X}_O$, $\dot{Y}_O$, $\dot{Z}_O$, $\dot{\psi}$, $\dot{\theta}$, $\dot{\varphi}$, $\dot{\xi}_1$, $\dot{\xi}_2$.

One successively determines:

- the matrices $[\boldsymbol{\psi}]$, $[\boldsymbol{\theta}]$, $[\boldsymbol{\varphi}]$, $[\mathbf{A}]$, $[\mathbf{A}_\psi]$, $[\mathbf{A}_\theta]$, $[\mathbf{A}_\varphi]$, $[\dot{\mathbf{A}}]$, $[\mathbf{Q}]$, $[\mathbf{Q}_\theta]$, $[\mathbf{Q}_\varphi]$, $[\dot{\mathbf{Q}}]$;
- the matrices $\{\mathbf{r}_i\}$, $\{\mathbf{r}_{ip}\}$, $\{\mathbf{r}_{is}\}$, $[\mathbf{r}_i]$, $[\mathbf{r}_{ip}]$, $i = 1, 2$;
- the matrices $\{\mathbf{F}_{ip}\}$, $[\mathbf{F}_{is}]$, $i = 1, 2$;
- the parameter ζ from the relation (7.243);
- the matrices $\{\widetilde{\mathbf{F}}_{\mathbf{s}}\}$, $\{\widetilde{\mathbf{F}}_{\boldsymbol{\beta}}\}$, $\{\widetilde{\mathbf{F}}\}$, $\{\mathbf{F}\}$, $[\mathbf{M}^*]$, $[\mathbf{B}_1]$, $[\mathbf{B}_2]$;

- the parameters λ_i, $i=\overline{1,6}$, and the new value of the parameter ζ;
- the new values of the matrices $\{\mathbf{R}_O\}$, $\{\boldsymbol{\beta}\}$;
- the new values of the parameters ξ_i, $i=1,2$,

and then one passes to the next step.

7.12 The Rolling without Sliding of an Axle with Wheels (Disks) with Angular Deviations on a Horizontal Plan

One considers the axle of length $2l$ (Fig. 7.51) and the wheels of radius r_0 symmetrically inclined with the angle α.

Choosing the reference systems in Fig. 7.51, one obtains for the two circles the equations

$$x_1=r_0\sin\xi_1, y_1=l-r_0\sin\alpha\cos\xi_1, z_1=-r_0\cos\alpha\cos\xi_1, \tag{7.267}$$

$$x_2=r_0\sin\xi_2, y_2=-l+r_0\sin\alpha\cos\xi_2, z_2=-r_0\cos\alpha\cos\xi_2, \tag{7.268}$$

and for the contact plan the equation

$$Z=0. \tag{7.269}$$

One writes the derivatives

$$\begin{aligned}&x_{ip}=r_0\cos\xi_i, y_{ip}=-(-1)^i r_0\sin\alpha\sin\xi_i, z_{ip}=r_0\cos\alpha\sin\xi_i, x_{is}=-r_0\sin\xi_i,\\&y_{is}=-(-1)^i r_0\sin\alpha\cos\xi_i, z_{is}=r_0\cos\alpha\cos\xi_i, i=1,2,\end{aligned} \tag{7.270}$$

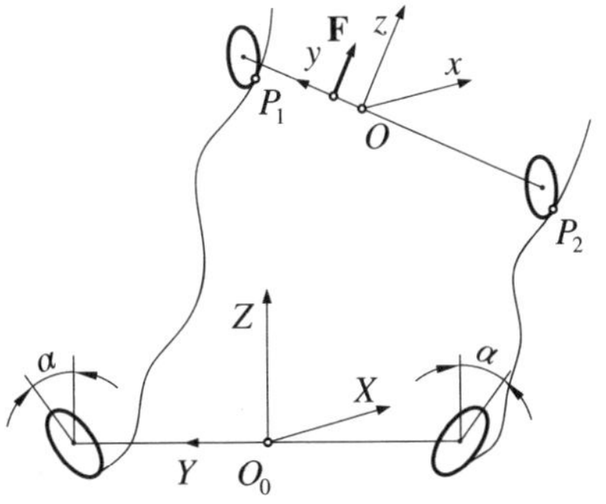

Figure 7.51 The axle with angular deviations rolling without sliding on the horizontal plan.

and calculates the matrices

$$\{\mathbf{r}_i\} = \begin{bmatrix} x_i \\ y_i \\ z_i \end{bmatrix}, \{\mathbf{r}_{ip}\} = \begin{bmatrix} x_{ip} \\ y_{ip} \\ z_{ip} \end{bmatrix}, \{\mathbf{r}_{is}\} = \begin{bmatrix} x_{is} \\ y_{is} \\ z_{is} \end{bmatrix}, i=1, 2, \tag{7.271}$$

$$[\mathbf{r}_i] = \begin{bmatrix} 0 & -z_i & y_i \\ z_i & 0 & -x_i \\ -y_i & x_i & 0 \end{bmatrix}, [\mathbf{r}_{ip}] = \begin{bmatrix} 0 & -z_{ip} & y_{ip} \\ z_{ip} & 0 & -x_{ip} \\ -y_{ip} & x_{ip} & 0 \end{bmatrix}, [\mathbf{r}_{is}] = \begin{bmatrix} 0 & -z_{is} & y_{is} \\ z_{is} & 0 & -x_{is} \\ -y_{is} & x_{is} & 0 \end{bmatrix}, i=1, 2, \tag{7.272}$$

$$\{\dot{\mathbf{r}}_i\} = \xi_i\{\mathbf{r}_{ip}\}, [\dot{\mathbf{r}}_i] = \xi_i[\mathbf{r}_{ip}], i=1, 2; \tag{7.273}$$

Further on, with the aid of the relations (7.259)–(7.263) one determines the matrices $\{\widetilde{\mathbf{F}}_\mathbf{s}\}$, $\{\widetilde{\mathbf{F}}_\boldsymbol{\beta}\}$, $\{\widetilde{\mathbf{F}}\}$, $[\mathbf{M}^*]$, $[\mathbf{B}_1]$, $[\mathbf{B}_2]$, the parameter ζ being given by the initial conditions.

By solving the equation (7.264), one determines the parameters λ_i, $i=\overline{1, 6}$, and the new value of the parameter ζ, while by the integration of the equations

$$[\mathbf{Q}]\{\dot{\boldsymbol{\beta}}\} = \zeta\{\{\mathbf{r}_1\}-\{\mathbf{r}_2\}\}, \{\dot{\mathbf{R}}_O\} = \zeta[\mathbf{A}][\mathbf{r}_2]^\mathrm{T}\{\mathbf{r}_1\} \tag{7.274}$$

one determines the new values of the matrices $\{\mathbf{R}_O\}$, $\{\boldsymbol{\beta}\}$.

Further on, one solves the equations (7.266) in which $\{\mathbf{F}_{ip}\} = [0 \;\; 0 \;\; 1]^\mathrm{T}$, $\{\mathbf{F}_{is}\} = [\mathbf{0}]$, and determines the new values of the parameters ξ_1, ξ_2.

In a numerical application one considers as known

- the inertial and geometric characteristics: $m=50\,\mathrm{kg}$, $J_x=16.5\,\mathrm{kgm}^2$, $J_y=3\,\mathrm{kgm}^2$, $J_z=17\,\mathrm{kgm}^2$, $J_{xy}=J_{yz}=J_{zx}=0\,\mathrm{kgm}^2$, $r_0=0.4\,\mathrm{m}$, $l=0.75\,\mathrm{m}$, $\alpha=10^0$;
- the forces: the own weight and a force $F=\begin{cases} F_0(1-t), & t\le 1\,\mathrm{s} \\ 0, & t>0 \end{cases}$, $F_0=100\,\mathrm{N}$, parallel to the axis O_0X and acting at the point E, $OE=0.1\,\mathrm{m}$;
- the initial conditions at $t=0\,\mathrm{s}$, $X=Y=0\,\mathrm{m}$, $Z=r_0\cos\alpha\,\mathrm{m}$, $Z=0.393923\,\mathrm{m}$, $\psi=\theta=\varphi=0\,\mathrm{rad}$, $\zeta=0\,\mathrm{m}^{-1}\mathrm{s}^{-1}$, $\xi_1=\xi_2=0\,\mathrm{rad}$, $\dot{X}_O=\dot{Y}_O=\dot{Z}_O=0\,\mathrm{m/s}$, $\dot{\psi}=\dot{\theta}=\dot{\varphi}=0\,\mathrm{m/s}$, $\dot{\xi}_1=\dot{\xi}_2=0\,\mathrm{rad/s}$.

The numerical results are represented in the following diagrams (Fig. 7.52 to Fig. 7.62). One easily observes that after a transitory regime of approximate 3.5 s, the parameters X_O, Y_O, Z_O (the coordinates of the center of axle), X_1, Y_1 (the coordinates of the contact point of the first wheel) and X_2, Y_2 (the coordinates of the contact

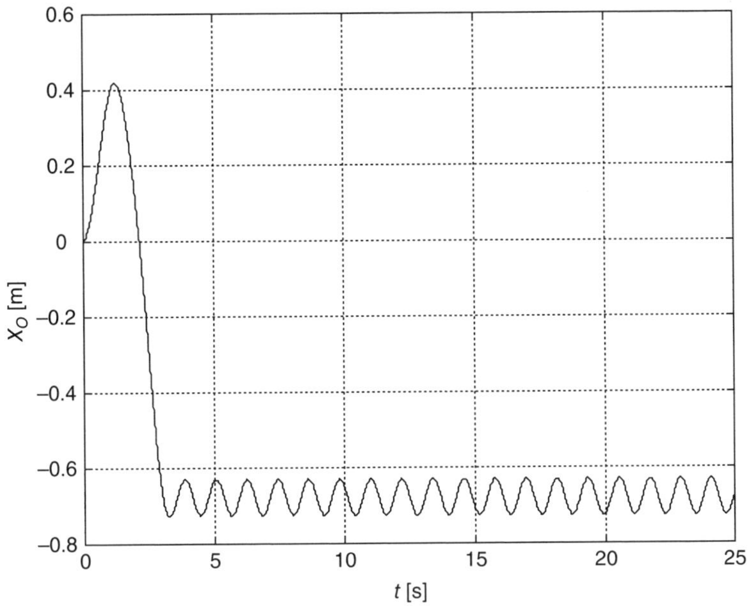

Figure 7.52 Time history $X_O = X_O(t)$ for $0 \le t \le 25$ s.

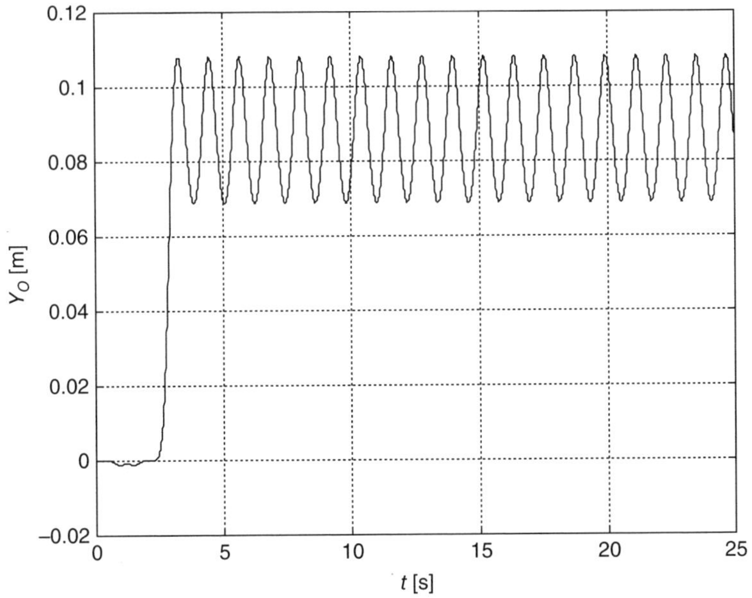

Figure 7.53 Time history $Y_O = Y_O(t)$ for $0 \le t \le 25$ s.

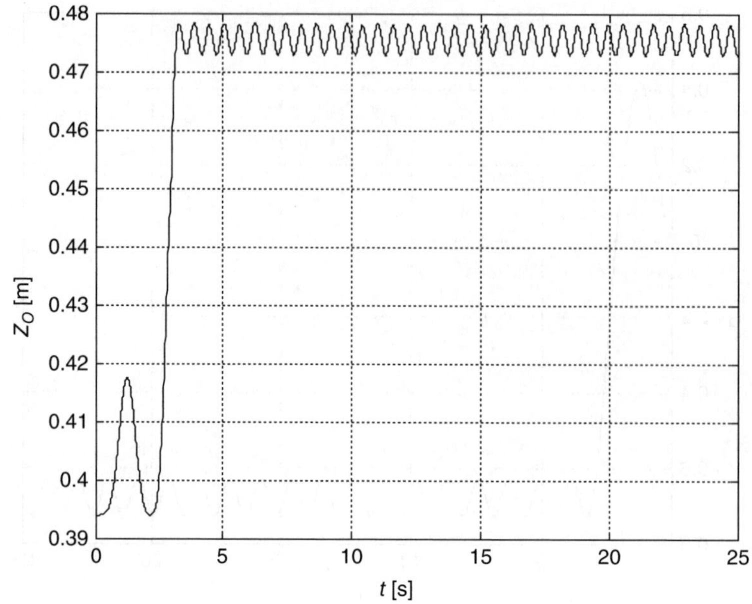

Figure 7.54 Time history $Z_O = Z_O(t)$ for $0 \le t \le 25$ s.

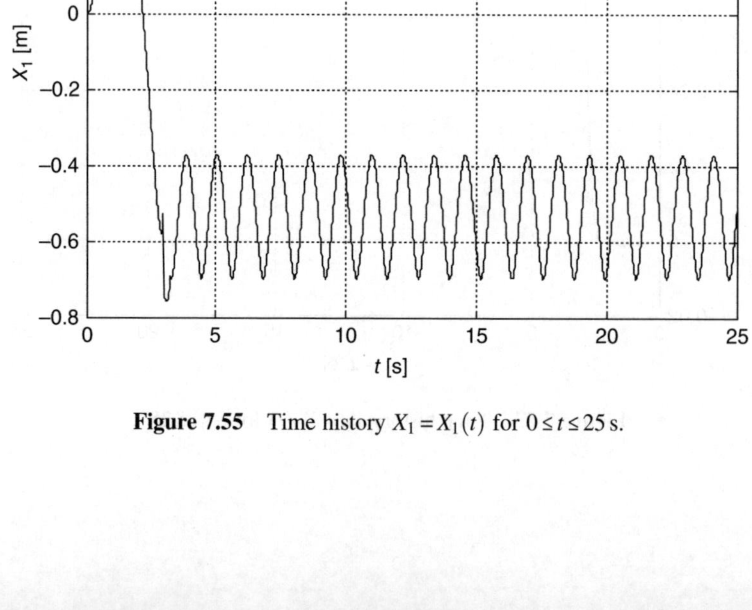

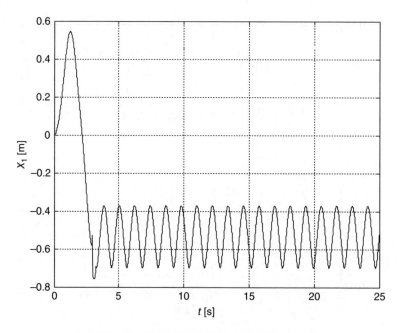

Figure 7.55 Time history $X_1 = X_1(t)$ for $0 \le t \le 25$ s.

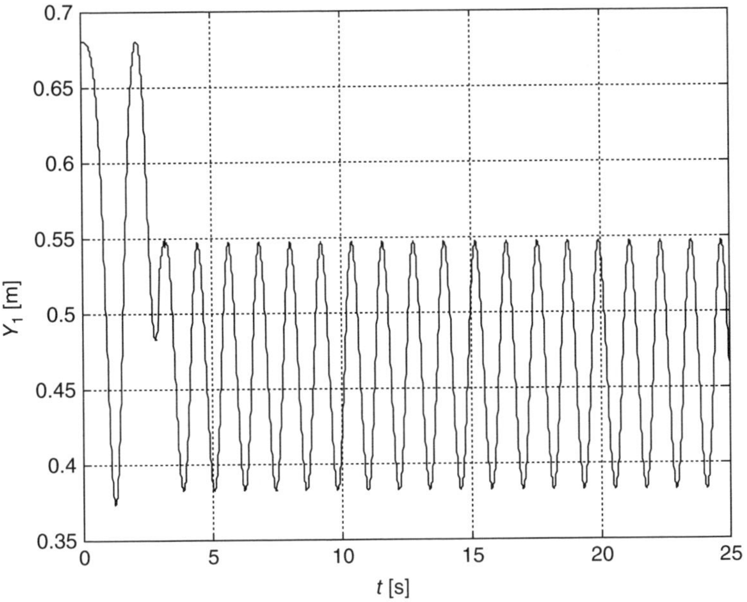

Figure 7.56 Time history $Y_1 = Y_1(t)$ for $0 \le t \le 25$ s.

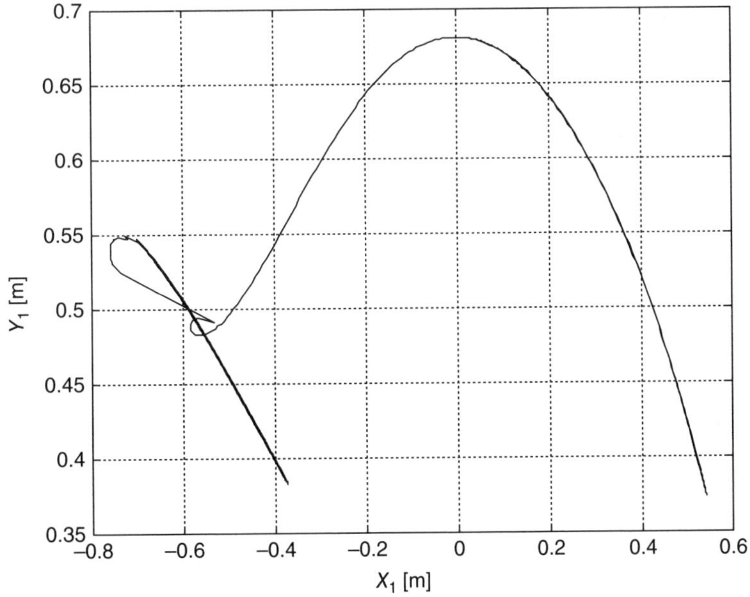

Figure 7.57 Trajectory of the first contact point in the first 25 seconds.

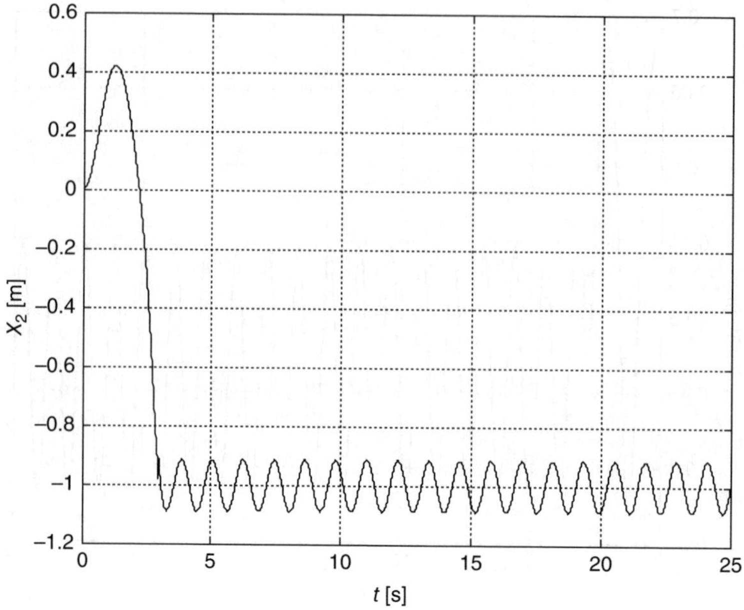

Figure 7.58 Time history $X_2 = X_2(t)$ for $0 \le t \le 25$ s.

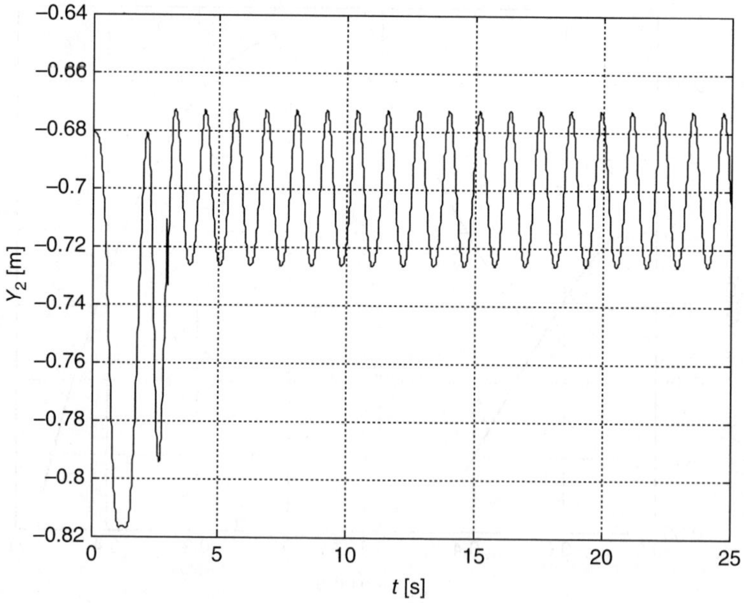

Figure 7.59 Time history $Y_2 = Y_2(t)$ for $0 \le t \le 25$ s.

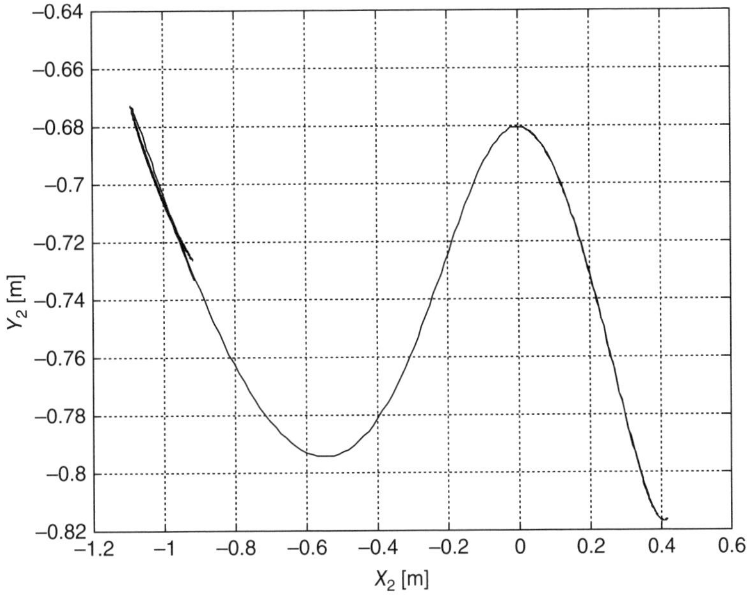

Figure 7.60 Trajectory of the second contact point in the first 25 seconds.

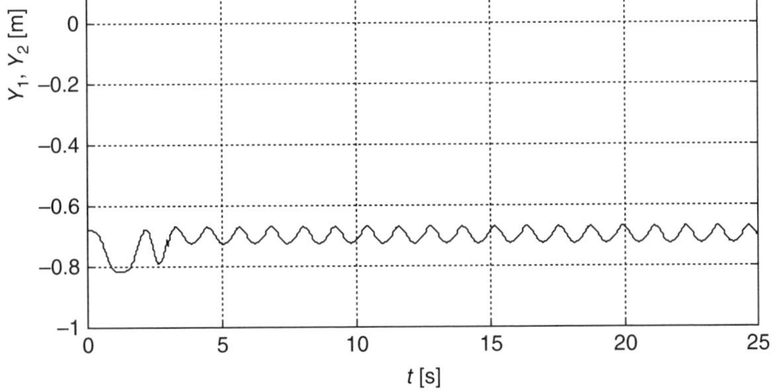

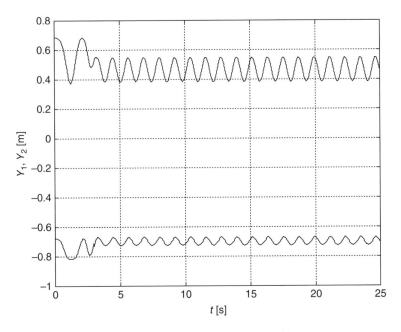

Figure 7.61 Time histories $Y_1 = Y_1(t)$ and $Y_2 = Y_2(t)$ for $0 \le t \le 25$ s.

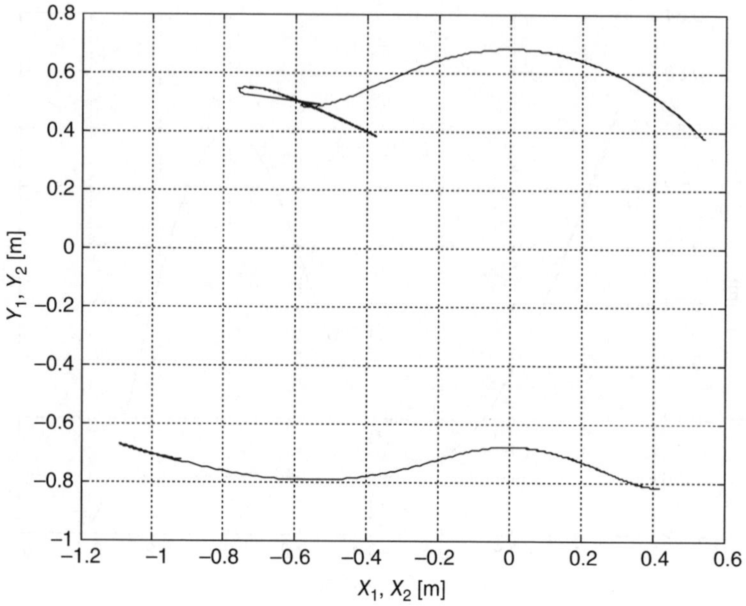

Figure 7.62 Trajectories of the two contact points in the first 25 seconds.

point of the second wheel) become periodic functions (but not of the same period; for all parameters the period is approximate 1.25 s, except for Z_O, for which the period is approximate 0.83 s). From Fig. 7.61 it results an oscillatory motion, the amplitudes of the contact points being not equal.

7.13 The Rolling without Sliding of an Axle with Disks on a Hyperbolic Paraboloid

7.13.1 General Aspects

We consider the axle of length $2l$ (Fig. 7.63) with two disks of radius r_0 and the hyperbolic surface on which the disks roll, of equation

$$kXY - Z = 0. \tag{7.275}$$

One chooses the local reference frame so that the axis Ox coincides to the axis O_1O_2 of the centers of disks, and the point O (the center of weight of the assemble) is situated at the middle of the segment O_1O_2.

The parametric equations of the circles that bound the disks read in the form

$$x_1 = -l, y_1 = r_0 \sin \xi_1,\ z_1 = -r_0 \cos \xi_1, \tag{7.276}$$

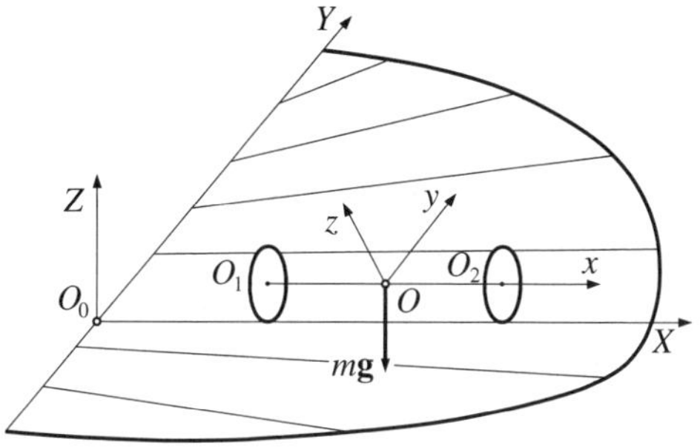

Figure 7.63 The axle rolling without sliding on a hyperbolic paraboloid.

$$x_2 = l, y_2 = r_0 \sin \xi_2, \ z_2 = -r_0 \cos \xi_2. \tag{7.277}$$

The contact points P_1, P_2 are defined in the local reference system $Oxyz$ by the column matrices

$$\{\mathbf{r}_1\} = \begin{bmatrix} -l \\ r_0 \sin \xi_1 \\ -r_0 \cos \xi_1 \end{bmatrix}, \{\mathbf{r}_2\} = \begin{bmatrix} l \\ r_0 \sin \xi_2 \\ -r_0 \cos \xi_2 \end{bmatrix}, \tag{7.278}$$

and in the general reference system O_0XYZ by the column matrices

$$\{\mathbf{R}_1\} = \begin{bmatrix} X_1 \\ Y_1 \\ kX_1Y_1 \end{bmatrix}, \{\mathbf{R}_2\} = \begin{bmatrix} X_2 \\ Y_2 \\ kX_2Y_2 \end{bmatrix}. \tag{7.279}$$

The contact and tangency conditions at the points P_1, P_2 lead to the equations

$$\{\mathbf{R}_i\} = \{\mathbf{R}_O\} + [\mathbf{A}]\{\mathbf{r}_i\}, i = 1, 2, \tag{7.280}$$

$$\{\mathbf{F}_{ip}\}^{\mathrm{T}}[\mathbf{A}]\{\mathbf{r}_{ip}\} = 0, i = 1, 2, \tag{7.281}$$

where

$$\{\mathbf{R}_O\} = [X_O \ \ Y_O \ \ Z_O]^{\mathrm{T}}, \tag{7.282}$$

$$\{\mathbf{r}_{ip}\} = [0 \ \ r_0 \cos \xi_i r_0 \sin \xi_i]^{\mathrm{T}}, \{\mathbf{F}_{ip}\} = [kY_i \ \ kX_i \ \ -1]^{\mathrm{T}}, i = 1, 2. \tag{7.283}$$

7.13.2 The Initial Position

If one chooses the initial position of the point P, on the hyperbolic paraboloid and on the disk, then it means that the parameters X_1, Y_1, ξ_1 are known and if one chooses the coordinate Y_2 too, then the system formed by the relations (7.277) and (7.280) has eight equations with the unknowns X_O, Y_O, Z_O, ψ, θ, φ, X_2, ξ_2. If one eliminates, by subtraction, the parameters X_O, Y_O, Z_O from the relation (7.280), then one obtains the following system of five equations with five unknowns

$$\{\mathbf{f}\} = \begin{bmatrix} [\mathbf{A}]\{\{\mathbf{r}_1\}-\{\mathbf{r}_2\}\}-\{\mathbf{R}_1\}+\{\mathbf{R}_2\} \\ \{\mathbf{F}_{1p}\}^{\mathrm{T}}[\mathbf{A}]\{\mathbf{r}_{1p}\} \\ \{\mathbf{F}_{2p}\}^{\mathrm{T}}[\mathbf{A}]\{\mathbf{r}_{2p}\} \end{bmatrix} = \{\mathbf{0}\}. \tag{7.284}$$

Using the notations

$$\{\mathbf{F}_{2s}\} = [0\ \ k\ \ 0]^{\mathrm{T}}, \{\mathbf{r}_{2s}\} = [0\ \ -r_0 \sin\xi_2 r_0 \cos\xi_2]^{\mathrm{T}}, \{\mathbf{R}_{2s}\} = [1\ \ 0\ \ kY_2]^{\mathrm{T}}, \tag{7.285}$$

$$\{\mathbf{f}_\psi\} = \begin{bmatrix} [\mathbf{A}_\psi]\{\{\mathbf{r}_1\}-\{\mathbf{r}_2\}\} \\ \{\mathbf{F}_{1p}\}^{\mathrm{T}}[\mathbf{A}_\psi]\{\mathbf{r}_{1p}\} \\ \{\mathbf{F}_{2p}\}^{\mathrm{T}}[\mathbf{A}_\psi]\{\mathbf{r}_{2p}\} \end{bmatrix}, \{\mathbf{f}_\theta\} = \begin{bmatrix} [\mathbf{A}_\theta]\{\{\mathbf{r}_1\}-\{\mathbf{r}_2\}\} \\ \{\mathbf{F}_{1p}\}^{\mathrm{T}}[\mathbf{A}_\theta]\{\mathbf{r}_{1p}\} \\ \{\mathbf{F}_{2p}\}^{\mathrm{T}}[\mathbf{A}_\theta]\{\mathbf{r}_{2p}\} \end{bmatrix}, \{\mathbf{f}_\varphi\} = \begin{bmatrix} [\mathbf{A}_\varphi]\{\{\mathbf{r}_1\}-\{\mathbf{r}_2\}\} \\ \{\mathbf{F}_{1p}\}^{\mathrm{T}}[\mathbf{A}_\varphi]\{\mathbf{r}_{1p}\} \\ \{\mathbf{F}_{2p}\}^{\mathrm{T}}[\mathbf{A}_\varphi]\{\mathbf{r}_{2p}\} \end{bmatrix},$$

$$\{\mathbf{f}_{X_2}\} = \begin{bmatrix} \{\mathbf{R}_{2s}\} \\ 0 \\ [\mathbf{F}_{2s}][\mathbf{A}]\{\mathbf{r}_{2p}\} \end{bmatrix}, \{\mathbf{f}_{\xi_2}\} = \begin{bmatrix} -[\mathbf{A}]\{\mathbf{r}_{2p}\} \\ 0 \\ \{\mathbf{F}_{2s}\}^{T}[\mathbf{A}]\{\mathbf{r}_{2s}\} \end{bmatrix}, \tag{7.286}$$

$$\{\mathbf{q}\} = [\psi\ \ \theta\ \ \varphi\ \ X_2\ \ \xi_2]^{\mathrm{T}}, \tag{7.287}$$

$$[\mathbf{J}] = [\{\mathbf{f}_\psi\}\ \ \{\mathbf{f}_\theta\}\ \ \{\mathbf{f}_\varphi\}\ \ \{\mathbf{f}_{X_2}\}\ \ \{\mathbf{f}_{\xi_2}\}]^{\mathrm{T}}, \tag{7.288}$$

one obtains the system

$$[\mathbf{J}]\{\Delta\mathbf{q}\} + \{\mathbf{f}\} = \{\mathbf{0}\} \tag{7.289}$$

for the deviation $\{\Delta\mathbf{q}\}$ relative to the considered approximate value.

7.13.3 The Differential Equations

Using the notations (7.259)–(7.261), with the initial values previously determined, it results from the system (7.264)

$$\dot{\zeta}=\frac{\{\mathbf{r}_1\}^{\mathrm{T}}[\mathbf{r}_2]^{\mathrm{T}}[\mathbf{A}]\left\{\{\mathbf{F}_s\}+\left\{\widetilde{\mathbf{F}}_s\right\}\right\}+\{\{\mathbf{r}_1\}-\{\mathbf{r}_2\}^{\mathrm{T}}\}\left\{\{\mathbf{F}_\beta\}+\left\{\widetilde{\mathbf{F}}_\beta\right\}\right\}}{m\left\{\{\mathbf{r}_1\}^{\mathrm{T}}[\mathbf{r}_2]^{\mathrm{T}}[\mathbf{r}_2]\{\mathbf{r}_1\}\right\}+\{\{\mathbf{r}_1\}-\{\mathbf{r}_2\}\}^{\mathrm{T}}[\mathbf{J}]\{\{\mathbf{r}_1\}-\{\mathbf{r}_2\}\}}, \tag{7.290}$$

$$\{\boldsymbol{\lambda}\}=\dot{\zeta}\begin{bmatrix}[\mathbf{m}][\mathbf{r}_2]\{\mathbf{r}_1\}\\ [\mathbf{J}]\{\{\mathbf{r}_1\}-\{\mathbf{r}_2\}\}\end{bmatrix}-\begin{bmatrix}[\mathbf{A}]^{\mathrm{T}}\left\{\{\mathbf{F}_\beta\}+\left\{\widetilde{\mathbf{F}}_\beta\right\}\right\}\\ \{\mathbf{F}_\beta\}+\left\{\widetilde{\mathbf{F}}_\beta\right\}\end{bmatrix}, \tag{7.291}$$

while from the equations

$$\begin{bmatrix}\dot{\psi}\\ \dot{\theta}\\ \dot{\varphi}\end{bmatrix}=\zeta[\mathbf{Q}]^{-1}\{\{\mathbf{r}_1\}-\{\mathbf{r}_2\}\},\ \begin{bmatrix}\dot{X}_O\\ \dot{Y}_O\\ \dot{Z}_O\end{bmatrix}=[\mathbf{A}][\mathbf{r}_2]\{\mathbf{r}_1\} \tag{7.292}$$

one determines the new values of the parameters ψ, θ, φ, X_O, Y_O, Z_O.

Further on, from the equations (7.266) one obtains the new values of the parameters ξ_i, $i=1, 2$.

In a numerical application one knows:

- the inertial and geometric characteristics: $m=50\,\mathrm{kg}$, $J_x=13\,\mathrm{kgm}^2$, $J_y=16.5\,\mathrm{kgm}^2$, $J_z=17\,\mathrm{kgm}^2$, $J_{xy}=J_{yz}=J_{zx}=0\,\mathrm{kgm}^2$, $r_0=0.4\,\mathrm{m}$, $l=0.75\,\mathrm{m}$, $k=0.01$;
- the forces: only the own weight, that is, $\{\mathbf{F}_s\}=[0\ \ 0\ \ -mg]^{\mathrm{T}}$, $\{\mathbf{F}_\beta\}=\{\mathbf{0}\}$;
- the initial position: one chooses the values $X_1=1\,\mathrm{m}$, $\psi=0\,\mathrm{rad}$, $\theta=0\,\mathrm{rad}$, $\varphi=0\,\mathrm{rad}$, $\dot{X}_O=0\,\mathrm{m/s}$, $\dot{Y}_O=0\,\mathrm{m/s}$, $\dot{Z}_O=0\,\mathrm{m/s}$, $\dot{\psi}=0\,\mathrm{rad/s}$, $\dot{\theta}=0\,\mathrm{rad/s}$, $\dot{\varphi}=0$ rad/s, $\dot{\xi}_1=0\,\mathrm{rad/s}$, $\dot{\xi}_2=0\,\mathrm{rad/s}$, $\dot{\zeta}=0\,\mathrm{m^{-1}s^{-2}}$ and it results $X_O=1.75\,\mathrm{m}$, $Y_O=-6.998732\cdot10^{-3}\,\mathrm{m}$, $Z_O=3.9995\cdot10^{-1}\,\mathrm{m}$, $\zeta=0\,\mathrm{m^{-1}s^{-1}}$, $\xi_1=9.999667\cdot10^{-3}\,\mathrm{rad}$, $\xi_2=2.499479\cdot10^{-2}\,\mathrm{rad}$, $Y_1=-2.998932\cdot10^{-3}\,\mathrm{m}$, $Z_1=-2.998932\cdot10^{-5}\approx0\,\mathrm{m}$, $X_2=2.5\,\mathrm{m}$, $Y_2=-2.998932\cdot10^{-3}\,\mathrm{m}$, $Z_2=7.495362\cdot10^{-5}\approx0\,\mathrm{m}$.

One obtains the diagrams captured in the Fig. 7.64 to Fig. 7.76.

From these diagrams one may deduce a certain pattern of periodicity for the shape of the trajectories both for the center of axle (point O), and for the trajectories of the two contact points. There also exists a synchronization for these three trajectories, that is, the same moments of time for which appear extremes in the trajectories. Moreover, one may state that the motion has a jerky characteristic (there exist close moments of time at which the diagrams of the characteristic parameters have a maximum and a

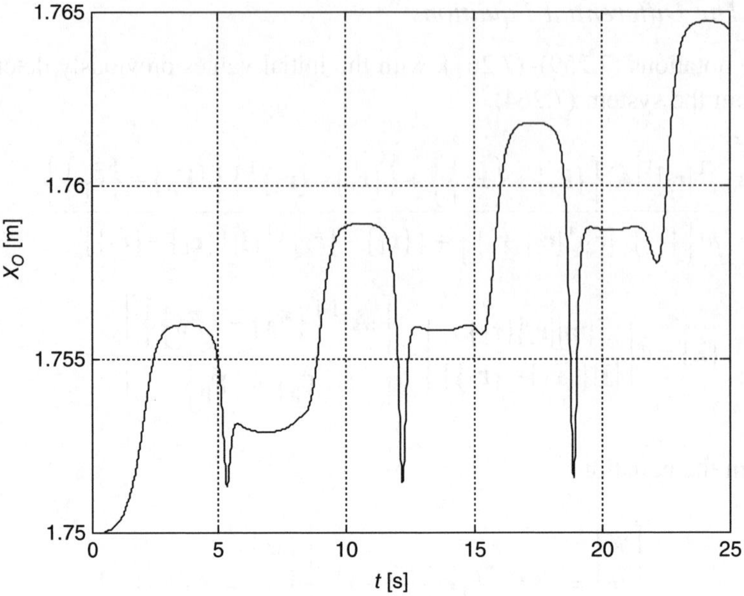

Figure 7.64 Time history $X_O = X_O(t)$ for $0 \le t \le 25$ s.

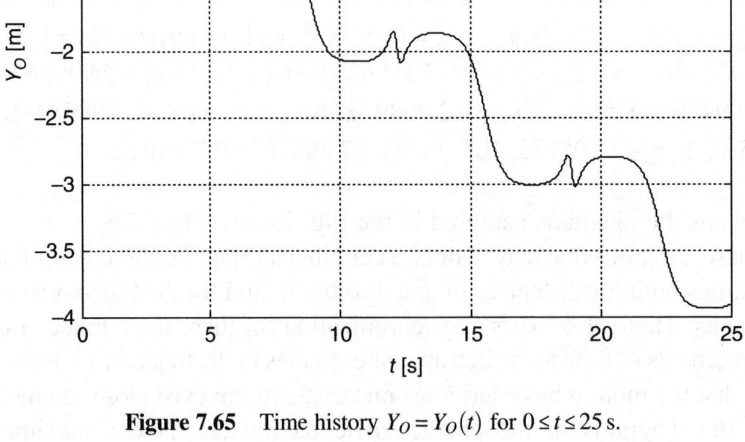

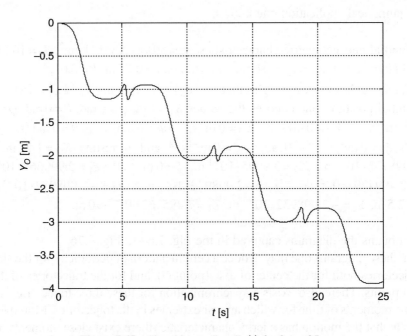

Figure 7.65 Time history $Y_O = Y_O(t)$ for $0 \le t \le 25$ s.

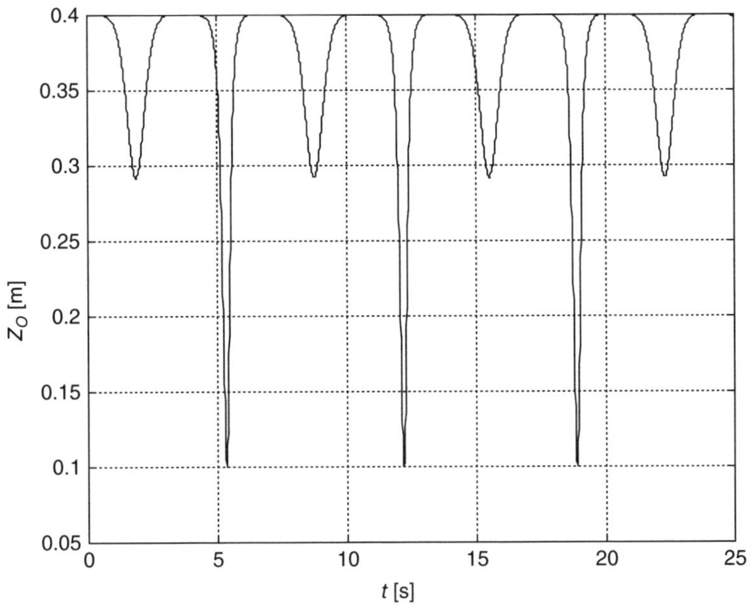

Figure 7.66 Time history $Z_O = Z_O(t)$ for $0 \le t \le 25$ s.

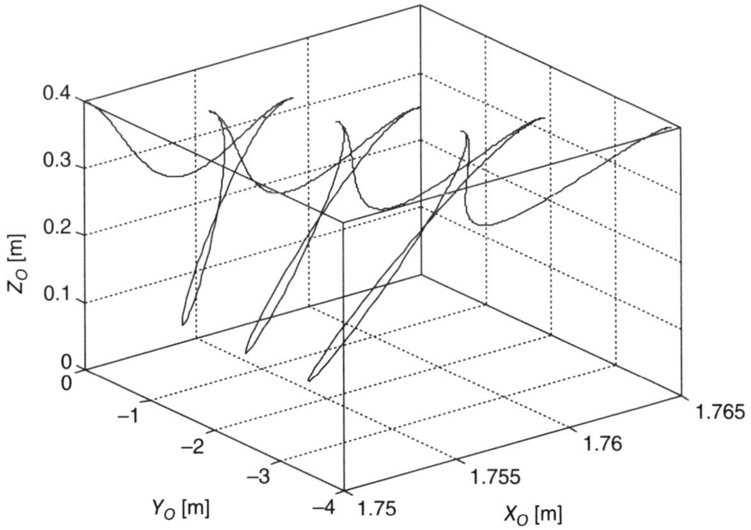

Figure 7.67 Trajectory of the center of the axle (point O) for the first 25 seconds.

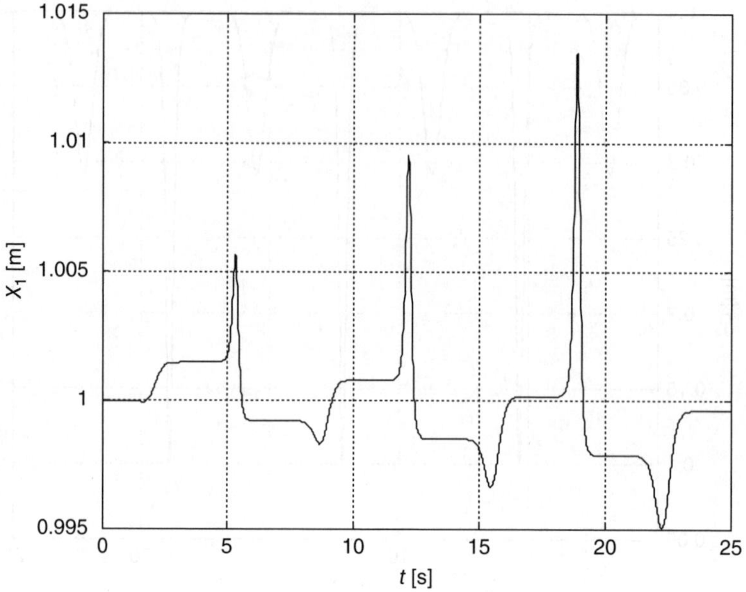

Figure 7.68 Time history $X_1 = X_1(t)$ for $0 \le t \le 25$ s.

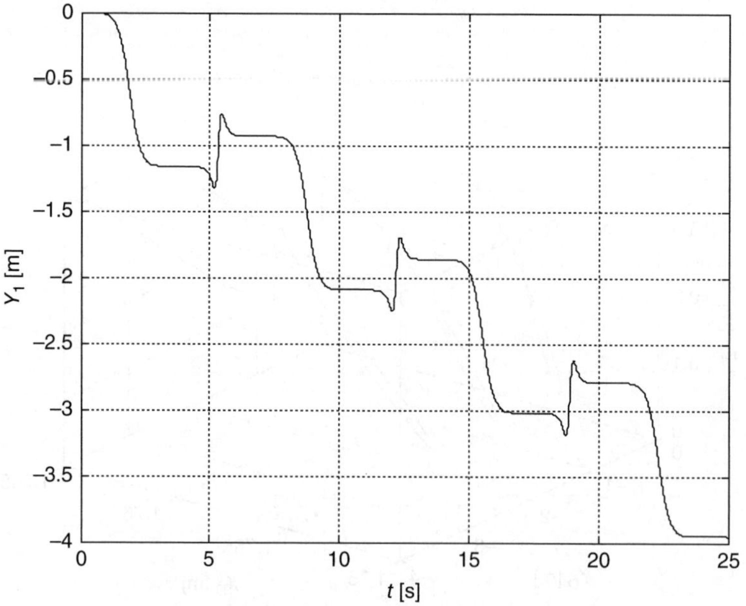

Figure 7.69 Time history $Y_1 = Y_1(t)$ for $0 \le t \le 25$ s.

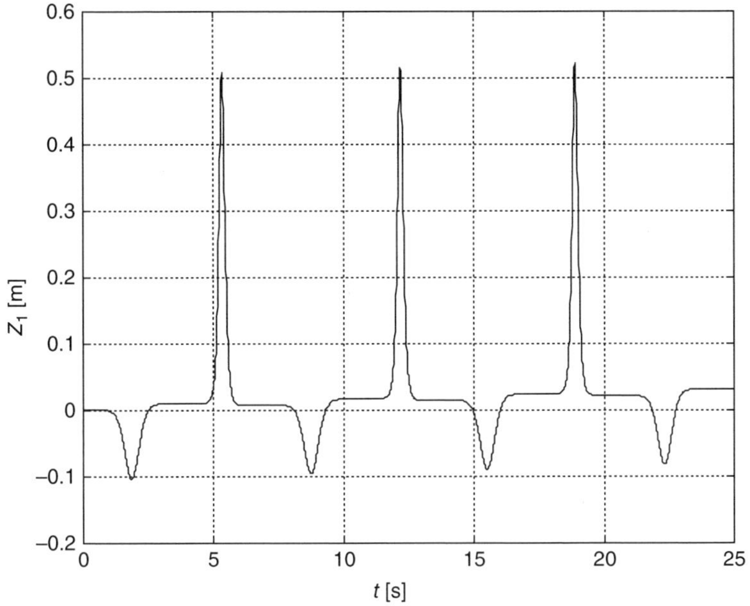

Figure 7.70 Time history $Z_1 = Z_1(t)$ for $0 \le t \le 25$ s.

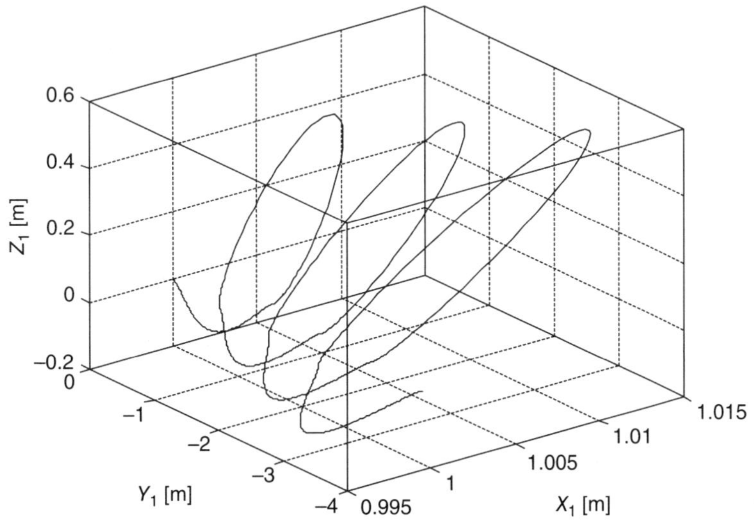

Figure 7.71 Trajectory of the first contact point for the first 25 seconds.

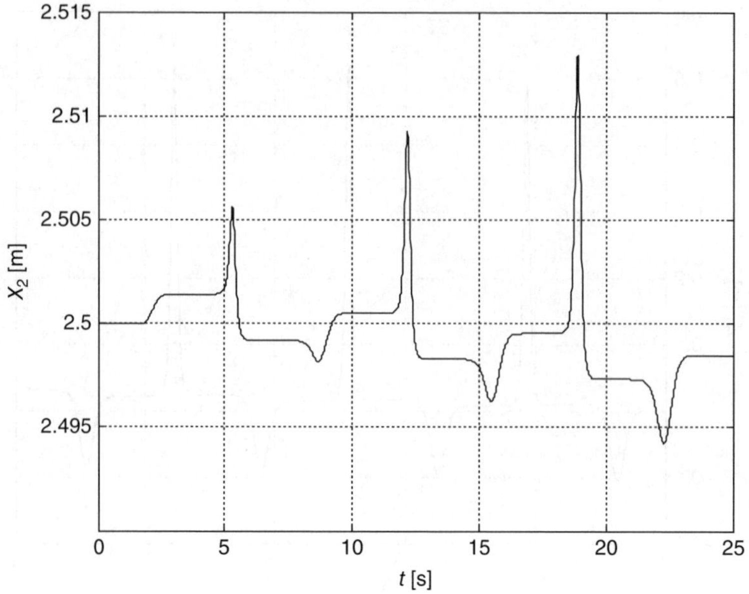

Figure 7.72 Time history $X_2 = X_2(t)$ for $0 \le t \le 25$ s.

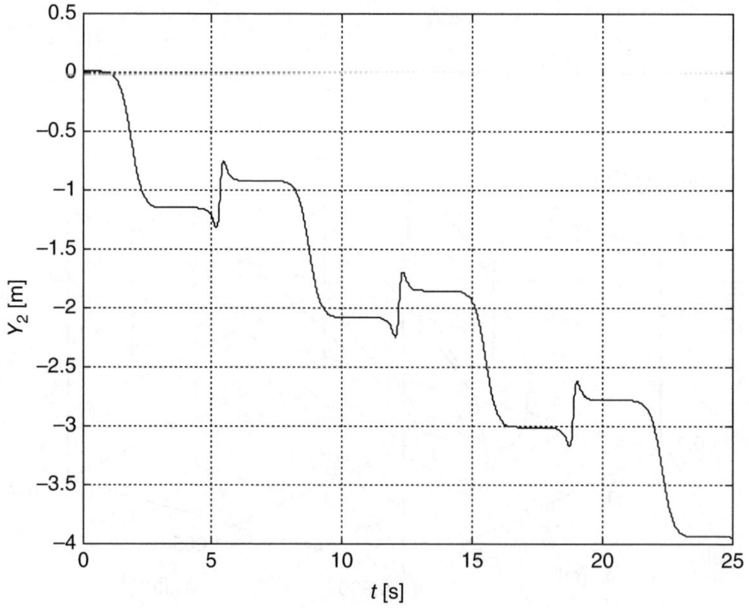

Figure 7.73 Time history $Y_2 = Y_2(t)$ for $0 \le t \le 25$ s.

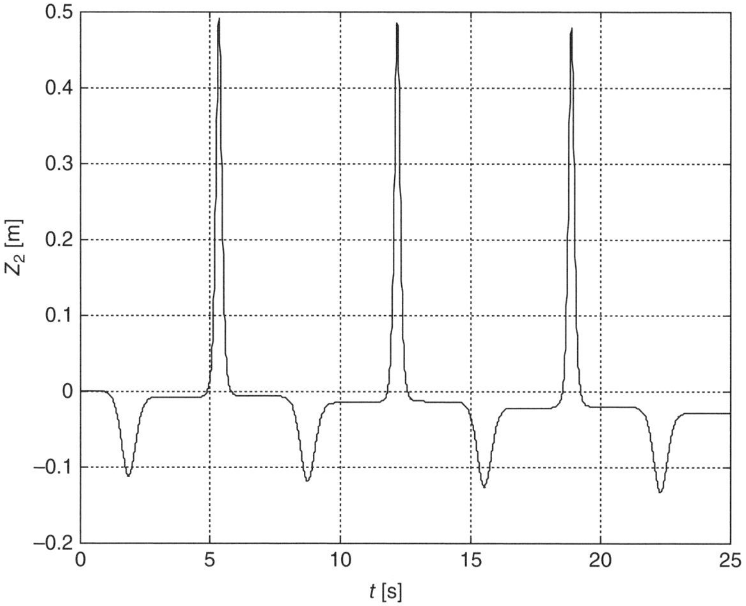

Figure 7.74 Time history $Z_2 = Z_2(t)$ for $0 \le t \le 25$ s.

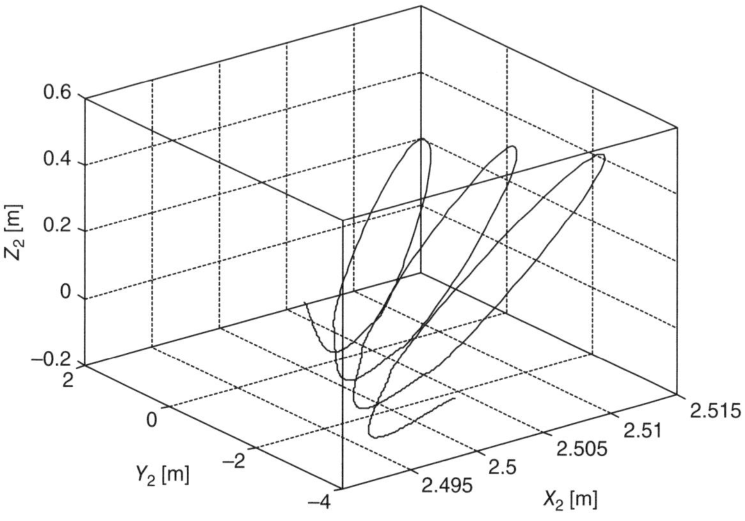

Figure 7.75 Trajectory of the second contact point for the first 25 seconds.

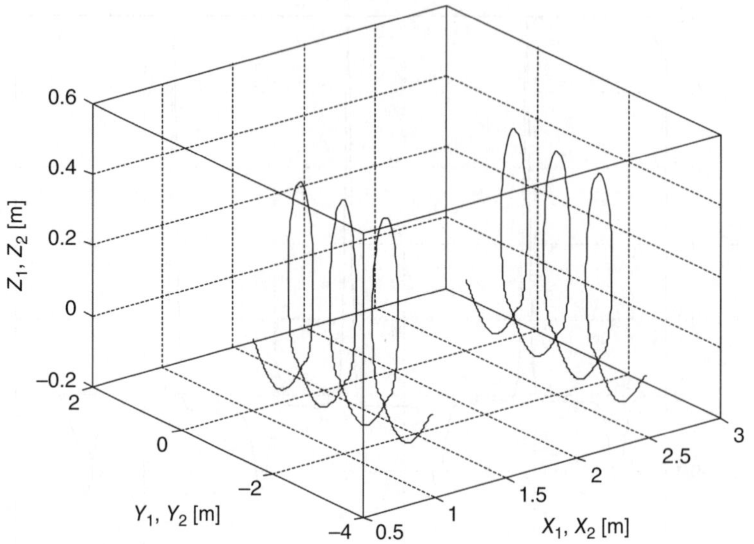

Figure 7.76 Trajectories of the two contact points for the first 25 seconds.

minimum, e.g. 5.5 s, 12 s etc.). It also results the existence of a quasi-period of the motion of approximate 6.5 s.

Further Reading

Amirouche FML (1992). Computational Methods in Multibody Dynamics. Englewood Cliffs: Prentice Hall.

Arnold VI (1997). Mathematical Methods of Classical Mechanics (Graduate Texts in Mathematics, Vol. 60). 2nd ed. New York: Springer.

Bloch AM, Baillieul J, Crouch P, Marsden J (2007). Nonholonomic Mechanics and Control (Interdisciplinary Applied Mathematics). Berlin: Springer.

Blundell M, Harty D (2004). The Multibody Systems Approach to Vehicle Dynamics. Amsterdam: Elsevier Butterworth-Heinemann.

Cortes J (2002). Geometric, Control and Numerical Aspects of Nonholonomic Systems. Berlin: Springer.

Coutinho MG (2001). Dynamic Simulations of Multibody Systems. New York: Springer.

de Jalón JG, Bayo E (2011). Kinematic and Dynamic Simulation of Multibody Systems: The Real-Time Challenge. Berlin: Springer Verlag.

Douglas Gregory R (2006). Classical Mechanics. Cambridge: Cambridge University Press.

Edelen DG (2010). Lagrangian Mechanics of Nonconservative Nonholonomic Systems. Leyden: Kluwer.

Eich-Soellner E, Führer C (2013). Numerical Methods in Multibody Dynamics. Wiesbaden: Springer.

Fasano A, Marmi S, Pelloni B (2006). Analytical Mechanics: An Introduction. Oxford: Oxford University Press.

Fearstone R (2007). Rigid body Dynamics Algorithms. New York: Springer.

Gattringer H, Gerstmayr J (2013). Multibody Systems Dynamics, Robotics and Control. Wien: Springer.
Greenwood DT (2006). Advanced Dynamics. Cambridge: Cambridge University Press.
Hahn H (2002). Rigid Body Dynamics of Mechanisms 1: Theoretical Basis. Berlin: Springer.
Hahn H (2010). Rigid Body Dynamics of Mechanisms 2: Applications. Berlin: Springer.
Hand LN, Finch JD (1998). Analytical Mechanics. Cambridge: Cambridge University Press.
Lagrange JL (1996). Analytical Mechanics. Dordrecht: Kluwer Academic Publishers,
Lurie AI (2002). Analytical Mechanics. Berlin: Springer.
Neimark IuI, Fufaev NA (1972). Dynamics of Nonholonomic Systems (Translations of Mathematical Monographs, V. 33). Providence, Rhode Island: American Mathematical Society.
Nikravesh PE (2007). Planar Multibody Dynamics: Formulation, Programming and Applications. Boca Raton: CRC Press.
Pandrea N, Stănescu ND (2002). Mecanica. Bucureşti: Editura Didactică şi Pedagogică (in Romanian).
Papastavridis JG (2002). Analytical Mechanics: A Comprehensive Treatise on the Dynamics of Constrained Systems; For Engineers, Physicists, and Mathematicians. Oxford: Oxford University Press.
Pfeiffer F, Glocker C (1996). Multibody Dynamics with Unilateral Contacts. New York: John Wiley & Sons, Inc.
Roberson RE, Schwertassek R (1988). Dynamics of Multibody Systems. Berlin: Springer Verlag.
Shabana AA (2008). Computational Dynamics. New York: John Wiley & Sons.
Shabana AA (2013). Dynamics of Multibody Systems. Cambridge: Cambridge University Press.
Soltakhanov ShKh, Yushkov MPP Zegzhda SA (2009). Mechanics of non-nolonomic systems: A New Class of control systems (Foundations of Engineering Mechanics). Berlin: Springer.
Stănescu ND (2013). Mecanica sistemelor. Bucureşti: Editura Didactică şi Pedagogică (inRomanian).
Stănescu ND, Munteanu L, Chiroiu V, Pandrea N (2007). Sisteme dinamice. Teorie şi aplicaţii vol. I. Bucureşti: Editura Academiei Române (in Romanian).
Stănescu ND, Munteanu L, Chiroiu V, Pandrea N (2011). Sisteme dinamice. Teorie şi aplicaţii vol. II. Bucureşti: Editura Academiei Române (in Romanian).
Teodorescu PP (2009). Mechanical Systems, Classical Models: Volume 3: Analytical Mechanics (Mathematical and Analytical Techniques with Applications to Engineering). Dordrecht: Springer.
Teodorescu PP, Stănescu ND, Pandrea N (2013). Numerical Analysis with Applications in Mechanics and Engineering. Hoboken: Wiley.
Udwadia FE, Kalaba RE (2007). Analytical Dynamics: A New Approach. Cambridge: Cambridge University Press.
Wittenburg J (2007). Dynamics of Multibody Systems. Berlin: Springer.
Woodhouse N (2010). Introduction to Analytical Dynamics. London: Springer.

8

The Motion of the Rigid Solid with Constraints on the Bounding Surface

This chapter is dedicated to the motion of the rigid solid with constraints on the bounded surfaces; in this case the rigid body is supported at fixed points, or it rolls on curves or surfaces.

8.1 General Aspects: Classification

There exist certain motions of the rigid solid at which the contact with other physical elements is realized at some points situated on the bounding surface of the rigid solid.

In the general analysis, one may distinguish contacts of the following types:

- mobile surface supported at fixed points (Fig. 8.1),
- mobile surface supported on fixed curves (Fig. 8.2),
- mobile surface supported on fixed surface (Fig. 8.3).

Dynamics of the Rigid Solid with General Constraints by a Multibody Approach, First Edition.
Nicolae Pandrea and Nicolae-Doru Stănescu.

Companion website: www.wiley.com/go/pandrea

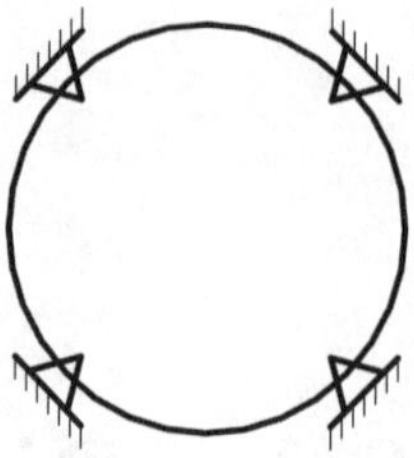

Figure 8.1 Mobile surface supported at fixed points.

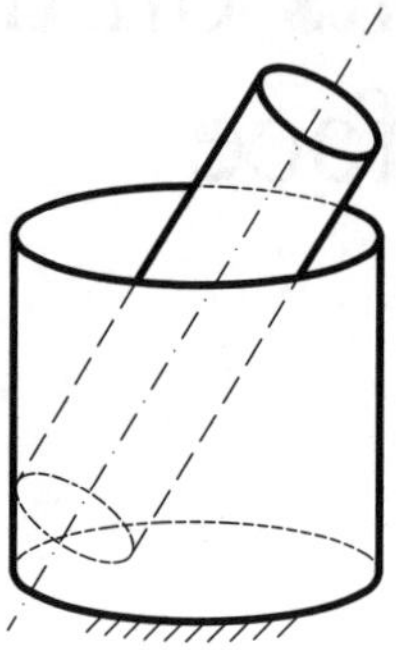

Figure 8.2 Contact mobile cylinder-fixed curve.

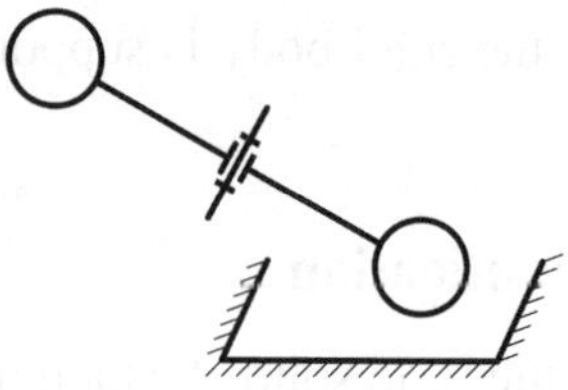

Figure 8.3 Contact mobile torus-fixed plan.

8.2 The Rigid Solid Supported at Fixed Points

8.2.1 The Matrix of Constraints

We consider that the fixed points have the coordinates X_i, Y_i, Z_i, $i=\overline{1,n}$, relative to the general fixed reference system, while the equation of the bounding surface of the rigid solid, relative to the local reference system $Oxyz$ is

$$f(x,y,z)=0. \tag{8.1}$$

With the known notations

$$\{\mathbf{R}_i\}=[X_i \;\; Y_i \;\; Z_i]^{\mathrm{T}}, \{\mathbf{R}_O\}=[X_O \;\; Y_O \;\; Z_O]^{\mathrm{T}}, \{\mathbf{r}_i\}=[x_i \;\; y_i \;\; z_i]^{\mathrm{T}} \tag{8.2}$$

we may write the contact conditions at the points A_i

$$\{\mathbf{R}_i\}=\{\mathbf{R}_O\}+[\mathbf{A}]\{\mathbf{r}_i\}, \; i=\overline{1,n}, \tag{8.3}$$

wherefrom it results

$$\{\mathbf{r}_i\} = [\mathbf{A}]^{\mathrm{T}}\{\mathbf{R}_i\} - [\mathbf{A}]^{\mathrm{T}}\{\mathbf{R}_O\},\ i = \overline{1, n}. \tag{8.4}$$

In the conditions in which the matrices $[\mathbf{A}]$, $\{\mathbf{R}_O\}$, $\{\mathbf{R}_i\}$ are known, from the relation (8.4) one may determine the coordinates x_i, y_i, z_i of the points A_i relative to the local reference system. Further on, using these values, one may determine the matrices

$$[\mathbf{r}_i] = \begin{bmatrix} 0 & -z_i & y_i \\ z_i & 0 & -x_i \\ -y_i & x_i & 0 \end{bmatrix},\ i = \overline{1, n}, \tag{8.5}$$

and the first and second order partial derivatives

$$\{\mathbf{f}_{ip}\} = \begin{bmatrix} f,_{x_i} \\ f,_{y_i} \\ f,_{z_i} \end{bmatrix},\ [\mathbf{f}_{is}] = \begin{bmatrix} f,_{x^2} & f,_{xy} & f,_{xz} \\ f,_{xy} & f,_{y^2} & f,_{yz} \\ f,_{xz} & f,_{yz} & f,_{z^2} \end{bmatrix},\ i = \overline{1, n}. \tag{8.6}$$

Deriving the relation (8.3) with respect to time and using the matrix $\{\dot{\boldsymbol{\beta}}\} = \begin{bmatrix} \dot{\psi} & \dot{\theta} & \dot{\varphi} \end{bmatrix}^{\mathrm{T}}$, we obtains the equalities

$$\{\dot{\mathbf{R}}_O\} + [\mathbf{A}][\mathbf{r}_i][\mathbf{Q}]\{\dot{\boldsymbol{\beta}}\} + [\mathbf{A}]\{\dot{\mathbf{r}}_i\} = \{\mathbf{0}\},\ i = \overline{1, n}, \tag{8.7}$$

wherefrom we may firstly calculate the matrices

$$\{\dot{\mathbf{r}}_i\} = -[\mathbf{A}]^{\mathrm{T}}\{\dot{\mathbf{R}}_O\} - [\mathbf{r}_i]^{\mathrm{T}}[\mathbf{Q}]\{\dot{\boldsymbol{\beta}}\},\ i = \overline{1,n}, \tag{8.8}$$

and then the matrices

$$[\dot{\mathbf{r}}_i] = \begin{bmatrix} 0 & -\dot{z}_i & \dot{y}_i \\ \dot{z}_i & 0 & -\dot{x}_i \\ -\dot{y}_i & \dot{x}_i & 0 \end{bmatrix},\ \{\dot{\mathbf{f}}_{ip}\} = [\mathbf{f}_{is}]\{\dot{\mathbf{r}}_i\},\ i = \overline{1, n}. \tag{8.9}$$

Deriving the equation (8.1) with respect to time, we get

$$\{\mathbf{f}_{ip}\}^{\mathrm{T}}\{\dot{\mathbf{r}}_i\} = 0,\ i = \overline{1, n}; \tag{8.10}$$

if we keep into account the expression (8.8), then we obtain the relations

$$\{\mathbf{f}_{ip}\}^{\mathrm{T}}[\mathbf{A}]^{\mathrm{T}}\{\dot{\mathbf{R}}_O\} + \{\mathbf{f}_{ip}\}^{\mathrm{T}}[\mathbf{r}_i]^{\mathrm{T}}[\mathbf{Q}]\{\dot{\boldsymbol{\beta}}\} = 0,\ i = \overline{1, n}. \tag{8.11}$$

Since the column matrix of the coordinates

$$\{\mathbf{q}\} = [X_O \; Y_O \; Z_O \; \psi \; \theta \; \varphi]^{\mathrm{T}} \tag{8.12}$$

has six components, it results that may exist at most five equations ($n \le 5$) of (8.11) type for a possible motion.

With the notations

$$\left[\mathbf{B}_1^{(i)}\right] = \{\mathbf{f}_{ip}\}^{\mathrm{T}}[\mathbf{A}]^{\mathrm{T}}, \; \left[\mathbf{B}_2^{(i)}\right] = \{\mathbf{f}_{ip}\}^{\mathrm{T}}[\mathbf{r}_i]^{\mathrm{T}}[\mathbf{Q}] \tag{8.13}$$

from the expression (8.11) one obtains the matrix of constraints

$$[\mathbf{B}] = \begin{bmatrix} \left[\mathbf{B}_1^{(1)}\right] & \left[\mathbf{B}_2^{(1)}\right] \\ \left[\mathbf{B}_1^{(2)}\right] & \left[\mathbf{B}_2^{(2)}\right] \\ \cdots & \cdots \\ \left[\mathbf{B}_1^{(n)}\right] & \left[\mathbf{B}_2^{(n)}\right] \end{bmatrix}. \tag{8.14}$$

For the calculation of the derivative of the matrix of constraints with respect to time we recall the relations (8.9) and the equalities

$$\left[\dot{\mathbf{B}}_1^{(i)}\right] = \{\dot{\mathbf{f}}_{ip}\}^{\mathrm{T}}[\mathbf{A}]^{\mathrm{T}} + \{\mathbf{f}_{ip}\}^{\mathrm{T}}[\dot{\mathbf{A}}]^{\mathrm{T}}, \; \left[\dot{\mathbf{B}}_2^{(i)}\right] = \{\dot{\mathbf{f}}_{ip}\}^{\mathrm{T}}[\mathbf{r}_i]^{\mathrm{T}}[\mathbf{Q}] + \{\mathbf{f}_{ip}\}^{\mathrm{T}}[\dot{\mathbf{r}}_i]^{\mathrm{T}}[\mathbf{Q}] + \{\mathbf{f}_{ip}\}^{\mathrm{T}}[\mathbf{r}_i]^{\mathrm{T}}[\dot{\mathbf{Q}}]; \tag{8.15}$$

it results

$$[\dot{\mathbf{B}}] = \begin{bmatrix} \left[\dot{\mathbf{B}}_1^{(1)}\right] & \left[\dot{\mathbf{B}}_2^{(1)}\right] \\ \left[\dot{\mathbf{B}}_1^{(2)}\right] & \left[\dot{\mathbf{B}}_2^{(2)}\right] \\ \cdots & \cdots \\ \left[\dot{\mathbf{B}}_1^{(n)}\right] & \left[\dot{\mathbf{B}}_2^{(n)}\right] \end{bmatrix}. \tag{8.16}$$

8.2.2 *The Matrix Differential Equation of Motion*

Recalling the relations

$$[\mathbf{M}] = \begin{bmatrix} [\mathbf{m}] & [\mathbf{A}][\mathbf{S}]^{\mathrm{T}}[\mathbf{Q}] \\ [\mathbf{Q}]^{\mathrm{T}}[\mathbf{S}][\mathbf{A}]^{\mathrm{T}} & [\mathbf{Q}]^{\mathrm{T}}[\mathbf{J}][\mathbf{Q}] \end{bmatrix}, \tag{8.17}$$

$$\left\{\widetilde{\mathbf{F}}_{\mathrm{s}}\right\} = -\left[[\mathbf{A}][\mathbf{S}]^{\mathrm{T}}[\dot{\mathbf{Q}}] + [\dot{\mathbf{A}}][\mathbf{S}]^{\mathrm{T}}[\mathbf{Q}]\right]\{\dot{\boldsymbol{\beta}}\},$$
$$\left\{\widetilde{\mathbf{F}}_{\beta}\right\} = -\left[[\mathbf{Q}]^{\mathrm{T}}[\mathbf{J}][\dot{\mathbf{Q}}] + [\mathbf{Q}]^{\mathrm{T}}[\boldsymbol{\omega}][\mathbf{J}][\mathbf{Q}]\right]\{\dot{\boldsymbol{\beta}}\}, \left\{\widetilde{\mathbf{F}}\right\} = \left[\left\{\widetilde{\mathbf{F}}_{\mathrm{s}}\right\}^{\mathrm{T}} \ \left\{\widetilde{\mathbf{F}}_{\beta}\right\}^{\mathrm{T}}\right]^{\mathrm{T}} \tag{8.18}$$

and knowing the expression $\{\mathbf{F}\}$ of the generalized forces, one obtains the matrix differential equation of motion

$$\begin{bmatrix} [\mathbf{M}] & -[\mathbf{B}]^{\mathrm{T}} \\ [\mathbf{B}] & [\mathbf{0}_{nn}] \end{bmatrix} \begin{bmatrix} \{\ddot{\mathbf{q}}\} \\ \{\boldsymbol{\lambda}\} \end{bmatrix} = \begin{bmatrix} \{\mathbf{F}\} + \left\{\widetilde{\mathbf{F}}\right\} \\ -[\dot{\mathbf{B}}]\{\dot{\mathbf{q}}\} \end{bmatrix}, \tag{8.19}$$

where

$$\{\boldsymbol{\lambda}\} = [\lambda_1 \ \ \lambda_2 \ \ \ldots \ \ \lambda_n]^{\mathrm{T}}. \tag{8.20}$$

The components of the reactions $\{\mathbf{N}_i\}$ at the points A_i onto the directions of the axes O_0X, O_0Y, O_0Z are given by the equalities

$$\{\mathbf{N}_i\} = \lambda_i[\mathbf{A}]\{\mathbf{f}_{ip}\}, i = \overline{1, n}, n \le 5. \tag{8.21}$$

8.2.3 *The Algorithm of Calculation*

One knows:

- the inertial matrices

$$[\mathbf{m}] = \begin{bmatrix} m & 0 & 0 \\ 0 & m & 0 \\ 0 & 0 & J \end{bmatrix}, [\mathbf{J}] = \begin{bmatrix} J_x & -J_{xy} & -J_{xz} \\ -J_{xy} & J_y & -J_{yz} \\ -J_{xz} & -J_{yz} & J_z \end{bmatrix}, [\mathbf{S}] = \begin{bmatrix} 0 & -mz_C & my_C \\ mz_C & 0 & -mx_C \\ -my_C & mx_C & 0 \end{bmatrix};$$

- the geometric characteristics

$$\{\mathbf{R}_i\} = \begin{bmatrix} X_i \\ Y_i \\ Z_i \end{bmatrix}, i = \overline{1, n}, \{\mathbf{f}_p\} = \begin{bmatrix} f_{,x} \\ f_{,y} \\ f_{,z} \end{bmatrix}, [\mathbf{f}_s] = \begin{bmatrix} f_{,x^2} & f_{,xy} & f_{,xz} \\ f_{,xy} & f_{,y^2} & f_{,yz} \\ f_{,xz} & f_{,yz} & f_{,z^2} \end{bmatrix};$$

- the initial conditions

$$\{\mathbf{R}_O\} = \begin{bmatrix} X_O \\ Y_O \\ Z_O \end{bmatrix}, \{\boldsymbol{\beta}\} = \begin{bmatrix} \psi \\ \theta \\ \varphi \end{bmatrix}, \{\mathbf{q}\} = [X_O \ Y_O \ Z_O \ \psi \ \theta \ \varphi]^{\mathrm{T}}, \{\dot{\mathbf{R}}_O\} = \begin{bmatrix} \dot{X}_O \\ \dot{Y}_O \\ \dot{Z}_O \end{bmatrix},$$

$$\{\dot{\boldsymbol{\beta}}\} = \begin{bmatrix} \dot{\psi} \\ \dot{\theta} \\ \dot{\varphi} \end{bmatrix}, \{\dot{\mathbf{q}}\} = \left[\dot{X}_O \ \dot{Y}_O \ \dot{Z}_O \ \dot{\psi} \ \dot{\theta} \ \dot{\varphi}\right]^{\mathrm{T}};$$

- the matrices $[\mathbf{U}_\psi]$, $[\mathbf{U}_\theta]$, $[\mathbf{U}_\varphi]$.

One successively calculates:

$$[\boldsymbol{\psi}], [\boldsymbol{\theta}], [\boldsymbol{\varphi}], [\mathbf{A}] = [\boldsymbol{\psi}][\boldsymbol{\theta}][\boldsymbol{\varphi}], [\mathbf{A}_\psi] = [\mathbf{U}_\psi][\mathbf{A}], [\mathbf{A}_\theta] = [\mathbf{A}][\boldsymbol{\varphi}]^{\mathrm{T}}[\mathbf{U}_\theta][\boldsymbol{\varphi}], [\mathbf{A}_\varphi] = [\mathbf{A}][\mathbf{U}_\varphi],$$

$$[\dot{\mathbf{A}}] = \dot{\psi}[\mathbf{A}_\psi] + \dot{\theta}[\mathbf{A}_\theta] + \dot{\varphi}[\mathbf{A}_\varphi], [\boldsymbol{\omega}] = [\mathbf{A}]^{\mathrm{T}}[\dot{\mathbf{A}}], [\mathbf{Q}], [\dot{\mathbf{Q}}], \{\widetilde{\mathbf{F}}_{\mathbf{s}}\}, [\mathbf{M}], \{\mathbf{r}_i\}, \{\mathbf{f}_{ip}\},$$

$$[\mathbf{f}_{is}], i = \overline{1, n}, \{\dot{\mathbf{r}}_i\}, [\dot{\mathbf{r}}_i], \{\dot{\mathbf{f}}_{ip}\}, i = \overline{1, n}, [\mathbf{B}_1^{(i)}], [\mathbf{B}_2^{(i)}], [\dot{\mathbf{B}}_1^{(i)}], [\dot{\mathbf{B}}_2^{(i)}], i = \overline{1, n}, [\mathbf{B}], [\dot{\mathbf{B}}].$$

One solves the equation (8.19).
One calculates the reactions with the aid of the relations (8.21).

Example 8.2.1 An ellipsoid of mass m and semi-axes a, b, c supports at the points $A_i(x_i, y_i, z_i)$, $i = 1, 2, 3$. Determine the laws of motion knowing the initial conditions at the moment $t = 0$.

Numerical application: $m = 100\,\mathrm{kg}$, $a = 0.4\,\mathrm{m}$, $b = 0.3\,\mathrm{m}$, $c = 0.2\,\mathrm{m}$, ω_0, $X_O = Y_O = Z_O = 0\,\mathrm{m}$, $\psi = \theta = \varphi = 0\,\mathrm{rad}$, $\dot{X}_O = \dot{Y}_O = \dot{Z}_O = 0\,\mathrm{m/s}$, $\dot{\psi} = 0\,\mathrm{rad/s}$, $\dot{\theta} = 0\,\mathrm{m/s}$, $\dot{\varphi} = \omega_0$, $X_1 = \frac{a}{2}$, $Y_1 = 0$, $Z_1 = -c\frac{\sqrt{3}}{2}$, $X_2 = -\frac{a}{2}$, $Y_2 = \frac{b}{2}$, $Z_2 = -\frac{c}{\sqrt{2}}$, $X_3 = -\frac{a}{2}$, $Y_3 = -\frac{b}{2}$, $Z_3 = -\frac{c}{\sqrt{2}}$.

Solution: It results $J_x = \frac{m(b^2 + c^2)}{5}$, $J_y = \frac{m(a^2 + c^2)}{5}$, $J_z = \frac{m(a^2 + b^2)}{5}$. Following the algorithm described above, one obtains the numerical results drawn in the next diagrams for $\omega_0 = 0\,\mathrm{rad/s}$, $\omega_0 = 0.01\,\mathrm{rad/s}$, and $\omega_0 = 1\,\mathrm{rad/s}$, respectively.

From the diagrams (Fig. 8.4 to Fig. 8.30) it results that for $\omega_0 = 0\,\mathrm{rad/s}$ the motion has a periodical character. The periods are of approximate 0.9 s for some parameters of motions and some components of the forces; the period is double, of approximate 1.8 s, for the rest of parameters and components of forces. For $\omega_0 = 0.01\,\mathrm{rad/s}$ and $\omega_0 = 1\,\mathrm{rad/s}$ this periodical character vanishes.

The case $\omega_0 = 0\,\mathrm{rad/s}$ leads to a maximum reactions of approximate 650 N, while for $\omega_0 = 0.01\,\mathrm{rad/s}$ this maximum is about 13000 N (an increasing of 20 times); for $\omega_0 = 1\,\mathrm{rad/s}$ the maximum is about 26000 N (that is, an increasing of 40 times).

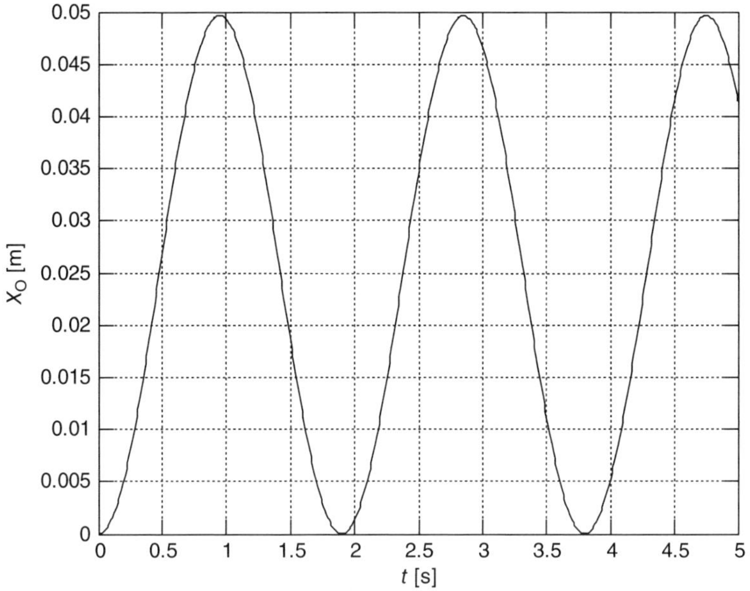

Figure 8.4 Time history $X_O = X_O(t)$ for $0 \le t \le 5$ s and $\omega_0 = 0$ rad/s.

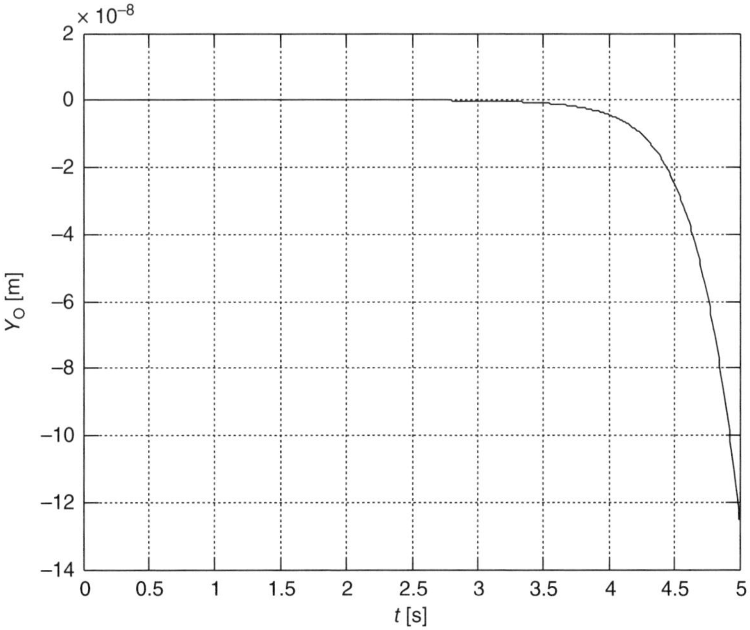

Figure 8.5 Time history $Y_O = Y_O(t)$ for $0 \le t \le 5$ s and $\omega_0 = 0$ rad/s.

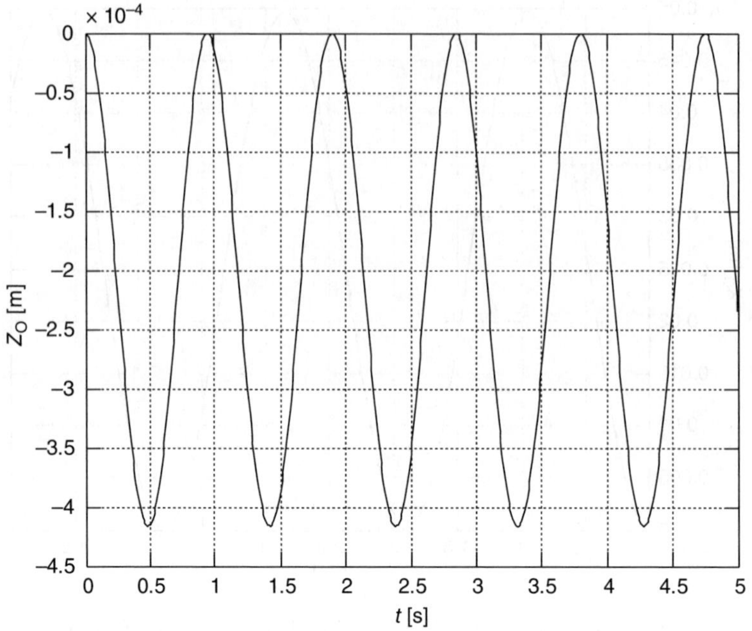

Figure 8.6 Time history $Z_O = Z_O(t)$ for $0 \le t \le 5$ s and $\omega_0 = 0$ rad/s.

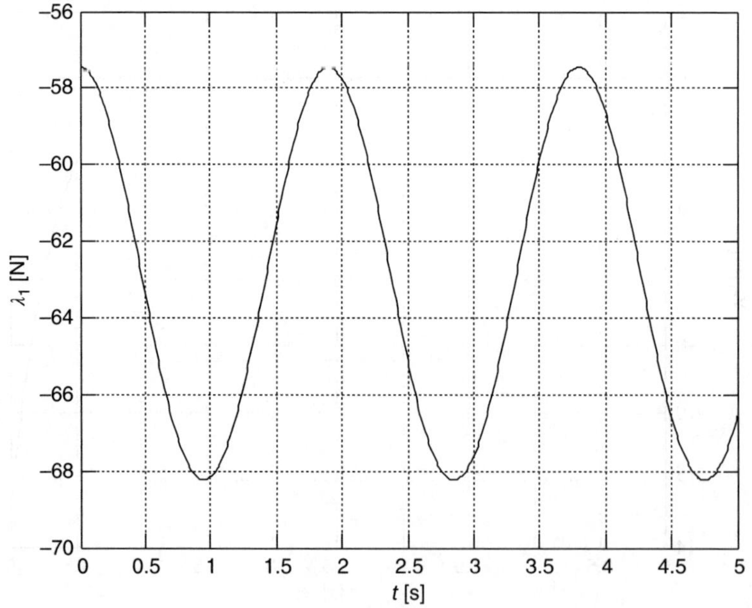

Figure 8.7 Time history $\lambda_1 = \lambda_1(t)$ for $0 \le t \le 5$ s and $\omega_0 = 0$ rad/s.

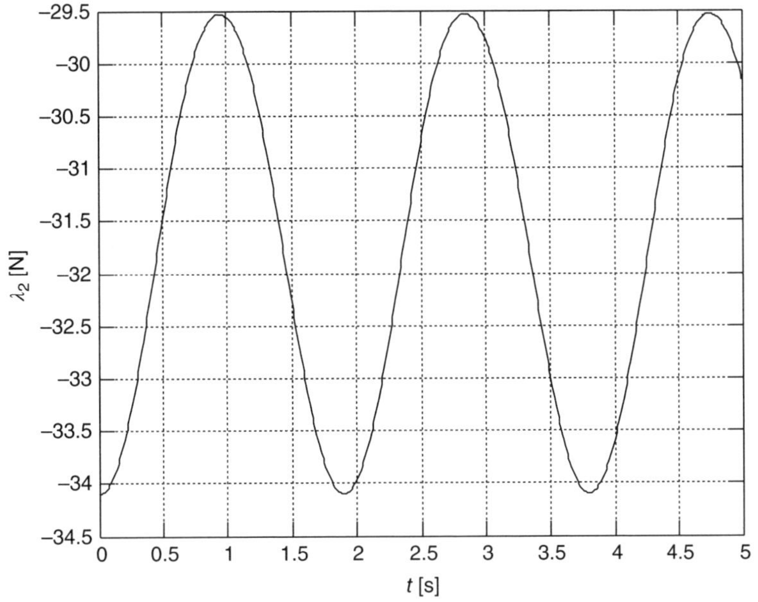

Figure 8.8 Time history $\lambda_2 = \lambda_2(t)$ for $0 \le t \le 5$ s and $\omega_0 = 0$ rad/s.

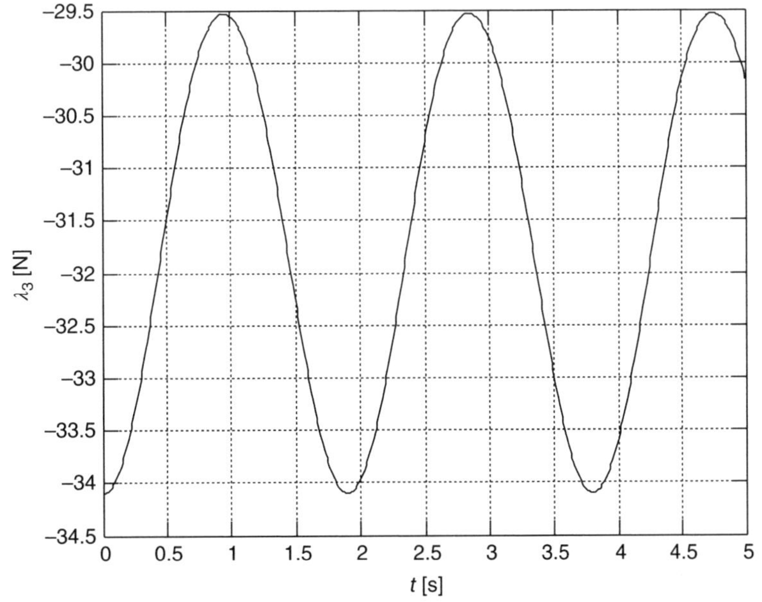

Figure 8.9 Time history $\lambda_3 = \lambda_3(t)$ for $0 \le t \le 5$ s and $\omega_0 = 0$ rad/s.

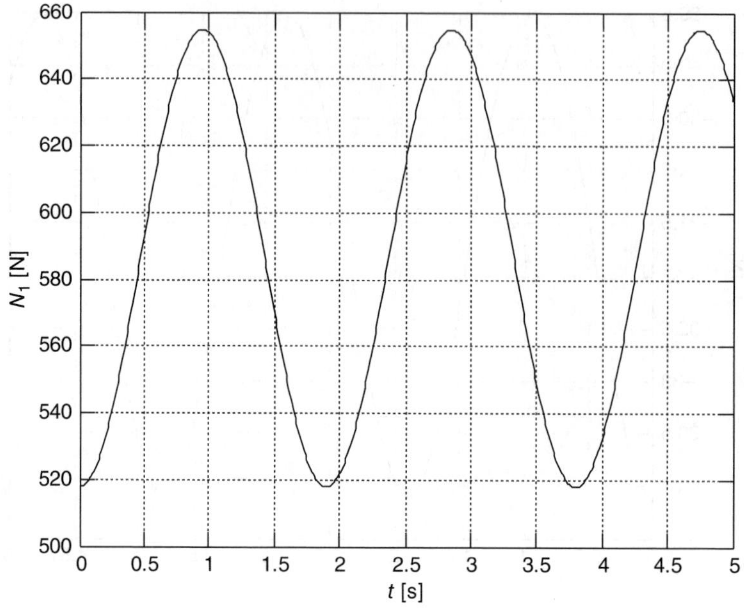

Figure 8.10 Time history $N_1 = N_1(t)$ for $0 \le t \le 5$ s and $\omega_0 = 0$ rad/s.

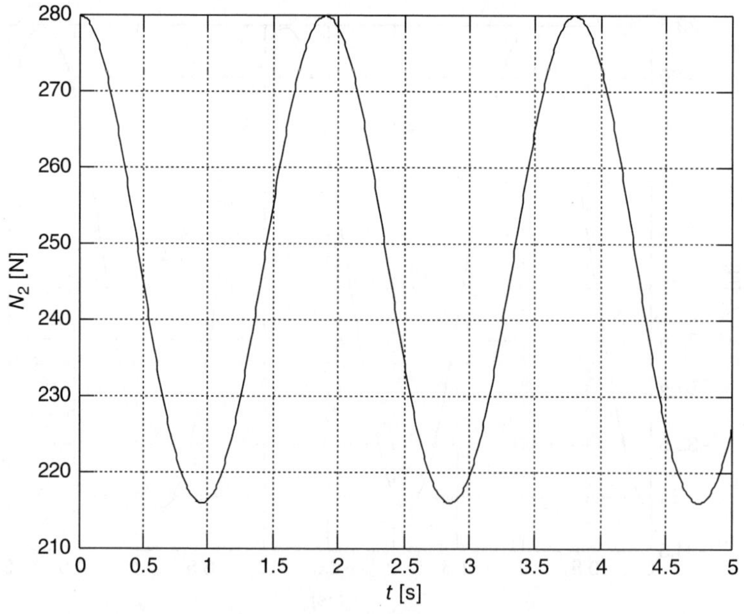

Figure 8.11 Time history $N_2 = N_2(t)$ for $0 \le t \le 5$ s and $\omega_0 = 0$ rad/s.

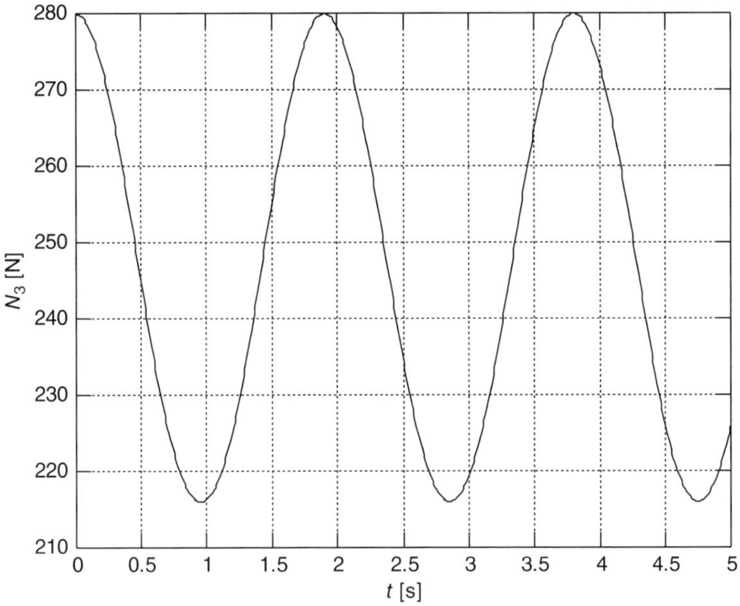

Figure 8.12 Time history $N_3 = N_3(t)$ for $0 \le t \le 5$ s and $\omega_0 = 0$ rad/s.

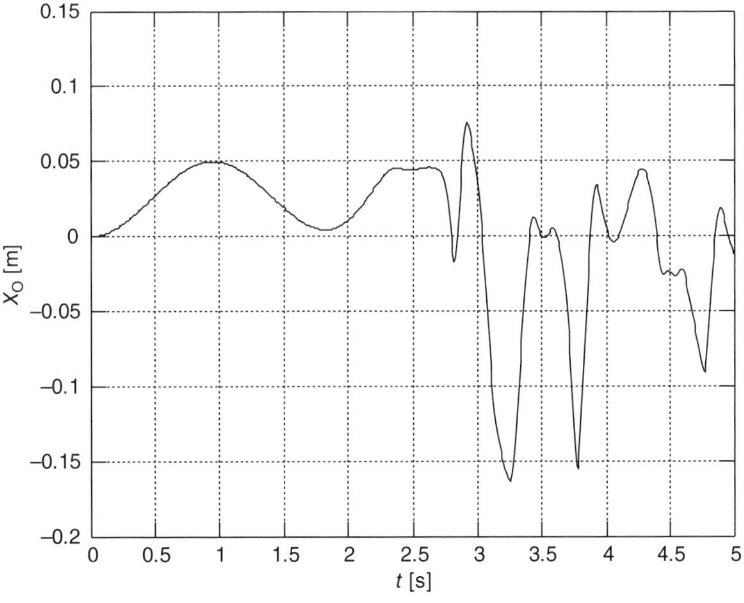

Figure 8.13 Time history $X_O = X_O(t)$ for $0 \le t \le 5$ s and $\omega_0 = 0.01$ rad/s.

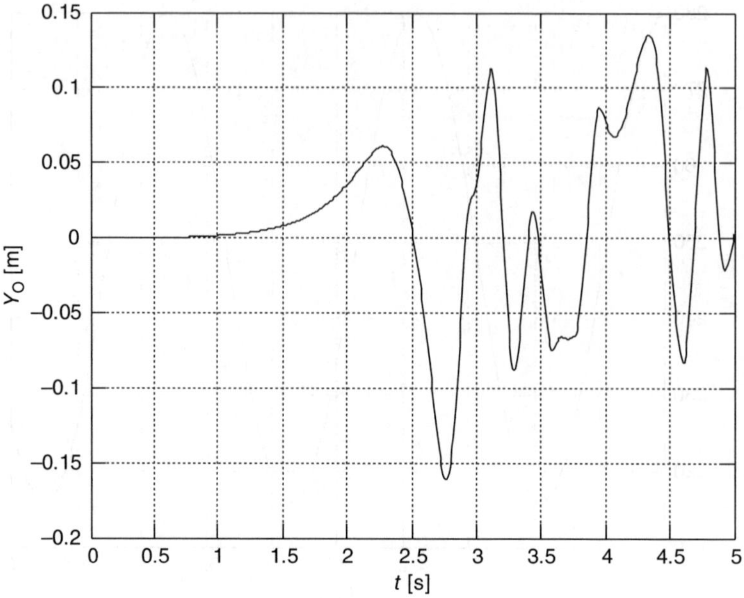

Figure 8.14 Time history $Y_O = Y_O(t)$ for $0 \leq t \leq 5$ s and $\omega_0 = 0.01$ rad/s.

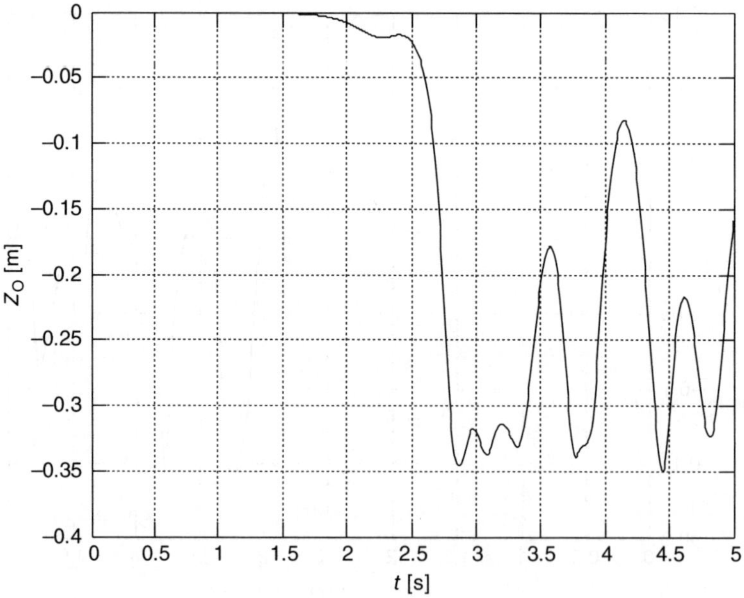

Figure 8.15 Time history $Z_O = Z_O(t)$ for $0 \leq t \leq 5$ s and $\omega_0 = 0.01$ rad/s.

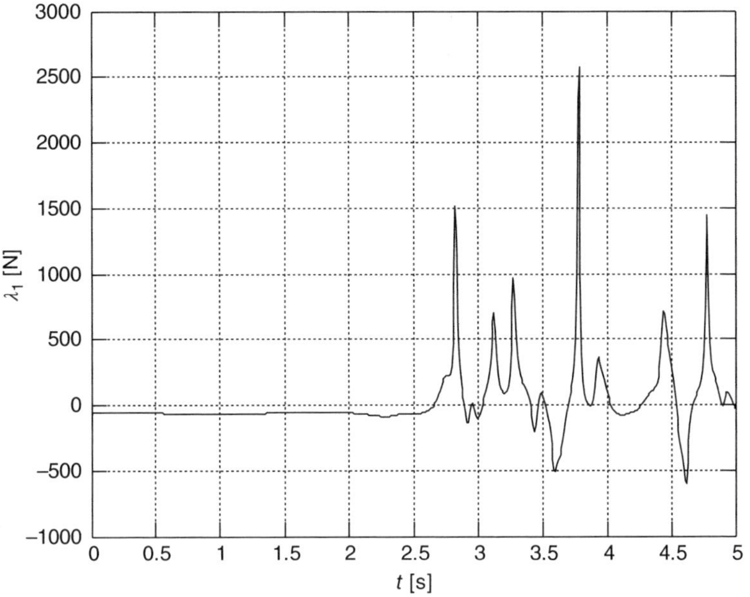

Figure 8.16 Time history $\lambda_1 = \lambda_1(t)$ for $0 \le t \le 5$ s and $\omega_0 = 0.01$ rad/s.

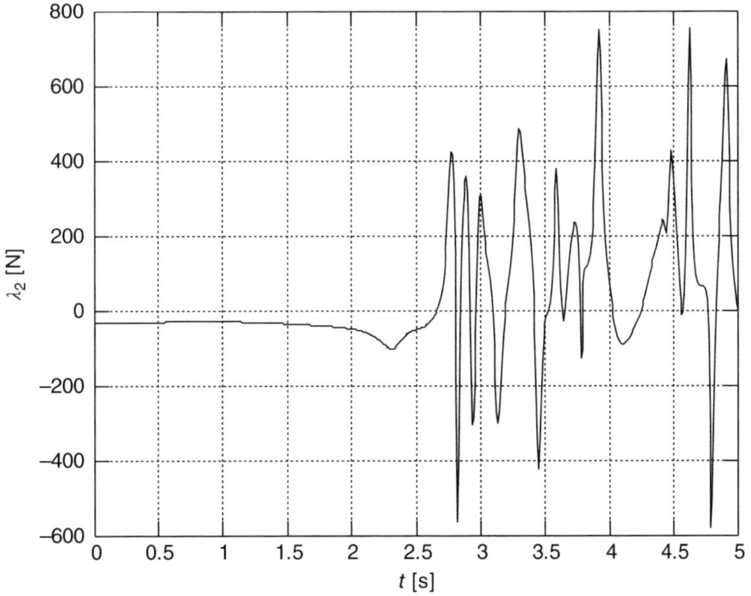

Figure 8.17 Time history $\lambda_2 = \lambda_2(t)$ for $0 \le t \le 5$ s and $\omega_0 = 0.01$ rad/s.

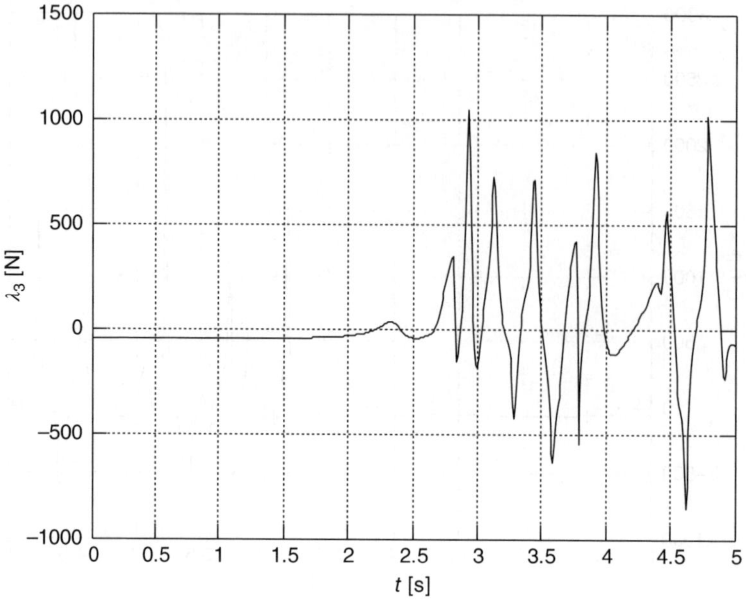

Figure 8.18 Time history $\lambda_3 = \lambda_3(t)$ for $0 \le t \le 5$ s and $\omega_0 = 0.01$ rad/s.

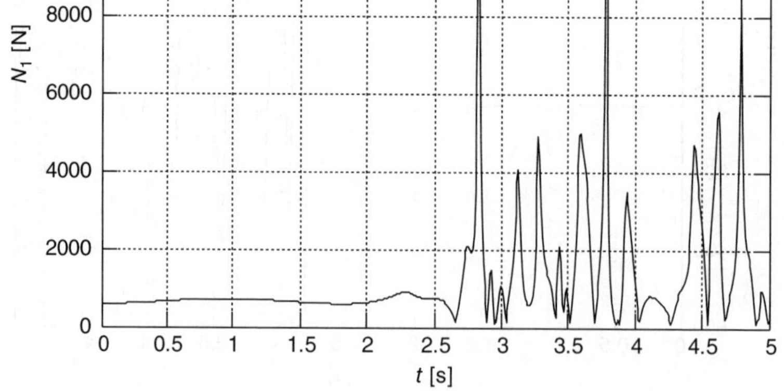

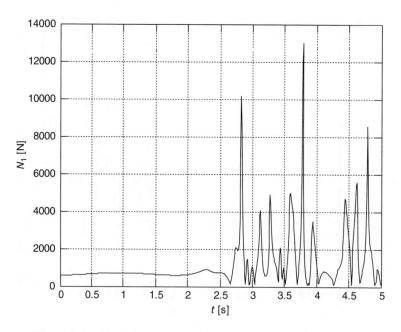

Figure 8.19 Time history $N_1 = N_1(t)$ for $0 \le t \le 5$ s and $\omega_0 = 0.01$ rad/s.

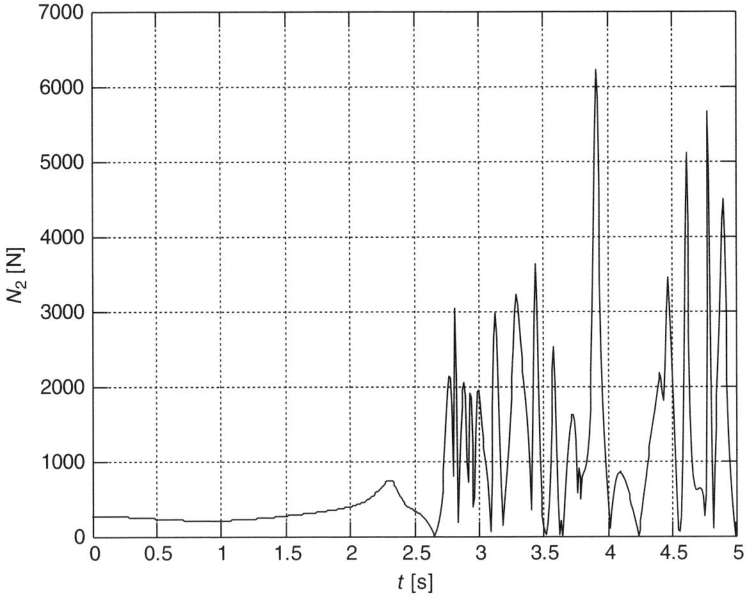

Figure 8.20 Time history $N_2 = N_2(t)$ for $0 \leq t \leq 5$ s and $\omega_0 = 0.01$ rad/s.

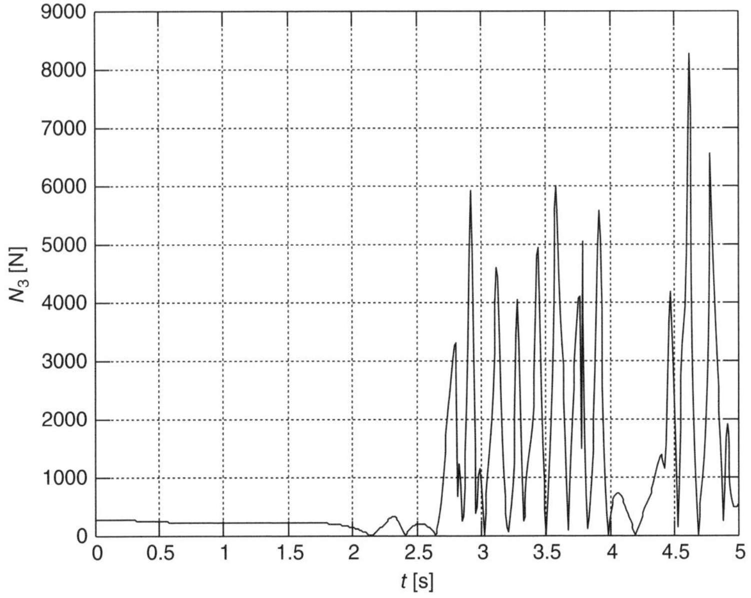

Figure 8.21 Time history $N_3 = N_3(t)$ for $0 \leq t \leq 5$ s and $\omega_0 = 0.01$ rad/s.

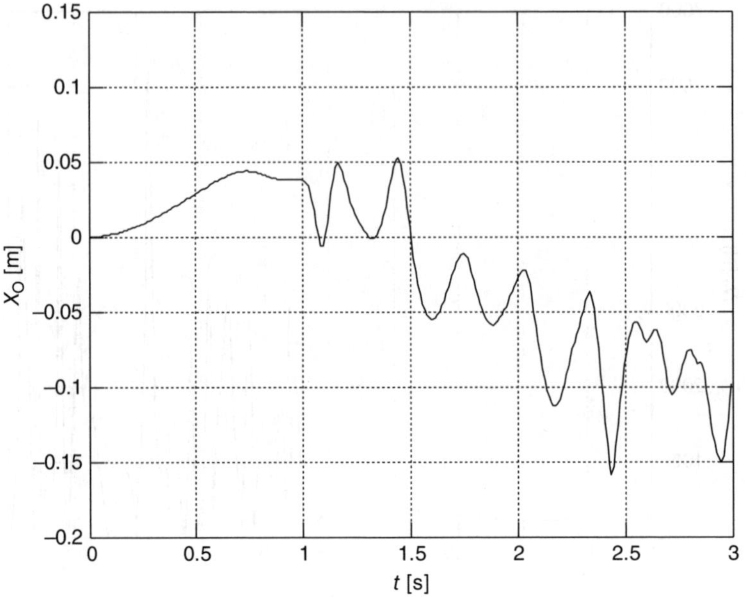

Figure 8.22 Time history $X_O = X_O(t)$ for $0 \le t \le 3$ s and $\omega_0 = 1$ rad/s.

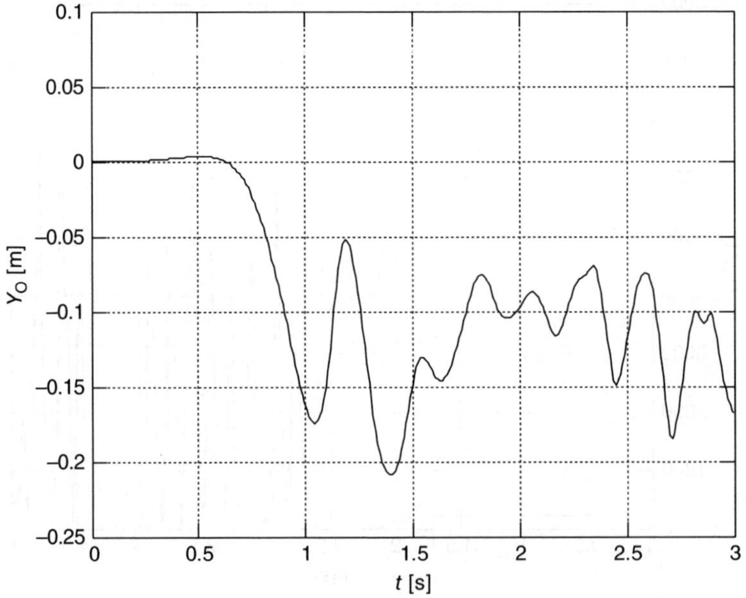

Figure 8.23 Time history $Y_O = Y_O(t)$ for $0 \le t \le 3$ s and $\omega_0 = 1$ rad/s.

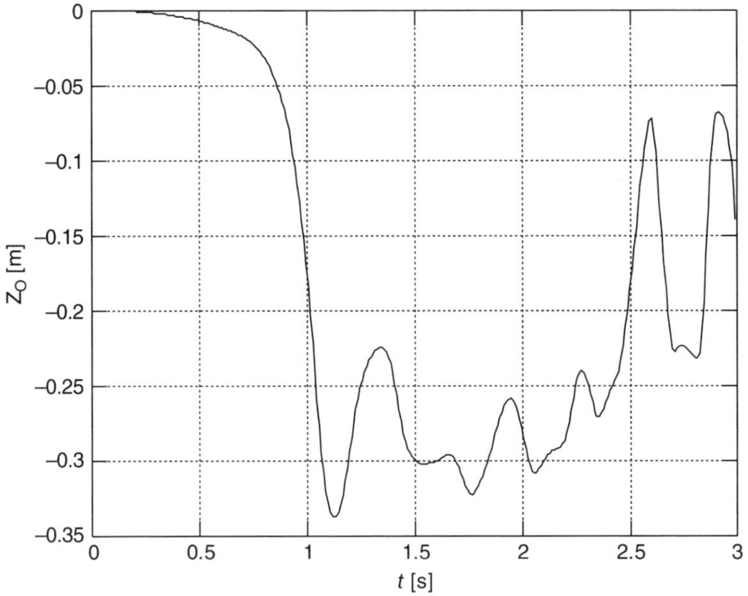

Figure 8.24 Time history $Z_O = Z_O(t)$ for $0 \leq t \leq 3$ s and $\omega_0 = 1$ rad/s.

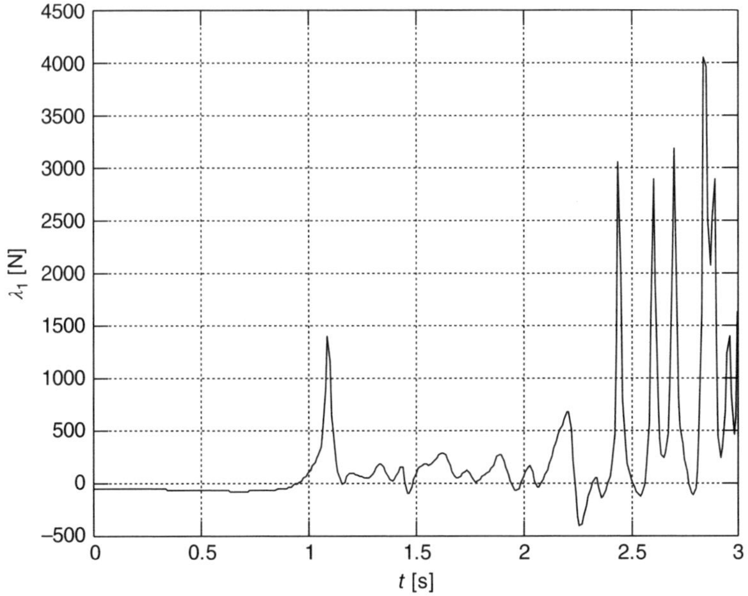

Figure 8.25 Time history $\lambda_1 = \lambda_1(t)$ for $0 \leq t \leq 3$ s and $\omega_0 = 1$ rad/s.

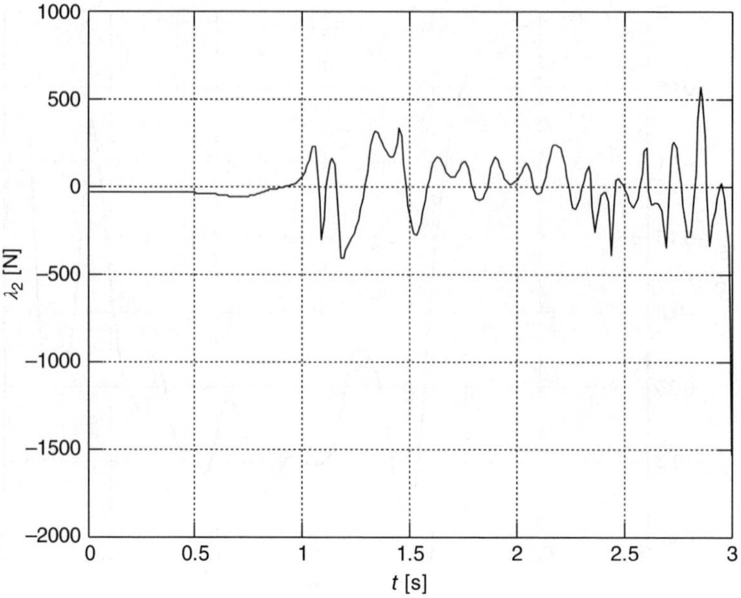

Figure 8.26 Time history $\lambda_2 = \lambda_2(t)$ for $0 \le t \le 3$ s and $\omega_0 = 1$ rad/s.

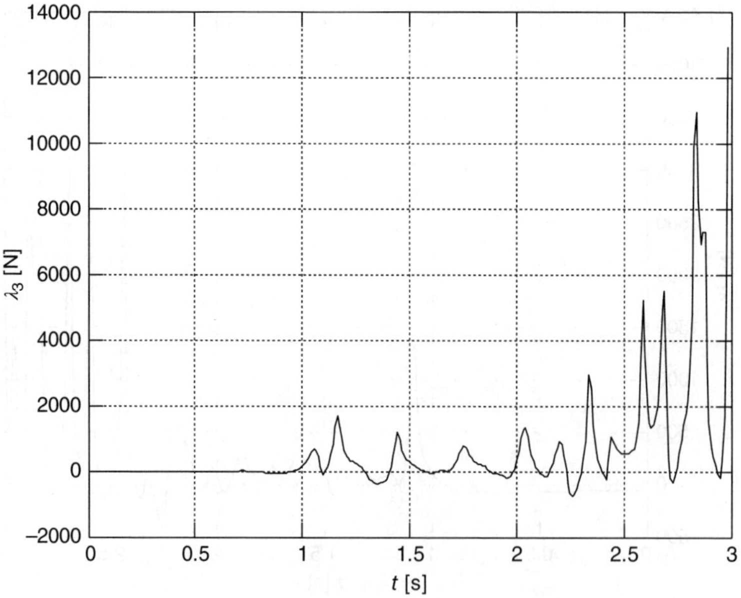

Figure 8.27 Time history $\lambda_3 = \lambda_3(t)$ for $0 \le t \le 3$ s and $\omega_0 = 1$ rad/s.

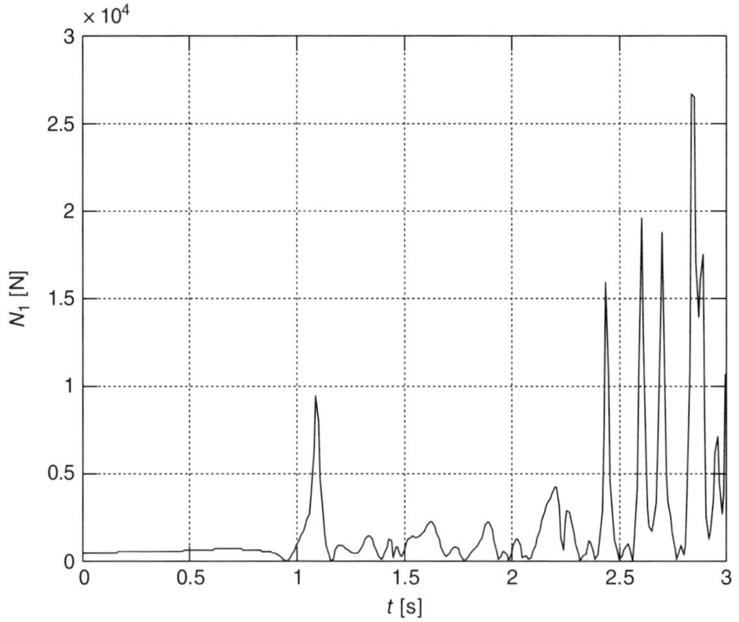

Figure 8.28 Time history $N_1 = N_1(t)$ for $0 \leq t \leq 3$ s and $\omega_0 = 1$ rad/s.

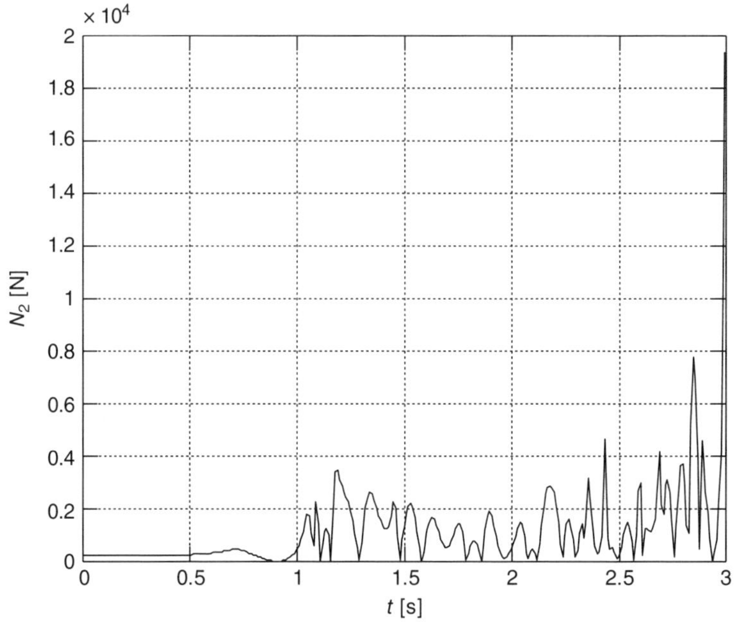

Figure 8.29 Time history $N_2 = N_2(t)$ for $0 \leq t \leq 3$ s and $\omega_0 = 1$ rad/s.

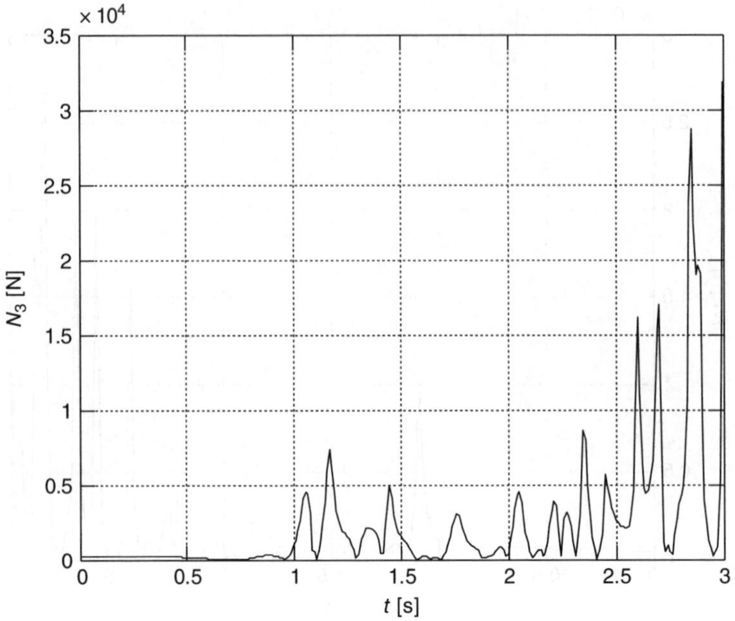

Figure 8.30 Time history $N_3 = N_3(t)$ for $0 \le t \le 3$ s and $\omega_0 = 1$ rad/s.

8.3 The Rigid Solid Supported with Sliding on Fixed Curves

8.3.1 The Matrix of Constraints

We consider that the equations of the fixed curves are given by the parametric equations

$$X_i = X_i(\eta_i), Y_i = Y_i(\eta_i), Z_i = Z_i(\eta_i), \tag{8.22}$$

while equation of the bounding surface of the rigid solid in the local reference system *Oxyz* is

$$f(x, y, z) = 0. \tag{8.23}$$

With the notations (8.2) the contact conditions between the surface and curves write in the form

$$\{\mathbf{R}_i\} = \{\mathbf{R}_O\} + [\mathbf{A}]\{\mathbf{r}_i\}, \ i = \overline{1, n}, \tag{8.24}$$

wherefrom it results

$$\{\mathbf{r}_i\} = [\mathbf{A}]^{\mathrm{T}}\{\mathbf{R}_i\} - [\mathbf{A}]^{\mathrm{T}}\{\mathbf{R}_O\}, \ i = \overline{1, n}. \tag{8.25}$$

In the conditions in which the matrices [**A**], {$\mathbf{R}_i$}, {$\mathbf{R}_O$} are known, from the relation (8.25) we determine the coordinates x_i, y_i, z_i of the contact points, relative to the local reference system, and, using these values, we may determine the matrices

$$[\mathbf{r}_i]=\begin{bmatrix}0 & -z_i & y_i\\ z_i & 0 & -x_i\\ -y_i & x_i & 0\end{bmatrix},\ \{\mathbf{f}_{ip}\}=\begin{bmatrix}f,_{x_i}\\ f,_{y_i}\\ f,_{z_i}\end{bmatrix},\ [\mathbf{f}_{ip}]=\begin{bmatrix}0 & -f,_{z_i} & f,_{y_i}\\ f,_{z_i} & 0 & -f,_{x_i}\\ -f,_{y_i} & f,_{x_i} & 0\end{bmatrix},$$
$$[\mathbf{f}_{is}]=\begin{bmatrix}f,_{x_i^2} & f,_{x_iy_i} & f,_{x_iz_i}\\ f,_{x_iy_i} & f,_{y_i^2} & f,_{y_iz_i}\\ f,_{x_iz_i} & f,_{y_iz_i} & f,_{z_i^2}\end{bmatrix}. \tag{8.26}$$

Deriving the relation (8.24) with respect to time and using the notation

$$\{\dot{\boldsymbol{\beta}}\}=\begin{bmatrix}\dot{\psi} & \dot{\theta} & \dot{\varphi}\end{bmatrix}^{\mathrm{T}}, \tag{8.27}$$

one obtains the equality

$$\dot{\eta}_i\{\mathbf{R}_{ip}\}=\{\dot{\mathbf{R}}_O\}+[\mathbf{A}][\mathbf{r}_i]^{\mathrm{T}}[\mathbf{Q}]\{\dot{\boldsymbol{\beta}}\}+[\mathbf{A}]\{\dot{\mathbf{r}}_i\}; \tag{8.28}$$

now one may calculate the matrices

$$\{\dot{\mathbf{r}}_i\}=-[\mathbf{A}]^{\mathrm{T}}\{\dot{\mathbf{R}}_O\}-[\mathbf{r}_i]^{\mathrm{T}}[\mathbf{Q}]\{\dot{\boldsymbol{\beta}}\}+\dot{\eta}_i[\mathbf{A}]^{\mathrm{T}}\{\mathbf{R}_{ip}\},\ i=\overline{1,n}, \tag{8.29}$$

and then the matrices

$$[\dot{\mathbf{r}}_i]=\begin{bmatrix}0 & -\dot{z}_i & \dot{y}_i\\ \dot{z}_i & 0 & -\dot{x}_i\\ -\dot{y}_i & \dot{x}_i & 0\end{bmatrix},\ \{\dot{\mathbf{f}}_{ip}\}=[\mathbf{f}_{is}]\{\dot{\mathbf{r}}_i\},\ i=\overline{1,n}. \tag{8.30}$$

By derivation of the relation (8.23) with respect to time one obtains

$$\{\mathbf{f}_{ip}\}^{\mathrm{T}}\{\dot{\mathbf{r}}_i\}=0; \tag{8.31}$$

keeping into account the tangency condition between the curve and surface

$$\{\mathbf{R}_{ip}\}^{\mathrm{T}}[\mathbf{A}]\{\mathbf{f}_{ip}\}=0, \tag{8.32}$$

from the relation (8.29) one gets the equality

$$\{\mathbf{f}_{ip}\}^{\mathrm{T}}[\dot{\mathbf{A}}]\{\dot{\mathbf{R}}_O\}+\{\mathbf{f}_{ip}\}^{\mathrm{T}}[\mathbf{r}_i]^{\mathrm{T}}[\mathbf{Q}]\{\dot{\boldsymbol{\beta}}\}=0. \tag{8.33}$$

From the second relation (8.30) and from the relation (8.29) one deduces the expression

$$\{\dot{\mathbf{f}}_{ip}\}^{\mathrm{T}} = -[\mathbf{f}_{is}][\mathbf{A}]^{\mathrm{T}}\{\dot{\mathbf{R}}_O\} - [\mathbf{f}_{is}][\mathbf{r}_i]^{\mathrm{T}}[\mathbf{Q}]\{\dot{\boldsymbol{\beta}}\} + \dot{\eta}_i[\mathbf{f}_{is}][\mathbf{A}]^{\mathrm{T}}\{\mathbf{R}_{ip}\}; \tag{8.34}$$

deriving the equation (8.32) with respect to time, one obtains

$$\begin{aligned} &\dot{\eta}_i\{\mathbf{R}_{is}\}^{\mathrm{T}}[\mathbf{A}]\{\mathbf{f}_{ip}\} - \{\mathbf{R}_{ip}\}^{\mathrm{T}}[\mathbf{A}][\mathbf{f}_{ip}]^{\mathrm{T}}[\mathbf{Q}]\{\dot{\boldsymbol{\beta}}\} - \{\mathbf{R}_{ip}\}^{\mathrm{T}}[\mathbf{A}][\mathbf{f}_{is}][\mathbf{A}]^{\mathrm{T}}\{\dot{\mathbf{R}}_0\} \\ &-\{\mathbf{R}_{ip}\}^{\mathrm{T}}[\mathbf{A}][\mathbf{f}_{is}][\mathbf{r}_i][\mathbf{Q}]\{\dot{\boldsymbol{\beta}}\} + \dot{\eta}_i\{\mathbf{R}_{ip}\}^{\mathrm{T}}[\mathbf{A}][\mathbf{f}_{is}][\mathbf{A}]^{\mathrm{T}}\{\mathbf{R}_{ip}\} = 0. \end{aligned} \tag{8.35}$$

Considering the notations

$$\left[\mathbf{B}_{11}^{(i)}\right] = \{\mathbf{f}_{ip}\}^{\mathrm{T}}[\mathbf{A}]^{\mathrm{T}},\ \left[\mathbf{B}_{12}^{(i)}\right] = \{\mathbf{f}_{ip}\}^{\mathrm{T}}[\mathbf{r}_i]^{\mathrm{T}}[\mathbf{Q}], \tag{8.36}$$

$$\begin{aligned} &\left[\mathbf{B}_{21}^{(i)}\right] = -\{\mathbf{R}_{ip}\}^{\mathrm{T}}[\mathbf{A}][\mathbf{f}_{is}][\mathbf{A}]^{\mathrm{T}},\ \left[\mathbf{B}_{22}^{(i)}\right] = \{\mathbf{R}_{ip}\}^{\mathrm{T}}[\mathbf{A}][\mathbf{f}_{ip}][\mathbf{Q}] - \{\mathbf{R}_{ip}\}^{\mathrm{T}}[\mathbf{A}][\mathbf{f}_{is}][\mathbf{r}_i][\mathbf{Q}], \\ &B_{23}^{(i)} = \{\mathbf{R}_{is}\}^{\mathrm{T}}[\mathbf{A}]\{\mathbf{f}_{ip}\} + \{\mathbf{R}_{ip}\}^{\mathrm{T}}[\mathbf{A}][\mathbf{f}_{is}][\mathbf{A}]^{\mathrm{T}}\{\mathbf{R}_{ip}\}, \end{aligned} \tag{8.37}$$

from the relations (8.33) and (8.35) one obtains the matrix of constraints

$$[\mathbf{B}] = \begin{bmatrix} \left[\mathbf{B}_{11}^{(1)}\right] & \left[\mathbf{B}_{12}^{(1)}\right] & 0 & 0 & 0 & \cdots & 0 \\ \left[\mathbf{B}_{21}^{(1)}\right] & \left[\mathbf{B}_{22}^{(1)}\right] & B_{23}^{(1)} & 0 & 0 & \cdots & 0 \\ \left[\mathbf{B}_{11}^{(2)}\right] & \left[\mathbf{B}_{12}^{(2)}\right] & 0 & 0 & 0 & \cdots & 0 \\ \left[\mathbf{B}_{21}^{(2)}\right] & \left[\mathbf{B}_{22}^{(2)}\right] & 0 & B_{23}^{(2)} & 0 & \cdots & 0 \\ \cdots & \cdots & \cdots & \cdots & \cdots & \cdots & \cdots \\ \left[\mathbf{B}_{11}^{(n)}\right] & \left[\mathbf{B}_{12}^{(n)}\right] & 0 & 0 & 0 & \cdots & 0 \\ \left[\mathbf{B}_{21}^{(n)}\right] & \left[\mathbf{B}_{22}^{(n)}\right] & 0 & 0 & 0 & \cdots & B_{23}^{(n)} \end{bmatrix}, \tag{8.38}$$

the matrix of coordinates being

$$\{\mathbf{q}^*\} = [X_O\ \ Y_O\ \ Z_O\ \ \psi\ \ \theta\ \ \varphi\ \ \eta_1\ \ \eta_2\ \ \ldots\ \ \eta_n]^{\mathrm{T}}. \tag{8.39}$$

We thus conclude that the maximum number of the sliding supports is equal to five.

8.3.2 The Matrix Differential Equation of Motion

With the aid of the notations (8.17) and (8.18), and constructing the modified matrix of inertia

$$[\widetilde{\mathbf{M}}] = \begin{bmatrix} [\mathbf{M}] & [\mathbf{0}_{6n}] \\ [\mathbf{0}_{n6}] & [\mathbf{0}_{nn}] \end{bmatrix}, \tag{8.40}$$

one obtains the matrix differential equation of motion

$$\begin{bmatrix} [\widetilde{\mathbf{M}}] & -[\mathbf{B}]^{\mathrm{T}} \\ [\mathbf{B}] & [\mathbf{0}_{2n,2n}] \end{bmatrix} \begin{bmatrix} \{\ddot{\mathbf{q}}^*\} \\ \{\boldsymbol{\lambda}\} \end{bmatrix} = \begin{bmatrix} \{\mathbf{F}\} + \{\widetilde{\mathbf{F}}\} \\ \{\mathbf{0}_{n,1}\} \\ -[\dot{\mathbf{B}}]\{\dot{\mathbf{q}}^*\} \end{bmatrix}, \tag{8.41}$$

where

$$\{\boldsymbol{\lambda}\} = [\lambda_1 \quad \lambda_2 \quad \ldots \quad \lambda_{2n}]^{\mathrm{T}}. \tag{8.42}$$

To be out to obtain the derivative matrix $[\dot{\mathbf{B}}]$ we recall the known relations, while the derivative $[\dot{\mathbf{f}}_{is}]$ is given by the expression

$$[\dot{\mathbf{f}}_{is}] = [\{\mathbf{A}_{i1}\} \quad \{\mathbf{A}_{i2}\} \quad \{\mathbf{A}_{i3}\}], \tag{8.43}$$

where

$$\{\mathbf{A}_{i1}\} = \begin{bmatrix} f_{,x_i^3} & f_{,x_i^2y_i} & f_{,x_i^2z_i} \\ f_{,x_i^2y_i} & f_{,x_iy_i^2} & f_{,x_iy_iz_i} \\ f_{,x_i^2z_i} & f_{,x_iy_iz_i} & f_{,x_iz_i^2} \end{bmatrix} \{\dot{\mathbf{r}}_i\},\ \{\mathbf{A}_{i2}\} = \begin{bmatrix} f_{,x_i^2y_i} & f_{,x_iy_i^2} & f_{,x_iy_iz_i} \\ f_{,x_iy_i^2} & f_{,y_i^3} & f_{,y_i^2z_i} \\ f_{,x_iy_iz_i} & f_{,y_i^2z_i} & f_{,y_iz_i^2} \end{bmatrix} \{\dot{\mathbf{r}}_i\},$$
$$\{\mathbf{A}_{i3}\} = \begin{bmatrix} f_{,x_i^2z_i} & f_{,x_iy_iz_i} & f_{,x_iz_i^2} \\ f_{,x_iy_iz_i} & f_{,y_i^2z_i} & f_{,y_iz_i^2} \\ f_{,x_iz_i^2} & f_{,y_iz_i^2} & f_{,z_i^3} \end{bmatrix} \{\dot{\mathbf{r}}_i\}. \tag{8.44}$$

8.3.3 The Reactions

The scalar equations resulted from the relation (8.41) and corresponding to the rows $7-6+2n$ lead to $\lambda_2 = \lambda_4 = \ldots = \lambda_{2n} = 0$; consequently, the reaction that corresponds to the index i has the components given by the relation $\lambda_{2i-1}[\mathbf{A}]\{\mathbf{f}_{ip}\}$, onto the axes of the reference system O_0XYZ, that is, as we expected, the reaction is normal to the surface.

8.3.4 *The Algorithm of Calculation*

The algorithm of calculation is identical to that given at the paragraph 8.2 with the particularities given by the matrix of coordinates $\{\mathbf{q}^*\}$ and the modified matrix of inertia $\{\widetilde{\mathbf{M}}\}$.

Example 8.3.1 Determine the matrix of constraints for the case in which a plan moves so that it remains in contact to an ellipse (Fig. 8.31) of semi-axes a, b.

Solution: Considering the axes in Fig. 8.31, it successively results the equalities

$$f(x,y,z)=z=0, X=a\cos\eta,\ Y=b\sin\eta, Z=0, \{\mathbf{f}_p\}=[0\ \ 0\ \ 1]^{\mathrm{T}},$$

$$[\mathbf{f}_p]=\begin{bmatrix}0&-1&0\\1&0&0\\0&0&0\end{bmatrix},\ [\mathbf{f}_s]=[\mathbf{0}],\ \{\mathbf{r}\}=[x\ \ y\ \ 0]^{\mathrm{T}},\ [\mathbf{r}]=\begin{bmatrix}0&0&y\\0&0&-x\\-y&x&0\end{bmatrix},$$

$$\{\mathbf{R}\}=[a\cos\eta\ \ b\sin\eta\ \ 0]^{\mathrm{T}},\ \{\mathbf{R}_p\}=[-a\sin\eta\ \ b\cos\eta\ \ 0]^{\mathrm{T}}; \tag{8.45}$$

if one knows the initial conditions, then one determines the numerical value of the matrix of constraints using the relations (8.25), (8.26), (8.27), (8.29), (8.30), (8.35), (8.36), and (8.41).

For the case considered in this example one may establish the analytical form of the matrix of constraints; it starts from the relation (8.25) which writes in the form

$$\begin{bmatrix}x\\y\\0\end{bmatrix}=[\mathbf{A}]^{\mathrm{T}}\begin{bmatrix}a\cos\eta\\b\sin\eta\\0\end{bmatrix}-[\mathbf{A}]^{\mathrm{T}}\begin{bmatrix}X_O\\Y_O\\Z_O\end{bmatrix}. \tag{8.46}$$

Choosing the Euler angles ψ, θ, φ as rotational angles, from the last row of the matrix equation (8.46) one obtains the equation

$$(a\sin\psi\cos\eta-b\cos\psi\sin\eta)\sin\theta-X_O\sin\psi\sin\theta+Y_O\cos\psi\sin\theta-Z_O\cos\theta=0. \tag{8.47}$$

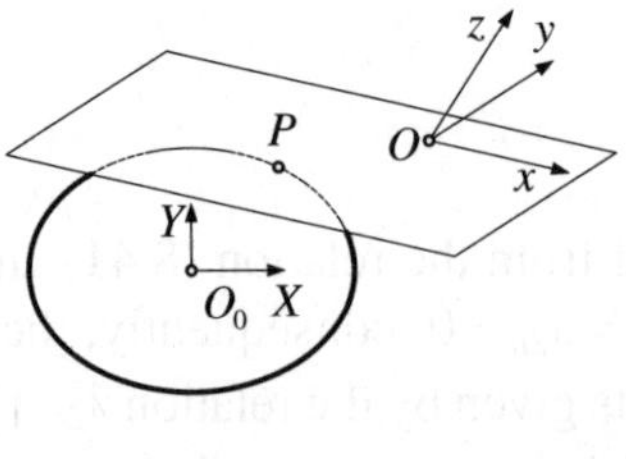

Figure 8.31 A plan moves remaining in contact to an ellipse in Example 8.3.1.

Further on, from the tangency condition (8.32) it results

$$a\cos\eta\sin\psi - b\sin\eta\cos\psi = 0. \tag{8.48}$$

By derivation of the equation (8.47) with respect to time one obtains the matrix of constraints

$$[\mathbf{B}] = \begin{bmatrix} -\sin\psi\sin\theta & \cos\psi\sin\theta & -\cos\theta & B_{1\psi} & B_{1\theta} & 0 & 0 \\ 0 & 0 & 0 & B_{2\psi} & 0 & 0 & B_{2\eta} \end{bmatrix}, \tag{8.49}$$

where

$$B_{1\psi} = (a\cos\psi\cos\eta + b\sin\psi\sin\eta)\sin\theta - X_O\cos\psi\sin\theta - Y_O\sin\psi\sin\theta,$$

$$B_{1\theta} = (a\sin\psi\cos\eta - b\cos\psi\sin\eta)\cos\theta - X_O\sin\psi\cos\theta + Y_O\cos\psi\cos\theta + Z_O\sin\theta,$$

$$B_{2\psi} = a\cos\eta\cos\psi + b\sin\eta\sin\psi, B_{2\eta} = -a\sin\eta\sin\psi - b\cos\eta\cos\psi. \tag{8.50}$$

Example 8.3.2 Determine the laws of motion for a sphere of radius r_0 and mass m, acted only by its own weight, and moving with sliding on two horizontal concurrent straight lines (Fig. 8.32), the angles between the straight lines being 2α.

Solution: We choose as origin O_0 the point of concurrence of the two straight lines, while the bisector of the angles between the two straight lines is chosen as the O_0Y axis.

The equations of the two support straight lines are

$$X = \eta_1\sin\alpha, Y = \eta_1\cos\alpha, Z = 0, \tag{8.51}$$

$$X = \eta_2\sin\alpha, Y = \eta_2\cos\alpha, Z = 0, \tag{8.52}$$

while the equation of the sphere relative to the local reference system $Oxyz$, with the origin O in the center of sphere, reads

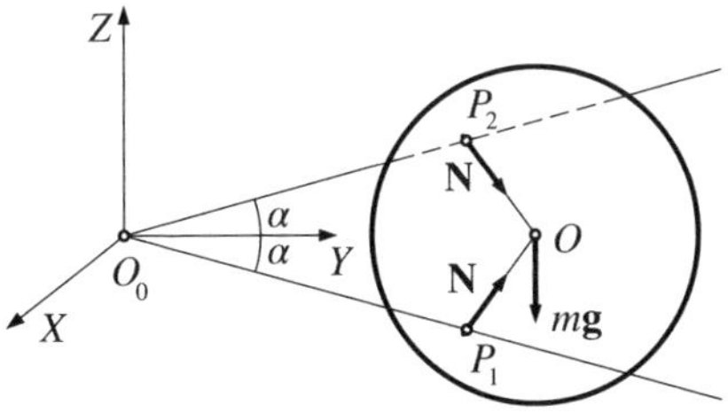

Figure 8.32 A sphere acted only by its own weight moving with sliding on two horizontal concurrent straight lines in Example 8.3.2.

$$f(x,y,z)=x^2+y^2+z^2-r_0^2=0. \tag{8.53}$$

The contact at the point P_1 leads to the equality

$$\begin{bmatrix}\eta_1\sin\alpha\\ \eta_1\cos\alpha\\ 0\end{bmatrix}=\begin{bmatrix}X_O\\ Y_O\\ Z_O\end{bmatrix}+[\mathbf{A}]\begin{bmatrix}x_1\\ y_1\\ z_1\end{bmatrix}, \tag{8.54}$$

while the tangency condition is

$$[\sin\alpha\cos\alpha\,0][\mathbf{A}]\begin{bmatrix}x_1\\ y_1\\ z_1\end{bmatrix}=0. \tag{8.55}$$

From the last equality one obtains

$$\eta_1=X_O\sin\alpha+Y_O\cos\alpha; \tag{8.56}$$

taking into account the equality

$$x_1^2+y_1^2+z_1^2=(\eta_1\sin\alpha-X_O)^2+(\eta_1\cos\alpha-Y_O)^2+Z_O^2=r_0^2, \tag{8.57}$$

we get the equation

$$X_O^2\cos^2\alpha+Y_O^2\sin^2\alpha+Z_O^2-2X_OY_O\sin\alpha\cos\alpha-r_0^2=0. \tag{8.58}$$

Analogically, for the contact at the point P_2 one obtains the equation

$$X_O^2\cos^2\alpha+Y_O^2\sin^2\alpha+Z_O^2+2X_OY_O\sin\alpha\cos\alpha-r_0^2=0; \tag{8.59}$$

it results that the point O (as we expected) remains in the plan $X_O=0$ and moves on the ellipse

$$Y_O^2\sin^2\alpha+Z_O^2-r_0^2=0. \tag{8.60}$$

Deriving the equation (8.60) with respect to time and considering the matrix of coordinates

$$\{\mathbf{q}\}=[X_O\;\;Y_O\;\;Z_O\;\;\psi\;\;\theta\;\;\varphi]^{\mathrm{T}}, \tag{8.61}$$

one obtains the matrix of constraints

$$[\mathbf{B}] = \left[Y_O \ \sin^2\alpha \ Z_O \ 0 \ 0 \ 0 \right]; \tag{8.62}$$

in this case the matrix differential equation of motion separates into the equations

$$m\ddot{Y}_O = \lambda Y_O \sin^2\alpha, \ m\ddot{Z}_O = \lambda Z_O - mg, \tag{8.63}$$

$$[\mathbf{Q}]^{\mathrm{T}}[\mathbf{J}][\mathbf{Q}]\{\ddot{\boldsymbol{\beta}}\} = \left\{\widetilde{\mathbf{F}}_{\boldsymbol{\beta}}\right\}, \tag{8.64}$$

where

$$\{\dot{\boldsymbol{\beta}}\} = \left[\dot{\psi} \ \dot{\theta} \ \dot{\varphi}\right]^{\mathrm{T}}, \tag{8.65}$$

$$[\mathbf{J}] = \begin{bmatrix} J & 0 & 0 \\ 0 & J & 0 \\ 0 & 0 & J \end{bmatrix}, \tag{8.66}$$

$$\left\{\widetilde{\mathbf{F}}_{\boldsymbol{\beta}}\right\} = -\left[[\mathbf{Q}]^{\mathrm{T}}[\mathbf{J}][\dot{\mathbf{Q}}] + [\mathbf{Q}]^{\mathrm{T}}[\boldsymbol{\omega}][\mathbf{J}][\mathbf{Q}]\right]\{\dot{\boldsymbol{\beta}}\}. \tag{8.67}$$

In these conditions, the term $[\mathbf{Q}]^{\mathrm{T}}[\boldsymbol{\omega}][\mathbf{J}][\mathbf{Q}]\{\dot{\boldsymbol{\beta}}\}$ vanishes, while the equation (8.64) becomes

$$[\mathbf{Q}]^{\mathrm{T}}\left[[\mathbf{Q}]\{\ddot{\boldsymbol{\beta}}\} + [\dot{\mathbf{Q}}]\{\dot{\boldsymbol{\beta}}\}\right] = \{\mathbf{0}\}. \tag{8.68}$$

Further on, based on the equality

$$\{\boldsymbol{\omega}\} = [\mathbf{Q}]\{\dot{\boldsymbol{\beta}}\}, \tag{8.69}$$

from the relation (8.68) one deduces the equation

$$\{\dot{\boldsymbol{\omega}}\} = \{\mathbf{0}\}; \tag{8.70}$$

considering that at the initial moment $t = 0\,\mathrm{s}$, $\psi = \theta = \varphi = 0\,\mathrm{rad}$, $\dot{\psi} = \dot{\theta} = \dot{\varphi} = 0\,\mathrm{rad/s}$, it results that the sphere has only translational motion.

Choosing the parametric representation

$$Y_O = \frac{r_0}{\sin\alpha}\sin\xi, \ Z_O = r_0\cos\xi \tag{8.71}$$

for the ellipse in the equation (8.60), from the relation (8.63) one obtains the differential equation

$$\left(\cos^2\xi + \sin^2\alpha\sin^2\xi\right)\ddot{\xi} - \dot{\xi}^2\sin\xi\cos\xi\cos^2\alpha - \frac{g}{r_0}\sin^2\alpha\sin\xi = 0. \tag{8.72}$$

8.4 The Rolling without Sliding of the Rigid Solid on Two Fixed Curves

8.4.1 General Considerations

We consider that the equations of the fixed curves are parametrically given by the equations

$$X_i = X_i(\eta_i), Y_i = Y_i(\eta_i), Z_i = Z_i(\eta_i),\ i = 1, 2, \tag{8.73}$$

while the bounding surface of the rigid solid, in the local reference systems $Oxyz$ is

$$f(x, y, z) = 0. \tag{8.74}$$

Maintaining the notations from paragraph 8.3 and taking into account that the rolling without sliding leads to the relations

$$\{\dot{\mathbf{R}}_O\} + [\mathbf{A}][\mathbf{r}_1]^{\mathrm{T}}[\mathbf{Q}]\{\dot{\boldsymbol{\beta}}\} = \{\mathbf{0}\}, \tag{8.75}$$

$$\{\dot{\mathbf{R}}_O\} + [\mathbf{A}][\mathbf{r}_2]^{\mathrm{T}}[\mathbf{Q}]\{\dot{\boldsymbol{\beta}}\} = \{\mathbf{0}\}, \tag{8.76}$$

from the equation (8.28) one obtains the equalities

$$\dot{\eta}_i\{\mathbf{R}_{ip}\} = [\mathbf{A}]\{\dot{\mathbf{r}}_i\},\ i = 1, 2. \tag{8.77}$$

From the equations (8.75) and (8.76), by subtraction and multiplication by $[\mathbf{A}]^{\mathrm{T}}$, one obtains the expression

$$[[\mathbf{r}_1] - [\mathbf{r}_2]][\mathbf{Q}]\{\dot{\boldsymbol{\beta}}\} = \{\mathbf{0}\}, \tag{8.78}$$

wherefrom one deduces the equality

$$[\mathbf{Q}]\{\dot{\boldsymbol{\beta}}\} = \zeta\{[\{\mathbf{r}_1\} - \{\mathbf{r}_2\}]\} \tag{8.79}$$

or

$$\{\boldsymbol{\omega}\} = \zeta\{[\{\mathbf{r}_1\} - \{\mathbf{r}_2\}]\}, \tag{8.80}$$

where ζ is a factor of proportionality. In these conditions, from the expressions (8.75) and (8.76) one obtains the relations

$$\{\dot{\mathbf{R}}_O\} = \zeta[\mathbf{A}][\mathbf{r}_2]\{\mathbf{r}_1\}. \tag{8.81}$$

In vector notation, the relations (8.80) and (8.81) are equivalent to the equalities

$$\boldsymbol{\omega}=\zeta(\mathbf{r}_1-\mathbf{r}_2), \tag{8.82}$$

$$\mathbf{v}_O=\zeta\mathbf{r}_2\times\mathbf{r}_1, \tag{8.83}$$

which show that, at the considered moment, the distribution of velocities is identical to that of the rotational motion about the straight line P_1P_2 (Fig. 8.33).

Denoting by $\mathbf{N}_1$, and $\mathbf{N}_2$ the reactions at the points P_1, and P_2, respectively, one may write the vector components $\mathbf{N}$, $\mathbf{C}_O$ of the torsor of reactions, reduced at the point O, in the form

$$\mathbf{N}=\mathbf{N}_1+\mathbf{N}_2,\ \mathbf{C}_O=-\mathbf{r}_1\times\mathbf{N}_1-\mathbf{r}_2\times\mathbf{N}_2. \tag{8.84}$$

This torsor has the particularity that its moment about the P_1P_2 axis is null; it results the relation

$$\mathbf{N}(\mathbf{r}_2\times\mathbf{r}_1)+\mathbf{C}_O(\mathbf{r}_1-\mathbf{r}_2)=0. \tag{8.85}$$

If we denote by λ_1, λ_2, λ_3, and λ_4, λ_5, λ_6 the projections of the vectors $\mathbf{N}$, and $\mathbf{C}_O$, respectively, onto the axes of the reference system $Oxyz$, then the relation (8.85) has the matrix correspondent

$$\left[\{\mathbf{r}_1\}^{\mathrm{T}}[\mathbf{r}_2]^{\mathrm{T}}\quad \{\{\mathbf{r}_1\}-\{\mathbf{r}_2\}\}^{\mathrm{T}}\right]\{\boldsymbol{\lambda}\}=0, \tag{8.86}$$

where

$$\{\boldsymbol{\lambda}\}=[\lambda_1\ \ \lambda_2\ \ \lambda_3\ \ \lambda_4\ \ \lambda_5\ \ \lambda_6]^{\mathrm{T}}. \tag{8.87}$$

A physical interpretation of the equation (8.86) states that the elementary work of the constraint forces is equal to zero, fact that leads to the equation

$$[\lambda_1\ \ \lambda_2\ \ \lambda_3]\{\mathbf{v}_O\}+[\lambda_4\ \ \lambda_5\ \ \lambda_6]\{\boldsymbol{\omega}\}=0; \tag{8.88}$$

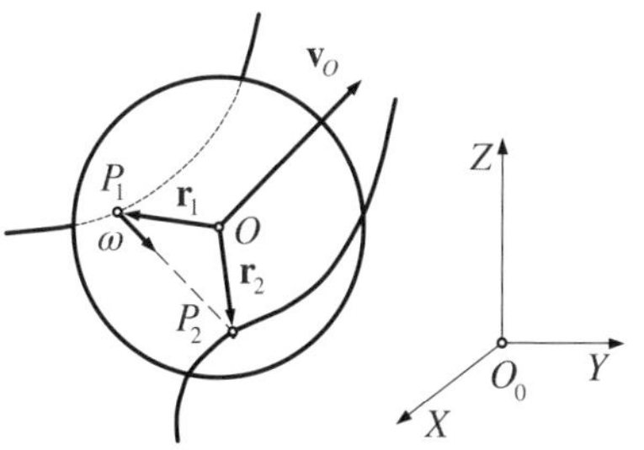

Figure 8.33 The rigid solid rolling without sliding on two fixed curves.

in the conditions of the relations (8.80) and (8.81) the last expression is equivalent to the equation (8.86).

8.4.2 *The Differential Equations of Motion*

To be out to obtain the differential equations of motion, we use the theorems of momentum and moment of momentum (3.61) and (3.69) which, for the case when the point O coincides to the center of weight, reads

$$[\mathbf{m}]\{\ddot{\mathbf{R}}_O\} = \{\mathbf{F}_s\} + \{\mathbf{F}_{Ls}\}, \tag{8.89}$$

$$[\mathbf{J}]\{\dot{\boldsymbol{\omega}}\} + [\boldsymbol{\omega}][\mathbf{J}]\{\boldsymbol{\omega}\} = \{\mathbf{F}_\beta\} + \{\mathbf{F}_{L\beta}\}, \tag{8.90}$$

where:

- $\{\mathbf{F}_s\}$, and $\{\mathbf{F}_{Ls}\}$ are the column matrices of the components of the resultants of the given forces, and constraint forces, respectively, onto the axes O_0X, O_0Y, O_0Z;
- $\{\mathbf{F}_\beta\}$, and $\{\mathbf{F}_{L\beta}\}$ are the column matrices of the components of the resultant moments of the given forces, and constraint forces, respectively, onto the axes Ox, Oy, Oz.

With the previous notations for the reactions, the equations (8.89) and (8.90) write in the form

$$[\mathbf{m}]\{\ddot{\mathbf{R}}_O\} = \{\mathbf{F}_s\} + [\mathbf{A}][\lambda_1 \ \ \lambda_2 \ \ \lambda_3]^{\mathrm{T}}, \tag{8.91}$$

$$[\mathbf{J}]\{\dot{\boldsymbol{\omega}}\} = \{\mathbf{F}_\beta\} + [\lambda_4 \ \ \lambda_5 \ \ \lambda_6]^{\mathrm{T}} - [\boldsymbol{\omega}][\mathbf{J}]\{\boldsymbol{\omega}\}. \tag{8.92}$$

Deriving the relations (8.80) and (8.81) with respect to time, one obtains the expressions

$$\{\ddot{\mathbf{R}}_O\} = \dot{\zeta}[\mathbf{A}][\mathbf{r}_2]\{\mathbf{r}_1\} + \zeta[\dot{\mathbf{A}}][\mathbf{r}_2]\{\mathbf{r}_1\} + \zeta[\mathbf{A}][\dot{\mathbf{r}}_2]\{\mathbf{r}_1\} + \zeta[\mathbf{A}][\mathbf{r}_2]\{\dot{\mathbf{r}}_1\}, \tag{8.93}$$

$$\{\dot{\boldsymbol{\omega}}\} = \dot{\zeta}\{\{\mathbf{r}_1\} - \{\mathbf{r}_2\}\} + \zeta\{\{\dot{\mathbf{r}}_1\} - \{\dot{\mathbf{r}}_2\}\}, \tag{8.94}$$

where $\{\dot{\mathbf{r}}_i\}$, $i = 1, 2$, are given by the relations (8.77), while $\{\mathbf{r}_i\}$, $i = 1, 2$, are given by the equalities

$$\{\mathbf{R}_i\} = \{\mathbf{R}_O\} + [\mathbf{A}]\{\mathbf{r}_i\},\ i = 1, 2. \tag{8.95}$$

Further on, using the notations

$$\{\widetilde{\mathbf{F}}_s\} = -\zeta[\mathbf{m}]\{[\dot{\mathbf{A}}][\mathbf{r}_2]\{\mathbf{r}_1\} + [\mathbf{A}][\dot{\mathbf{r}}_2]\{\mathbf{r}_1\} + [\mathbf{A}][\mathbf{r}_2]\{\dot{\mathbf{r}}_1\}\}, \tag{8.96}$$

$$\left\{\widetilde{\mathbf{F}}_{\boldsymbol{\beta}}\right\} = -\zeta^2[[\mathbf{r}_1]-[\mathbf{r}_2]][\mathbf{J}]\{\{\mathbf{r}_1\}-\{\mathbf{r}_2\}\}-\zeta[\mathbf{J}]\{\{\dot{\mathbf{r}}_1\}-\{\dot{\mathbf{r}}_2\}\}, \tag{8.97}$$

$$\{\mathbf{F}\} = \begin{bmatrix}\{\mathbf{F}_s\}\\ \{\mathbf{F}_{\boldsymbol{\beta}}\}\end{bmatrix}, \left\{\widetilde{\mathbf{F}}\right\} = \begin{bmatrix}\left\{\widetilde{\mathbf{F}}_s\right\}\\ \left\{\widetilde{\mathbf{F}}_{\boldsymbol{\beta}}\right\}\end{bmatrix}, \tag{8.98}$$

$$\{\mathbf{M}^*\} = \begin{bmatrix}[\mathbf{m}][\mathbf{A}][\mathbf{r}_2]\{\mathbf{r}_1\}\\ [\mathbf{J}]\{\{\mathbf{r}_1\}-\{\mathbf{r}_2\}\}\end{bmatrix}, \tag{8.99}$$

$$[\mathbf{B}_1] = \begin{bmatrix}[\mathbf{A}] & [\mathbf{0}]\\ [\mathbf{0}] & [\mathbf{I}]\end{bmatrix}, [\mathbf{B}_2] = \begin{bmatrix}\{\mathbf{r}_1\}^{\mathrm{T}}[\mathbf{r}_2]^{\mathrm{T}} & \{\{\mathbf{r}_1\}-\{\mathbf{r}_2\}\}^{\mathrm{T}}\end{bmatrix}, \tag{8.100}$$

from the relations (8.86), (8.91), (8.92), (8.93), and (8.94) one obtains the matrix equation

$$\begin{bmatrix}\{\mathbf{M}^*\} & -[\mathbf{B}_1]\\ 0 & [\mathbf{B}_2]\end{bmatrix}\begin{bmatrix}\dot{\zeta}\\ \{\boldsymbol{\lambda}\}\end{bmatrix} = \begin{bmatrix}\{\mathbf{F}\}+\left\{\widetilde{\mathbf{F}}\right\}\\ 0\end{bmatrix}, \tag{8.101}$$

wherefrom one deduces the unknowns $\{\boldsymbol{\lambda}\}$ and $\dot{\boldsymbol{\zeta}}$.

Indeed, the system (8.101) may be written as

$$[\mathbf{m}][\mathbf{A}][\mathbf{r}_2]\{\mathbf{r}_1\}\dot{\zeta}-[\mathbf{A}]\begin{bmatrix}\lambda_1\\ \lambda_2\\ \lambda_3\end{bmatrix} = \{\mathbf{F}_s\}+\left\{\widetilde{\mathbf{F}}_s\right\}, [\mathbf{J}]\{\{\mathbf{r}_1\}-\{\mathbf{r}_2\}\}\dot{\zeta}-\begin{bmatrix}\lambda_4\\ \lambda_5\\ \lambda_6\end{bmatrix} = \{\mathbf{F}_{\boldsymbol{\beta}}\}+\left\{\widetilde{\mathbf{F}}_{\boldsymbol{\beta}}\right\},$$

$$\{\mathbf{r}_1\}^{\mathrm{T}}[\mathbf{r}_2]^{\mathrm{T}}\begin{bmatrix}\lambda_1\\ \lambda_2\\ \lambda_3\end{bmatrix} + \{\{\mathbf{r}_1\}^{\mathrm{T}}-\{\mathbf{r}_2\}^{\mathrm{T}}\}\begin{bmatrix}\lambda_4\\ \lambda_5\\ \lambda_6\end{bmatrix} = 0 \tag{8.102}$$

or

$$\begin{bmatrix}\lambda_1\\ \lambda_2\\ \lambda_3\end{bmatrix} = [\mathbf{m}][\mathbf{r}_2]\{\mathbf{r}_1\}\dot{\zeta}-[\mathbf{A}]^{\mathrm{T}}\left\{\{\mathbf{F}_s\}+\left\{\widetilde{\mathbf{F}}_s\right\}\right\}, \begin{bmatrix}\lambda_4\\ \lambda_5\\ \lambda_6\end{bmatrix}[\mathbf{J}]\{\{\mathbf{r}_1\}-\{\mathbf{r}_2\}\}\dot{\zeta}-\{\mathbf{F}_{\boldsymbol{\beta}}\}-\left\{\widetilde{\mathbf{F}}_{\boldsymbol{\beta}}\right\},$$

$$[\mathbf{m}]\{\mathbf{r}_1\}^{\mathrm{T}}[\mathbf{r}_2]^{\mathrm{T}}[\mathbf{r}_2]\{\mathbf{r}_1\}\dot{\zeta}-\{\mathbf{r}_1\}^{\mathrm{T}}[\mathbf{r}_2]^{\mathrm{T}}[\mathbf{A}]^{\mathrm{T}}\left\{\{\mathbf{F}_s\}+\left\{\widetilde{\mathbf{F}}_s\right\}\right\}+$$

$$\{\{\mathbf{r}_1\}-\{\mathbf{r}_2\}\}^{\mathrm{T}}[\mathbf{J}]\{\{\mathbf{r}_1\}-\{\mathbf{r}_2\}\}\dot{\zeta}-\{\{\mathbf{r}_1\}-\{\mathbf{r}_2\}\}^{\mathrm{T}}\left\{\{\mathbf{F}_{\boldsymbol{\beta}}\}+\left\{\widetilde{\mathbf{F}}_{\boldsymbol{\beta}}\right\}\right\}=0; \tag{8.103}$$

it results

$$\dot{\zeta}=\frac{\{\mathbf{r}_1\}^{\mathrm{T}}[\mathbf{r}_2]^{\mathrm{T}}[\mathbf{A}]^{\mathrm{T}}\left\{\{\mathbf{F}_s\}+\left\{\widetilde{\mathbf{F}}_s\right\}\right\}+\{\{\mathbf{r}_1\}-\{\mathbf{r}_2\}\}^{\mathrm{T}}\left\{\{\mathbf{F}_\beta\}+\left\{\widetilde{\mathbf{F}}_\beta\right\}\right\}}{[\mathbf{m}]\{\mathbf{r}_1\}^{\mathrm{T}}[\mathbf{r}_2]^{\mathrm{T}}[\mathbf{r}_2]\{\mathbf{r}_1\}+\{\{\mathbf{r}_1\}-\{\mathbf{r}_2\}\}^{\mathrm{T}}[\mathbf{J}]\{\{\mathbf{r}_1\}-\{\mathbf{r}_2\}\}}. \tag{8.104}$$

We then determine the parameter ζ by numerical integration; it permits the determination of the parameters X_O, Y_O, Z_O, ψ, θ, φ by numerical integration using the relations (8.79) and (8.81).

To be out to determine the parameters η_i, $i=1$, 2, one considers the tangency equations

$$\{\mathbf{R}_{ip}\}^{\mathrm{T}}[\mathbf{A}]\{\mathbf{f}_{ip}\}=0; \tag{8.105}$$

by derivation with respect to time these relations become

$$\dot{\eta}_i\left[\{\mathbf{R}_{ip}\}^{\mathrm{T}}[\mathbf{A}][\mathbf{f}_s][\mathbf{A}]^{\mathrm{T}}\{\mathbf{R}_{ip}\}+\{\mathbf{R}_{is}\}^{\mathrm{T}}[\mathbf{A}]\{\mathbf{f}_{ip}\}\right]+\zeta\{\mathbf{R}_{ip}\}^{\mathrm{T}}[\mathbf{A}]\left[\mathbf{f}_{ip}\right]^{\mathrm{T}}\{\{\mathbf{r}_1\}-\{\mathbf{r}_2\}\}=0,\ i=1,2, \tag{8.106}$$

wherefrom, with the previous calculated values of the parameters, one may determine, by integration, the new values of the parameters η_i, $i=1$, 2.

8.4.3 The Algorithm for the Numerical Calculation

At the initial moment $t=0$ s one consider as known the parameters X_O, Y_O, Z_O, ψ, θ, φ, η_1, η_2, $\dot{X}_O$, $\dot{Y}_O$, $\dot{Z}_O$, $\dot{\psi}$, $\dot{\theta}$, $\dot{\varphi}$, $\dot{\eta}_1$, $\dot{\eta}_2$.

One successively determines:

- the matrices $[\boldsymbol{\psi}]$, $[\boldsymbol{\theta}]$, $[\boldsymbol{\varphi}]$, $[\mathbf{A}]$, $[\mathbf{A}_\psi]$, $[\mathbf{A}_\theta]$, $[\mathbf{A}_\varphi]$, $[\dot{\mathbf{A}}]$, $[\boldsymbol{\omega}]$, $[\mathbf{Q}]$, $[\mathbf{Q}_\theta]$, $[\mathbf{Q}_\varphi]$, $[\dot{\mathbf{Q}}]$;
- the matrices $\{\mathbf{R}_i\}$, $\{\mathbf{R}_{ip}\}$, $\{\mathbf{R}_{is}\}$;
- the matrices $\{\mathbf{r}_i\}$, $\{\dot{\mathbf{r}}_i\}$ from the relations (8.95) and (8.77), and then the matrices $[\mathbf{r}_i]$, $[\dot{\mathbf{r}}_i]$, $\{\mathbf{f}_{ip}\}$, $[\mathbf{f}_{ip}]$, $[\mathbf{f}_{is}]$;
- the parameter ζ from the relation (8.80);
- the matrices $\left\{\widetilde{\mathbf{F}}_s\right\}$, $\left\{\widetilde{\mathbf{F}}_\beta\right\}$, $\left\{\widetilde{\mathbf{F}}\right\}$, $\{\mathbf{M}^*\}$, $[\mathbf{B}_1]$, $[\mathbf{B}_2]$;
- the parameters $\dot{\zeta}$, λ_i;
- the new parameter ζ by integration;
- the new values of the matrices $\{\dot{\boldsymbol{\beta}}\}$, $\{\dot{\mathbf{R}}_O\}$ from the relations (8.79), (8.80);
- the new values of the matrices $\{\boldsymbol{\beta}\}$, $\{\mathbf{R}_O\}$ by integration;
- the new values of the parameters $\dot{\eta}_i$ from the relations (8.106);
- the new values of the parameters η_i by integration.

One passes to the next step in the interval of time.

Example 8.4.1 Determine the laws of motion for a sphere of radius r_0 and mass m, which rolls without sliding on two horizontal concurrent straight, the angle between the two straight lines being 2α (Fig. 8.32). The sphere is acted only by its own weight. The following parameters are known $m=32\,\text{kg}$, $r=0.1\,\text{m}$, $\alpha=50$, $J_x=J_y=J_z=\frac{2m}{5}r_0^2=0.048\,\text{kgm}^2$. One chooses as rotational angles ψ, θ, φ the Bryan angles. The initial conditions at $t=0\,\text{s}$ are $X_O=0\,\text{m}$, $Y_O=0.1r_0=0.01\,\text{m}$, $Z_O=r_0\sqrt{1-0{,}01\sin^2\alpha}=0.099996\,\text{m}$, $\psi=\theta=\varphi=0\,\text{rad}$, $\eta_1=\eta_2=Y_O\cos\alpha=0.009962\,\text{m}$, $\dot{X}_O=\dot{Y}_O=\dot{Z}_O=0\,\text{m/s}$, $\dot{\psi}=\dot{\theta}=\dot{\varphi}=0\,\text{rad/s}$, $\dot{\eta}_1=\dot{\eta}_2=0\,\text{rad/s}$, $\zeta=0\,\text{m}^{-1}\text{s}^{-1}$.

Solution: We successively have

$$[\boldsymbol{\psi}]=\begin{bmatrix}1 & 0 & 0\\ 0 & \cos\psi & -\sin\psi\\ 0 & \sin\psi & \cos\psi\end{bmatrix},\ [\boldsymbol{\theta}]=\begin{bmatrix}\cos\theta & 0 & \sin\theta\\ 0 & 1 & 0\\ -\sin\theta & 0 & \cos\theta\end{bmatrix},\ [\boldsymbol{\varphi}]=\begin{bmatrix}\cos\varphi & -\sin\varphi & 0\\ \sin\varphi & \cos\varphi & 0\\ 0 & 0 & 1\end{bmatrix},$$

$$[\mathbf{A}]=[\boldsymbol{\psi}][\boldsymbol{\theta}][\boldsymbol{\varphi}],\ [\mathbf{A}_\psi]=[\mathbf{U}_\psi][\mathbf{A}],[\mathbf{A}_\theta]=[\mathbf{A}][\boldsymbol{\varphi}]^{\mathrm{T}}[\mathbf{U}_\theta][\boldsymbol{\varphi}],\ [\mathbf{A}_\varphi]=[\mathbf{A}][\mathbf{U}_\varphi],$$

$$[\dot{\mathbf{A}}]=\dot{\psi}[\mathbf{A}_\psi]+\dot{\theta}[\mathbf{A}_\theta]+\dot{\varphi}[\mathbf{A}_\varphi],[\mathbf{Q}]=[\boldsymbol{\varphi}]^{\mathrm{T}}\left[[\boldsymbol{\theta}]^{\mathrm{T}}\{\mathbf{u}_\psi\}\ \{\mathbf{u}_\theta\}\ \{\mathbf{u}_\varphi\}\right]$$
$$=\begin{bmatrix}\cos\varphi\cos\theta & \sin\varphi & 0\\ -\sin\varphi\cos\theta & \cos\varphi & 0\\ \sin\theta & 0 & 1\end{bmatrix},$$

$$[\mathbf{Q}_\theta]=\begin{bmatrix}-\cos\varphi\sin\theta & 0 & 0\\ \sin\varphi\sin\theta & 0 & 0\\ \cos\theta & 0 & 0\end{bmatrix},\ [\mathbf{Q}_\varphi]=\begin{bmatrix}-\sin\varphi\cos\theta & \cos\varphi & 0\\ -\cos\varphi\cos\theta & -\sin\varphi & 0\\ 0 & 0 & 0\end{bmatrix},\ [\dot{\mathbf{Q}}]=\dot{\varphi}[\mathbf{Q}_\varphi]+\dot{\theta}[\mathbf{Q}_\theta],$$

$$\{\mathbf{R}_O\}=[X_O\ Y_O\ Z_O]^{\mathrm{T}},\{\boldsymbol{\beta}\}=[\psi\ \theta\ \varphi]^{\mathrm{T}},\{\dot{\mathbf{R}}_O\}=[\dot{X}_O\ \dot{Y}_O\ \dot{Z}_O]^{\mathrm{T}},\{\dot{\boldsymbol{\beta}}\}=[\dot{\psi}\ \dot{\theta}\ \dot{\varphi}]^{\mathrm{T}},$$

$$\{\mathbf{R}_1\}=\eta_1\begin{bmatrix}\sin\alpha\\ \cos\alpha\\ 0\end{bmatrix},\{\mathbf{R}_{1p}\}=\begin{bmatrix}\sin\alpha\\ \cos\alpha\\ 0\end{bmatrix},\{\mathbf{R}_{1s}\}=\begin{bmatrix}0\\ 0\\ 0\end{bmatrix},\{\mathbf{R}_2\}=\eta_2\begin{bmatrix}-\sin\alpha\\ \cos\alpha\\ 0\end{bmatrix},$$

$$\{\mathbf{R}_2\}=\begin{bmatrix}-\sin\alpha\\ \cos\alpha\\ 0\end{bmatrix},\{\mathbf{R}_{2s}\}=\begin{bmatrix}0\\0\\0\end{bmatrix},\{\mathbf{r}_i\}=\begin{bmatrix}x_i\\y_i\\z_i\end{bmatrix}=[\mathbf{A}]^{\mathrm{T}}\{\{\mathbf{R}_i\}-\{\mathbf{R}_O\}\},\ i=1,2,$$

$$[\mathbf{r}_i]=\begin{bmatrix}0 & -z_i & y_i\\ z_i & 0 & -x_i\\ -y_i & x_i & 0\end{bmatrix},\ i=1,2,\{\dot{\mathbf{r}}_i\}=\dot{\eta}_i[\mathbf{A}]^{\mathrm{T}}\{\mathbf{R}_{ip}\},\ i=1,2,[\dot{\mathbf{r}}_i]=\begin{bmatrix}0 & -\dot{z}_i & \dot{y}_i\\ \dot{z}_i & 0 & -\dot{x}_i\\ -\dot{y}_i & \dot{x}_i & 0\end{bmatrix},$$

$$i=1,2,\{\mathbf{f}_{ip}\}=\begin{bmatrix}2x_i\\2y_i\\2z_i\end{bmatrix}=2\{\mathbf{r}_i\},\ [\mathbf{f}_{ip}]=2[\mathbf{r}_i],\ i=1,2.$$

The obtained numerical results are represented in Fig. 8.34 to Fig. 8.40.

One may observe in these diagrams that the point O (the center of sphere) moves on a parabola. The diagrams are correct only for $Z_O \geq 0$ m, which is also the condition to stop the iterations.

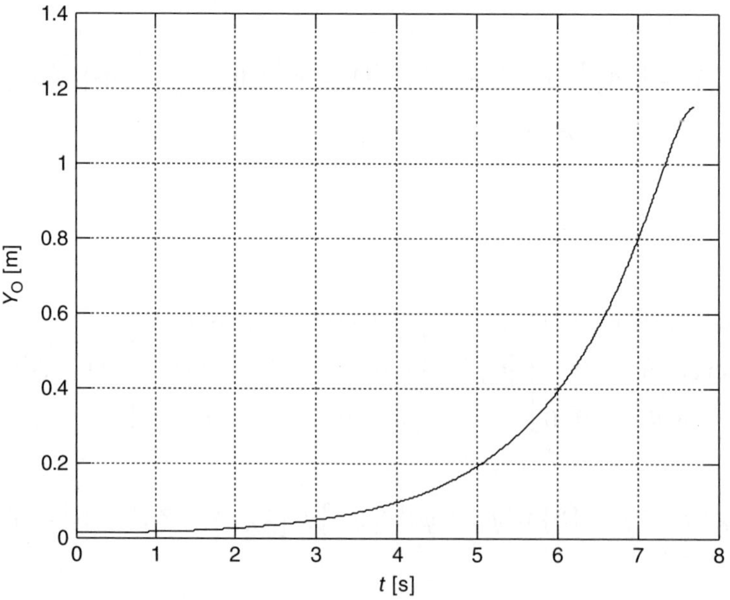

Figure 8.34 Time history $Y_O = Y_O(t)$.

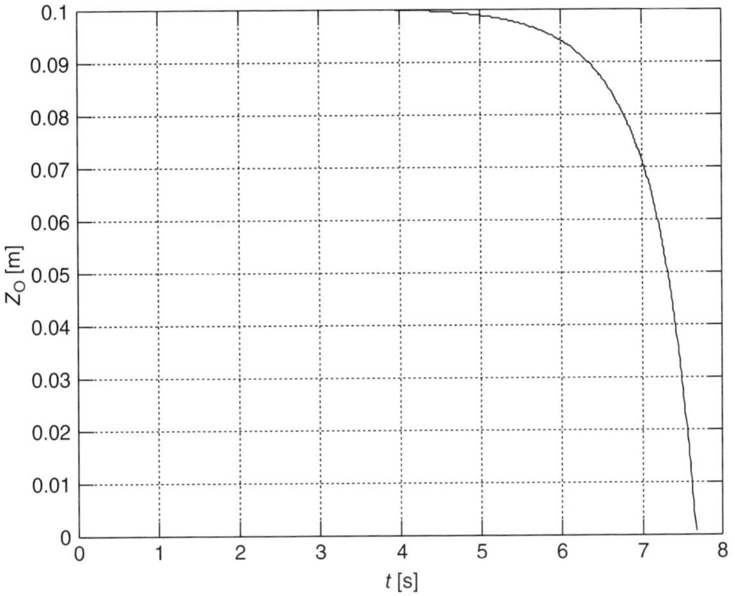

Figure 8.35 Time history $Z_O = Z_O(t)$.

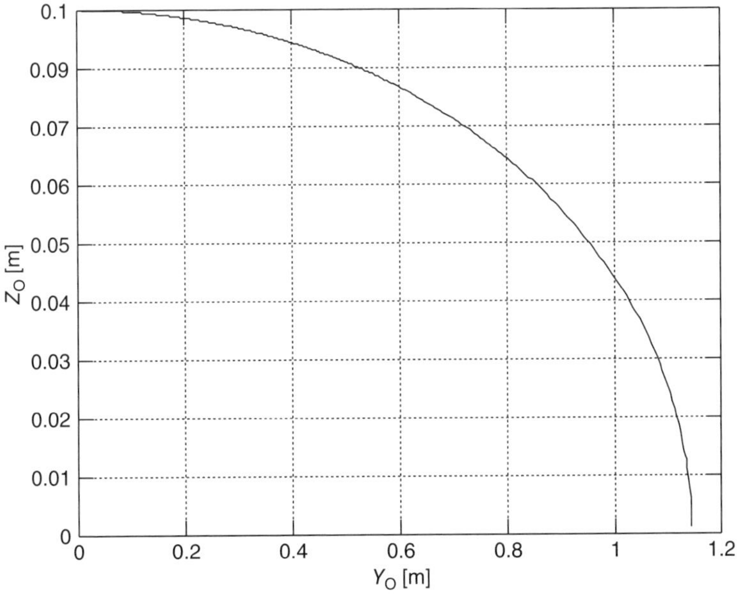

Figure 8.36 Time history $Z_O = Z_O(Y_O)$.

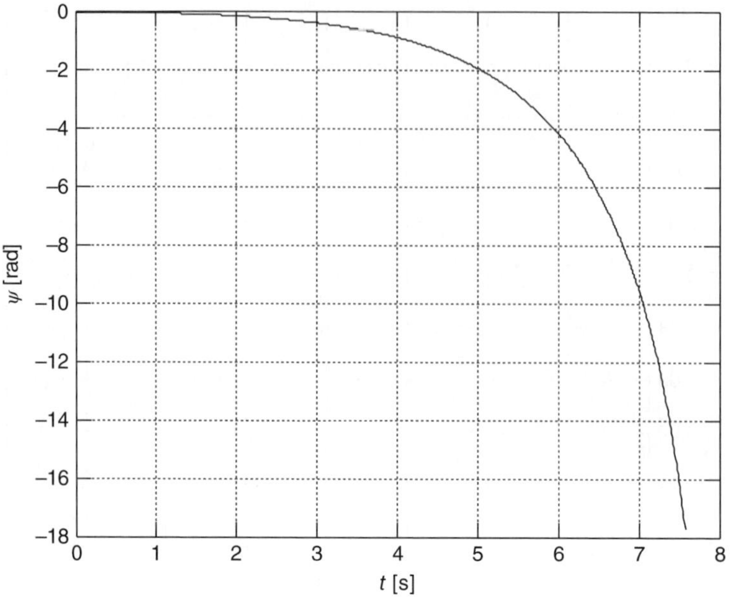

Figure 8.37 Time history $\psi = \psi(t)$.

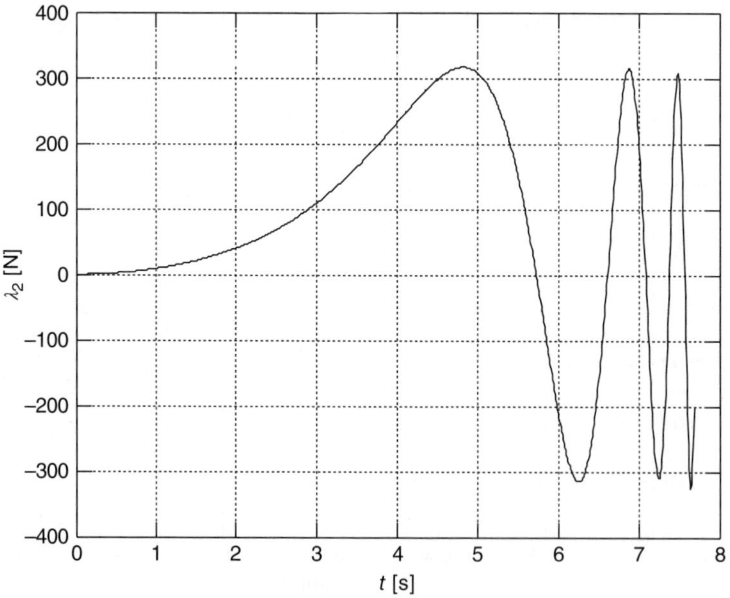

Figure 8.38 Time history $\lambda_2 = \lambda_2(t)$.

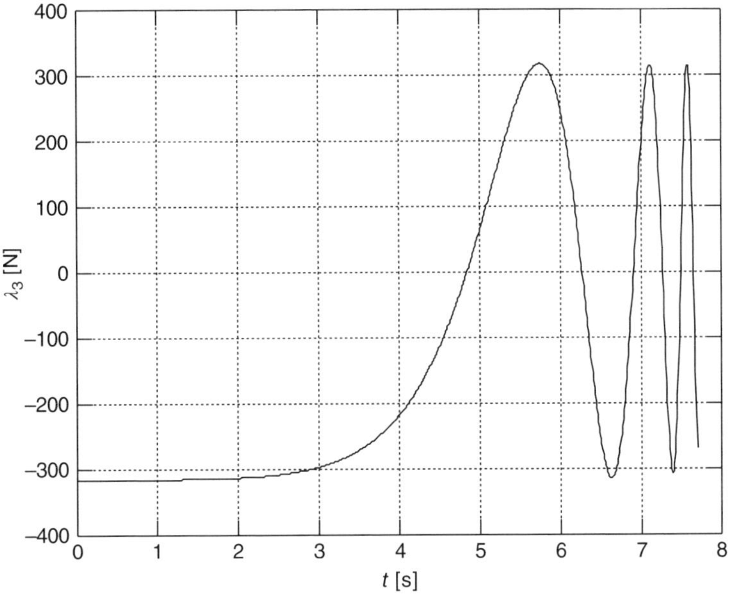

Figure 8.39 Time history $\lambda_3 = \lambda_3(t)$.

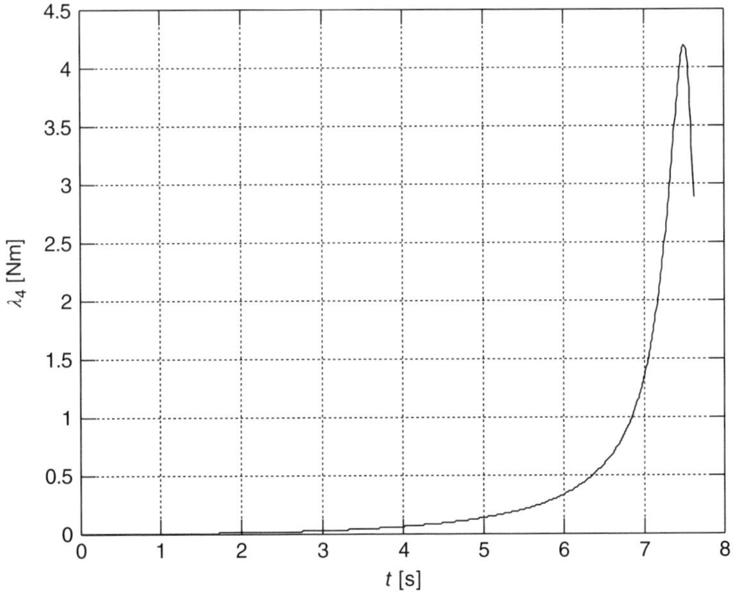

Figure 8.40 Time history $\lambda_4 = \lambda_4(t)$.

8.5 The Rolling without Sliding of a Rigid Solid on a Fixed Surface

8.5.1 The Matrix of Constraints

We consider that in the reference system O_0XYZ the equation of the fixed surface is given in implicit form

$$F(X,Y,Z)=0, \tag{8.107}$$

while in the local reference system $Oxyz$ the equations of the bounding surface of the rigid solid are given in parametric form

$$x=x(\xi,\eta),\ y=y(\xi,\eta),\ z=z(\xi,\eta). \tag{8.108}$$

Using the notations

$$\{\mathbf{R}\}=[X\ \ Y\ \ Z]^{\mathrm{T}},\{\mathbf{r}\}=[x\ \ y\ \ z]^{\mathrm{T}},\{\mathbf{R}_O\}=[X_O\ \ Y_O\ \ Z_O]^{\mathrm{T}}, \tag{8.109}$$

the contact condition between the surfaces reads

$$\{\mathbf{R}\}=\{\mathbf{R}_O\}+[\mathbf{A}]\{\mathbf{r}\}; \tag{8.110}$$

the tangency conditions are

$$\{\mathbf{F}_p\}^{\mathrm{T}}[\mathbf{A}]\{\mathbf{r}_\xi\}=0,\{\mathbf{F}_p\}^{\mathrm{T}}[\mathbf{A}]\{\mathbf{r}_\eta\}=0, \tag{8.111}$$

where $\{\mathbf{F}_p\}$, $\{\mathbf{r}_\xi\}$, $\{\mathbf{r}_\eta\}$ are the column matrices of the partial derivatives

$$\{\mathbf{F}_p\}=[F_{,X}\ \ F_{,Y}\ \ F_{,Z}]^{\mathrm{T}},\{\mathbf{r}_\xi\}=[x_{,\xi}\ \ y_{,\xi}\ \ z_{,\xi}]^{\mathrm{T}},\{\mathbf{r}_\eta\}=[x_{,\eta}\ \ y_{,\eta}\ \ z_{,\eta}]^{\mathrm{T}}. \tag{8.112}$$

The condition of rolling without sliding pretends the relative velocity at the contact point to be null, that is

$$\{\dot{\mathbf{R}}_O\}+[\mathbf{A}][\mathbf{r}]^{\mathrm{T}}[\mathbf{Q}]\{\dot{\boldsymbol{\beta}}\}=\{\mathbf{0}\}, \tag{8.113}$$

where

$$\{\dot{\boldsymbol{\beta}}\}=[\dot{\psi}\ \ \dot{\theta}\ \ \dot{\varphi}]^{\mathrm{T}},[\mathbf{r}]=\begin{bmatrix}0 & -z & y\\ z & 0 & -x\\ -y & x & 0\end{bmatrix}. \tag{8.114}$$

Deriving the relation (8.110) with respect to time and taking into account the relation (8.113), one obtains the expression

$$\{\dot{\mathbf{R}}\}=\dot{\xi}[\mathbf{A}]\{\mathbf{r}_\xi\}+\dot{\eta}[\mathbf{A}]\{\mathbf{r}_\eta\}, \tag{8.115}$$

Considering the notations

$$[\mathbf{F}_s]=\begin{bmatrix} f,_{X^2} & f,_{XY} & f,_{XZ} \\ f,_{XY} & f,_{Y^2} & f,_{YZ} \\ f,_{XZ} & f,_{YZ} & f,_{Z^2} \end{bmatrix},\left\{\mathbf{r}_{\xi^2}\right\}=\begin{bmatrix} x,_{\xi^2} \\ y,_{\xi^2} \\ z,_{\xi^2} \end{bmatrix},\left\{\mathbf{r}_{\xi\eta}\right\}=\begin{bmatrix} x,_{\xi\eta} \\ y,_{\xi\eta} \\ z,_{\xi\eta} \end{bmatrix},\left\{\mathbf{r}_{\eta^2}\right\}=\begin{bmatrix} x,_{\eta^2} \\ y,_{\eta^2} \\ z,_{\eta^2} \end{bmatrix},$$
$$[\mathbf{r}_\xi]=\begin{bmatrix} 0 & -z,_\xi & y,_\xi \\ z,_\xi & 0 & -x,_\xi \\ -y,_\xi & x,_\xi & 0 \end{bmatrix},[\mathbf{r}_\eta]=\begin{bmatrix} 0 & -z,_\eta & y,_\eta \\ z,_\eta & 0 & -x,_\eta \\ -y,_\eta & x,_\eta & 0 \end{bmatrix}, \tag{8.116}$$

$$\begin{gathered}[\mathbf{B}_{1\boldsymbol{\beta}}]=\{\mathbf{F}_p\}^{\mathrm{T}}[\mathbf{A}][\mathbf{r}_\xi]^{\mathrm{T}}[\mathbf{Q}],B_{1\xi}=\{\mathbf{F}_p\}^{\mathrm{T}}[\mathbf{A}]\left\{\mathbf{r}_{\xi^2}\right\}+\{\mathbf{r}_\xi\}^{\mathrm{T}}[\mathbf{A}]^{\mathrm{T}}[\mathbf{F}_s][\mathbf{A}]\{\mathbf{r}_\xi\},\\ B_{1\eta}=\{\mathbf{F}_p\}^{\mathrm{T}}[\mathbf{A}]\{\mathbf{r}_{\xi\eta}\}+\{\mathbf{r}_\xi\}^{\mathrm{T}}[\mathbf{A}]^{\mathrm{T}}[\mathbf{F}_s][\mathbf{A}]\{\mathbf{r}_\eta\},\end{gathered} \tag{8.117}$$

$$\begin{gathered}[\mathbf{B}_{2\boldsymbol{\beta}}]=\{\mathbf{F}_p\}^{\mathrm{T}}[\mathbf{A}][\mathbf{r}_\eta]^{\mathrm{T}}[\mathbf{Q}],B_{2\xi}=\{\mathbf{F}_p\}^{\mathrm{T}}[\mathbf{A}]\{\mathbf{r}_{\xi\eta}\}+\{\mathbf{r}_\eta\}^{\mathrm{T}}[\mathbf{A}]^{\mathrm{T}}[\mathbf{F}_s][\mathbf{A}]\{\mathbf{r}_\xi\},\\ B_{2\eta}=\{\mathbf{F}_p\}^{\mathrm{T}}[\mathbf{A}]\{\mathbf{r}_{\xi\eta}\}+\{\mathbf{r}_\eta\}^{\mathrm{T}}[\mathbf{A}]^{\mathrm{T}}[\mathbf{F}_s][\mathbf{A}]\{\mathbf{r}_\eta\},\end{gathered} \tag{8.118}$$

taking into account the equality (8.115) and the relation

$$\{\dot{\mathbf{F}}_p\}=[\mathbf{F}_s]\{\dot{\mathbf{R}}\}, \tag{8.119}$$

by derivation of the equations (8.111) with respect to time one obtains the expressions

$$[\mathbf{B}_{1\boldsymbol{\beta}}]\{\dot{\boldsymbol{\beta}}\}+B_{1\xi}\dot{\xi}+B_{1\eta}\dot{\eta}=0,\ [\mathbf{B}_{2\boldsymbol{\beta}}]\{\dot{\boldsymbol{\beta}}\}+B_{2\xi}\dot{\xi}+B_{2\eta}\dot{\eta}=0. \tag{8.120}$$

With the aid of the equation (8.113) one deduces the matrix of constraints

$$[\mathbf{B}]=\begin{bmatrix} [\mathbf{I}] & [\mathbf{A}][\mathbf{r}]^{\mathrm{T}}[\mathbf{Q}] & \{\mathbf{0}\} & \{\mathbf{0}\} \\ [\mathbf{0}] & [\mathbf{B}_{1\boldsymbol{\beta}}] & B_{1\xi} & B_{1\eta} \\ [\mathbf{0}] & [\mathbf{B}_{2\boldsymbol{\beta}}] & B_{2\xi} & B_{2\eta} \end{bmatrix}, \tag{8.121}$$

the column matrix of the coordinates being

$$\{\mathbf{q}^*\}=\left[\{\mathbf{q}\}^{\mathrm{T}}\ \ \{\widetilde{\mathbf{q}}\}^{\mathrm{T}}\right]^{\mathrm{T}}, \tag{8.122}$$

where

$$\{\mathbf{q}\}=[X_O\ \ Y_O\ \ Z_O\ \ \psi\ \ \theta\ \ \varphi]^{\mathrm{T}},\{\widetilde{\mathbf{q}}\}=[\xi\ \ \eta]^{\mathrm{T}}. \tag{8.123}$$

To be out to obtain the derivative matrix $[\dot{\mathbf{B}}]$ one has to consider the relations

$$[\dot{\mathbf{r}}]=\dot{\xi}[\mathbf{r}_\xi]+\dot{\eta}[\mathbf{r}_\eta],\ [\dot{\mathbf{r}}_\xi]=\dot{\xi}\left[\mathbf{r}_{\xi^2}\right]+\dot{\eta}[\mathbf{r}_{\xi\eta}],\ [\dot{\mathbf{r}}_\eta]=\dot{\xi}[\mathbf{r}_{\xi\eta}]+\dot{\eta}[\mathbf{r}_{\eta^2}],\ \{\dot{\mathbf{r}}_\xi\}=\dot{\xi}\left\{\mathbf{r}_{\xi^2}\right\}+\dot{\eta}\{\mathbf{r}_{\xi\eta}\},$$
$$\{\dot{\mathbf{r}}_\eta\}=\dot{\xi}\{\mathbf{r}_{\xi\eta}\}+\dot{\eta}\{\mathbf{r}_{\eta^2}\},\ \left\{\dot{\mathbf{r}}_{\xi^2}\right\}=\dot{\xi}\left\{\mathbf{r}_{\xi^3}\right\}+\dot{\eta}\left\{\mathbf{r}_{\xi^2\eta}\right\},\ \{\dot{\mathbf{r}}_{\xi\eta}\}=\dot{\xi}\left\{\mathbf{r}_{\xi^2\eta}\right\}+\dot{\eta}\{\mathbf{r}_{\xi\eta^2}\},$$
$$\{\dot{\mathbf{r}}_{\eta^2}\}=\dot{\xi}\{\mathbf{r}_{\xi\eta^2}\}+\dot{\eta}\{\mathbf{r}_{\eta^3}\},\ [\dot{\mathbf{F}}_s]=\left[[\mathbf{F}_{1X}]\{\dot{\mathbf{R}}\}\ \ [\mathbf{F}_{1Y}]\{\dot{\mathbf{R}}\}\ \ [\mathbf{F}_{1Z}]\{\dot{\mathbf{R}}\}\right], \tag{8.124}$$

where

$$[\mathbf{F}_{1X}]=\begin{bmatrix} F_{,X^3} & F_{,X^2Y} & F_{,X^2Z} \\ F_{,X^2Y} & F_{,XY^2} & F_{,XYZ} \\ F_{,X^2Z} & F_{,XYZ} & F_{,XZ^2}\end{bmatrix},\ [\mathbf{F}_{1Y}]=\begin{bmatrix} F_{,X^2Y} & F_{,XY^2} & F_{,XYZ} \\ F_{,XY^2} & F_{,Y^3} & F_{,Y^2Z} \\ F_{,XYZ} & F_{,Y^2Z} & F_{,YZ^2}\end{bmatrix},$$
$$[\mathbf{F}_{1Z}]=\begin{bmatrix} F_{,X^2Z} & F_{,XYZ} & F_{,XZ^2} \\ F_{,XYZ} & F_{,Y^2Z} & F_{,YZ^2} \\ F_{,XZ^2} & F_{,YZ^2} & F_{,Z^3}\end{bmatrix}. \tag{8.125}$$

8.5.2 *The Matrix Differential Equation of Motion*

Starting from the expression of the matrix of inertia

$$[\mathbf{M}]=\begin{bmatrix} [\mathbf{m}] & [\mathbf{A}][\mathbf{S}]^{\mathrm{T}}[\mathbf{Q}] \\ [\mathbf{Q}]^{\mathrm{T}}[\mathbf{S}][\mathbf{A}]^{\mathrm{T}} & [\mathbf{Q}]^{\mathrm{T}}[\mathbf{J}][\mathbf{Q}]\end{bmatrix}, \tag{8.126}$$

one deduces the modified matrix of inertia

$$\left[\widetilde{\mathbf{M}}\right]=\begin{bmatrix} [\mathbf{M}] & [\mathbf{0}_{62}] \\ [\mathbf{0}_{26}] & [\mathbf{0}_{22}]\end{bmatrix}, \tag{8.127}$$

the differential equation of motion being

$$\begin{bmatrix} \left[\widetilde{\mathbf{M}}\right] & -[\mathbf{B}]^{\mathrm{T}} \\ [\mathbf{B}] & [\mathbf{0}_{55}]\end{bmatrix}\begin{bmatrix} \{\ddot{\mathbf{q}}^*\} \\ \{\boldsymbol{\lambda}\}\end{bmatrix}=\begin{bmatrix} \{\mathbf{F}\}+\left\{\widetilde{\mathbf{F}}\right\} \\ \{\mathbf{0}_{21}\} \\ -[\dot{\mathbf{B}}]\{\dot{\mathbf{q}}^*\}\end{bmatrix}, \tag{8.128}$$

where

$$\{\boldsymbol{\lambda}\} = [\lambda_1 \quad \lambda_2 \quad \lambda_3 \quad \lambda_4 \quad \lambda_5]^{\mathrm{T}}. \tag{8.129}$$

By solving the equation (8.128) one obtains the solution $\{\mathbf{q}^*\} = \{\mathbf{q}^*(t)\}$, the reaction λ_1, λ_2, λ_3 at the contact point, and the values $\lambda_4 = \lambda_5 = 0$.

8.6 The Rolling without Sliding of a Toroidal Wheel on a Horizontal Plan

8.6.1 The Equations of Torus

Let $Oxyz$ be the fixed reference system (Fig. 8.41) and let $Ox'y'z'$ the mobile reference system obtained by rotating the system $Oxyz$ by the angle η about the axis Oz.

In the plan $Oy'z$ we consider the circle of radius r and center C ($OC = r_0$), situated on the Oy' axis. In the reference system $Ox'y'z'$ the parametric equations of this circle are

$$x' = 0,\ y' = r_0 + r\cos\xi,\ z' = z = r\sin\xi; \tag{8.130}$$

by rotation of angle η one obtains the parametric equations of the torus

$$\begin{bmatrix} x \\ y \\ z \end{bmatrix} = \begin{bmatrix} \cos\eta & -\sin\eta & 0 \\ \sin\eta & \cos\eta & 0 \\ 0 & 0 & 1 \end{bmatrix} \begin{bmatrix} 0 \\ -(r_0 + r\cos\xi) \\ r\sin\xi \end{bmatrix} \tag{8.131}$$

or

$$x = (r_0 + r\cos\xi)\sin\eta,\ y = -(r_0 + r\cos\xi)\cos\eta,\ z = r\sin\xi. \tag{8.132}$$

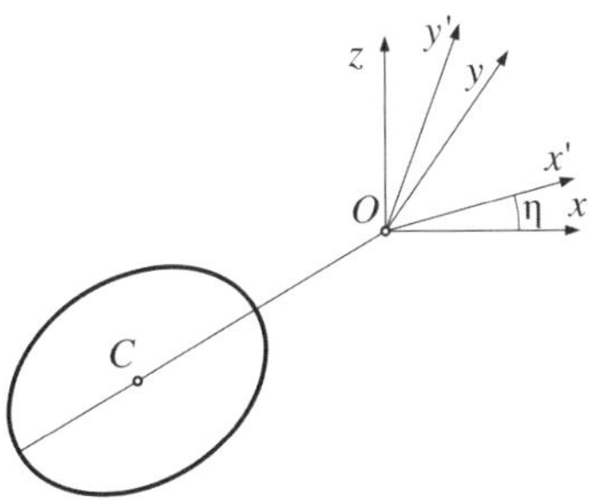

Figure 8.41 The equations of torus.

8.6.2 *The Tangency Conditions*

With the aid of the equations (8.132) we get the column matrices

$$\{\mathbf{r}\}=\begin{bmatrix}(r_0+r\cos\xi)\sin\eta\\-(r_0+r\cos\xi)\cos\eta\\r\sin\xi\end{bmatrix},\{\mathbf{r}_\eta\}=\begin{bmatrix}(r_0+r\cos\xi)\cos\eta\\(r_0+r\cos\xi)\sin\eta\\0\end{bmatrix},\{\mathbf{r}_\xi\}=\begin{bmatrix}-r\sin\xi\cos\eta\\r\sin\xi\sin\eta\\r\cos\xi\end{bmatrix}. \tag{8.133}$$

If the rotational angles ψ, θ, φ are the Euler angles, then the equations (8.111) write in the form

$$[0\ \ 0\ \ 1][\mathbf{A}]\{\mathbf{r}_\eta\}=0,[0\ \ 0\ \ 1][\mathbf{A}]\{\mathbf{r}_\xi\}=0 \tag{8.134}$$

or

$$\sin(\eta+\varphi)=0,\ \cos(\xi-\theta)=0, \tag{8.135}$$

wherefrom it results

$$\eta=-\varphi,\ \xi=-\frac{\pi}{2}+\theta, \tag{8.136}$$

so that the column matrix $\{\mathbf{r}\}$ becomes

$$\{\mathbf{r}\}=\begin{bmatrix}(r_0+r\sin\theta)\sin\varphi\\-(r_0+r\sin\theta)\cos\varphi\\-r\cos\theta\end{bmatrix}. \tag{8.137}$$

8.6.3 *The Initial Conditions*

If, at the initial moment, the Oy axis of torus (Fig. 8.42) is situated in the plan $Y=0$, and the torus is tangent to the plan $Z=0$ at the point O_0, then, in this position, $\psi=\frac{\pi}{2}$, $\varphi=0$, and the matrix of rotation becomes

$$[\mathbf{A}]=\begin{bmatrix}0&-1&0\\1&0&0\\0&0&1\end{bmatrix}\begin{bmatrix}1&0&0\\0&\cos\theta&-\sin\theta\\0&\sin\theta&\cos\theta\end{bmatrix}=\begin{bmatrix}0&-\cos\theta&\sin\theta\\1&0&0\\0&\sin\theta&\cos\theta\end{bmatrix}. \tag{8.138}$$

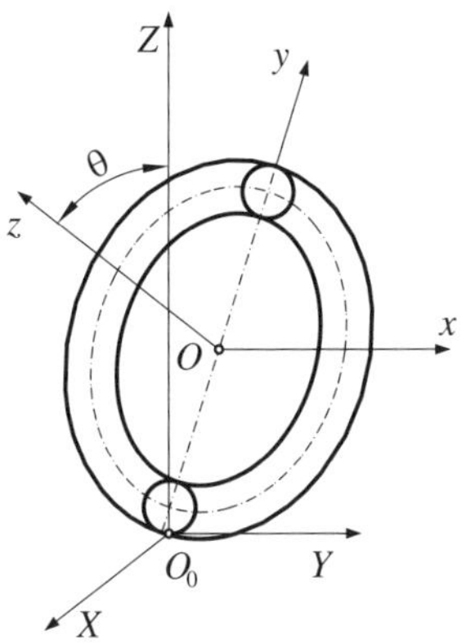

Figure 8.42 Establishing of the initial conditions.

From the contact equation

$$\begin{bmatrix} 0 \\ 0 \\ 0 \end{bmatrix} = \begin{bmatrix} X_O \\ Y_O \\ Z_O \end{bmatrix} + [\mathbf{A}] \begin{bmatrix} 0 \\ -(r_0 + r\cos\theta) \\ -r\cos\theta \end{bmatrix} \tag{8.139}$$

we get the initial conditions

$$X_O = -r_0\cos\theta,\ Y_O = 0,\ Z_O = r_0\sin\theta + r. \tag{8.140}$$

Considering that at the initial moment $\dot{\psi} = \dot{\theta} = 0\,\text{rad/s}$, $\dot{\varphi} \neq 0\,\text{rad/s}$, from the condition of rolling without sliding

$$\begin{bmatrix} \dot{X}_O \\ \dot{Y}_O \\ \dot{Z}_O \end{bmatrix} + [\mathbf{A}][\mathbf{r}]^{\mathrm{T}}[\mathbf{Q}] \begin{bmatrix} \dot{\psi} \\ \dot{\theta} \\ \dot{\varphi} \end{bmatrix} = \{\mathbf{0}\} \tag{8.141}$$

or

$$\begin{bmatrix} \dot{X}_O \\ \dot{Y}_O \\ \dot{Z}_O \end{bmatrix} + \begin{bmatrix} 0 & -\cos\theta & \sin\theta \\ 1 & 0 & 0 \\ 0 & \sin\theta & \cos\theta \end{bmatrix} \begin{bmatrix} 0 & r\cos\theta & -(r_0 + r\sin\theta) \\ -r\cos\theta & 0 & 0 \\ r_0 + r\sin\theta & 0 & 0 \end{bmatrix} \begin{bmatrix} 0 & 1 & 0 \\ \sin\theta & 0 & 0 \\ \cos\theta & 0 & 1 \end{bmatrix} \begin{bmatrix} 0 \\ 0 \\ \dot{\varphi} \end{bmatrix} = \{\mathbf{0}\} \tag{8.142}$$

one obtains the initial velocities

$$\dot{X}_O = \dot{Z}_O = 0\,\mathrm{m/s},\ \dot{Y}_O = -(r_0 + \sin\theta)\dot{\varphi}\,\mathrm{m/s}. \tag{8.143}$$

8.6.4 *The Differential Equations of Motion*

Denoting by $\lambda_1, \lambda_2, \lambda_3$ the components of the reaction at the contact point, onto the axes of the trihedron O_0XYZ, applying the general theorems, in the absence of the rolling friction and considering that the point O coincides to the center of weight, one obtains the equations

$$[\mathbf{m}]\begin{bmatrix}\ddot{X}_O\\ \ddot{Y}_O\\ \ddot{Z}_O\end{bmatrix} = \begin{bmatrix}0\\ 0\\ -mg\end{bmatrix} + \begin{bmatrix}\lambda_1\\ \lambda_2\\ \lambda_3\end{bmatrix}, [\mathbf{J}]\{\dot{\boldsymbol{\omega}}\} + [\boldsymbol{\omega}][\mathbf{J}]\{\boldsymbol{\omega}\} = [\mathbf{r}][\mathbf{A}]^{\mathrm{T}}\begin{bmatrix}\lambda_1\\ \lambda_2\\ \lambda_3\end{bmatrix}. \tag{8.144}$$

Keeping into account the relation (8.113), the equalities

$$\{\boldsymbol{\beta}\} = [\psi\ \ \theta\ \ \varphi]^{\mathrm{T}}, \{\boldsymbol{\omega}\} = [\mathbf{Q}]\{\dot{\boldsymbol{\beta}}\}, \tag{8.145}$$

and the notations

$$\begin{gathered}\left\{\widetilde{\mathbf{F}}_{\boldsymbol{\beta}}\right\} = -\left[[\mathbf{J}][\dot{\mathbf{Q}}] + [\mathbf{A}]^{\mathrm{T}}[\dot{\mathbf{A}}][\mathbf{J}][\mathbf{Q}]\right]\{\dot{\boldsymbol{\beta}}\}, \left\{\widetilde{\mathbf{F}}\right\} = \left[0\ \ 0\ \ 0\ \ \left\{\widetilde{\mathbf{F}}_{\boldsymbol{\beta}}\right\}^{\mathrm{T}}\right]^{\mathrm{T}},\\ \{\mathbf{F}\} = [0\ \ 0\ \ -mg\ \ 0\ \ 0\ \ 0]^{\mathrm{T}}, [\mathbf{B}] = \left[[\mathbf{I}]\ \ [\mathbf{A}][\mathbf{r}]^{\mathrm{T}}[\mathbf{Q}]\right], \left[\widetilde{\mathbf{B}}\right] = \left[[\mathbf{I}]\ \ [\mathbf{A}][\mathbf{r}]^{\mathrm{T}}\right],\\ \{\boldsymbol{\lambda}\} = [\lambda_1\ \ \lambda_2\ \ \lambda_3]^{\mathrm{T}}, [\mathbf{M}] = \begin{bmatrix}[\mathbf{m}] & [\mathbf{0}]\\ [\mathbf{0}] & [\mathbf{J}][\mathbf{Q}]\end{bmatrix}, \{\mathbf{q}\} = [X_O\ \ Y_O\ \ Z_O\ \ \psi\ \ \theta\ \ \varphi]^{\mathrm{T}},\end{gathered} \tag{8.146}$$

the equations (8.144) reunite in the matrix equation

$$\begin{bmatrix}[\mathbf{M}] & -\left[\widetilde{\mathbf{B}}\right]^{\mathrm{T}}\\ [\mathbf{B}] & [\mathbf{0}_{nn}]\end{bmatrix}\begin{bmatrix}\{\ddot{\mathbf{q}}\}\\ \{\boldsymbol{\lambda}\}\end{bmatrix} = \begin{bmatrix}\{\mathbf{F}\} + \left\{\widetilde{\mathbf{F}}\right\}\\ -[\dot{\mathbf{B}}]\{\dot{\mathbf{q}}\}\end{bmatrix}. \tag{8.147}$$

If there exists rolling friction of coefficient $s = s^* r_0$, then appears the rolling moment oriented along the Oz axis, and equal to $s\lambda_3$; it can be expressed in matrix form by the relation

$$\{\mathbf{M}_r\} = [\mathbf{B}^*]\{\boldsymbol{\lambda}\}, \tag{8.148}$$

where

$$[\mathbf{B}^*] = \begin{bmatrix} 0 & 0 & 0 \\ 0 & 0 & 0 \\ 0 & 0 & -s\dfrac{\omega_z}{|\boldsymbol{\omega}|} \end{bmatrix}. \tag{8.149}$$

In these conditions the matrix $\left[\widetilde{\mathbf{B}}\right]$ in the equation (8.147) becomes

$$\left[\widetilde{\mathbf{B}}\right] = \left[[\mathbf{I}]\ \ [\mathbf{A}][\mathbf{r}]^{\mathrm{T}} + [\mathbf{B}^*]\right]. \tag{8.150}$$

Example 8.6.1 Considering as known the values $r_0 = 0.3\,\mathrm{m}$, $r = 0.05\,\mathrm{m}$, $m = 20\,\mathrm{kg}$, $J_x = J_y = \dfrac{mr_0^2}{2} = 0.9\,\mathrm{kgm}^2$, $J_z = mr_0^2 = 1.8\,\mathrm{kgm}^2$, the initial conditions at $t = 0\,\mathrm{s}$, $\psi = \dfrac{\pi}{2}\,\mathrm{rad}$, $\theta = \dfrac{5\pi}{12}\,\mathrm{rad}$, $\varphi = 0\,\mathrm{rad}$, $X_O = -r_0\cos\theta = -0.077646\,\mathrm{m}$, $Y_O = 0\,\mathrm{m}$, $Z_O = r_0\sin\theta + r = 0.339778\,\mathrm{m}$, $\dot{\psi} = \dot{\theta} = 0\,\mathrm{rad/s}$, $\dot{\varphi} = -2\,\mathrm{rad/s}$, $\dot{X}_O = \dot{Z}_O = 0\,\mathrm{m/s}$, $\dot{Y}_O = -(r_0 + r\sin\theta)\dot{\varphi} = 0.696593\,\mathrm{m/s}$, and taking $s^* = 0$, we obtained the numerical results plotted in the next diagrams (Fig. 8.43 to Fig. 8.50).

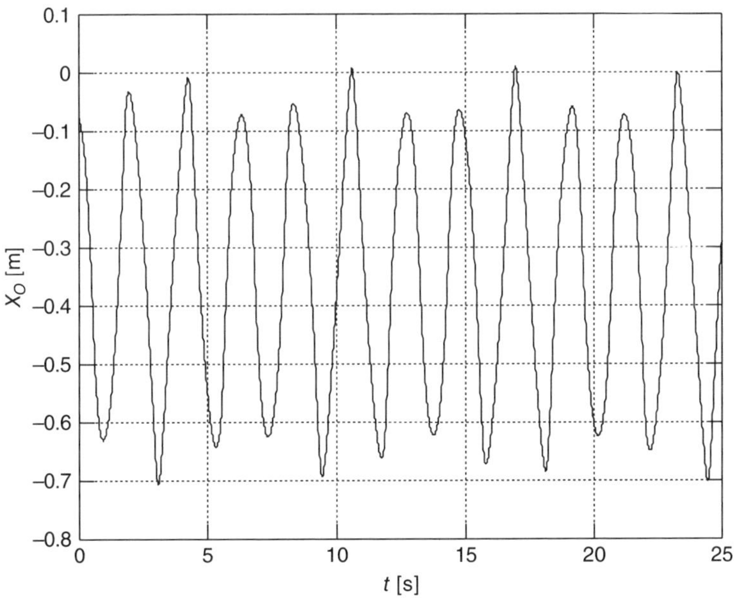

Figure 8.43 Time history $X_O = X_O(t)$ for $0 \le t \le 25\,\mathrm{s}$.

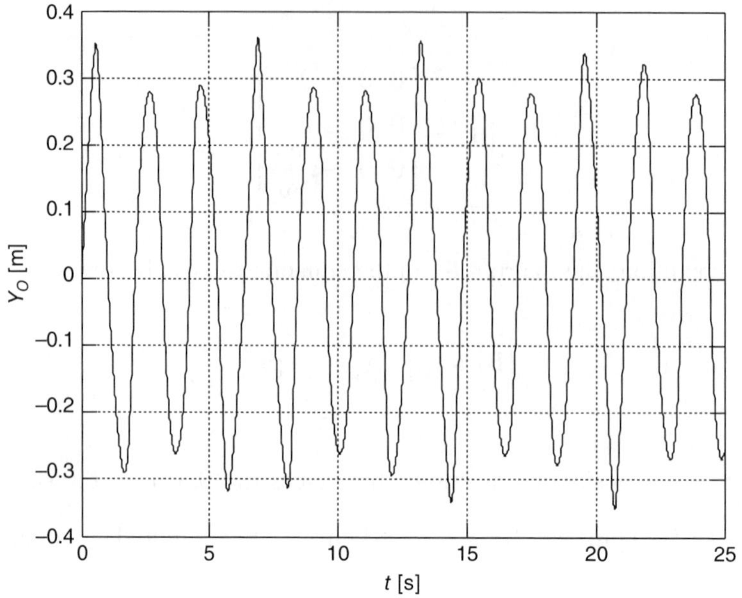

Figure 8.44 Time history $Y_O = Y_O(t)$ for $0 \le t \le 25$ s.

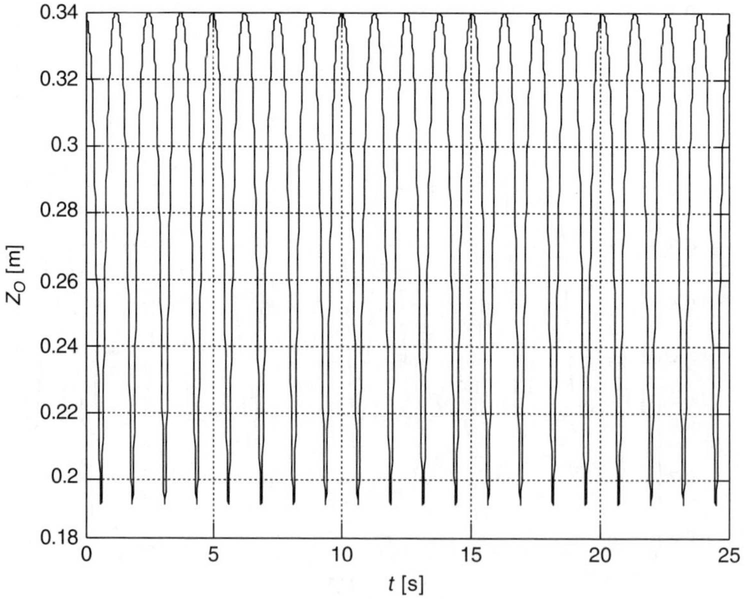

Figure 8.45 Time history $Z_O = Z_O(t)$ for $0 \le t \le 25$ s.

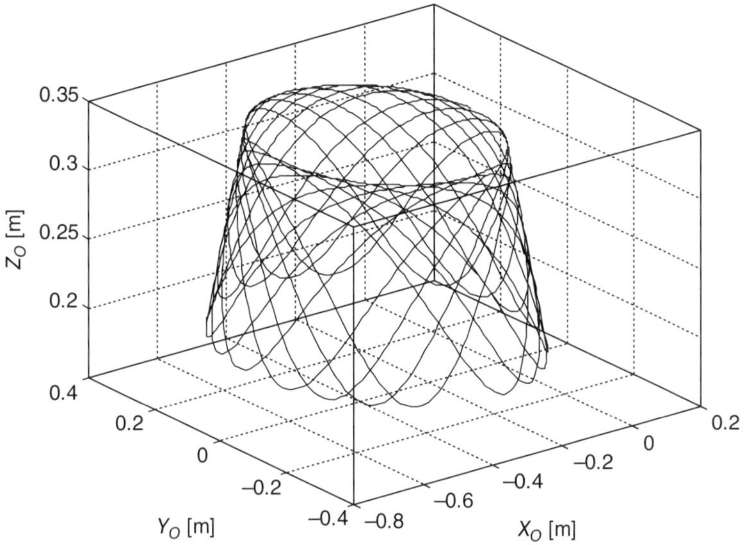

Figure 8.46 Trajectory of the point O (the center of the toroidal wheel) for the first 25 seconds.

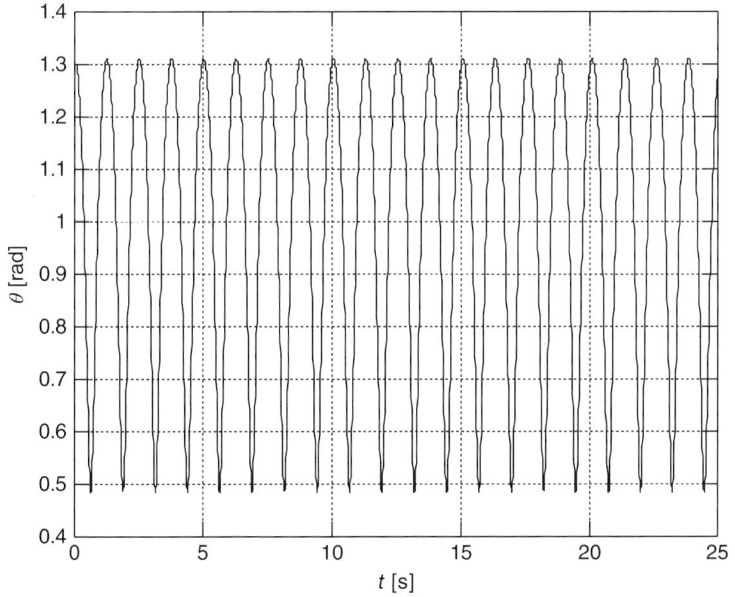

Figure 8.47 Time history $\theta = \theta(t)$ for $0 \le t \le 25$ s.

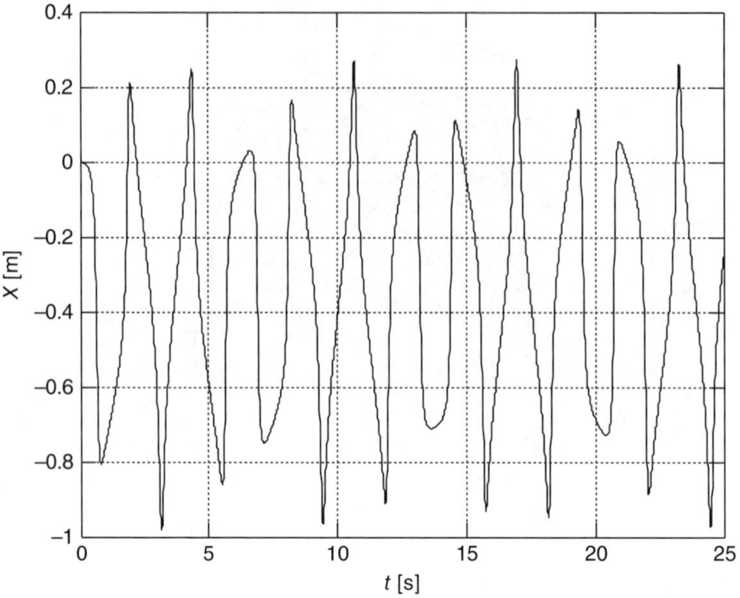

Figure 8.48 Time history $X = X(t)$ for $0 \le t \le 25$ s.

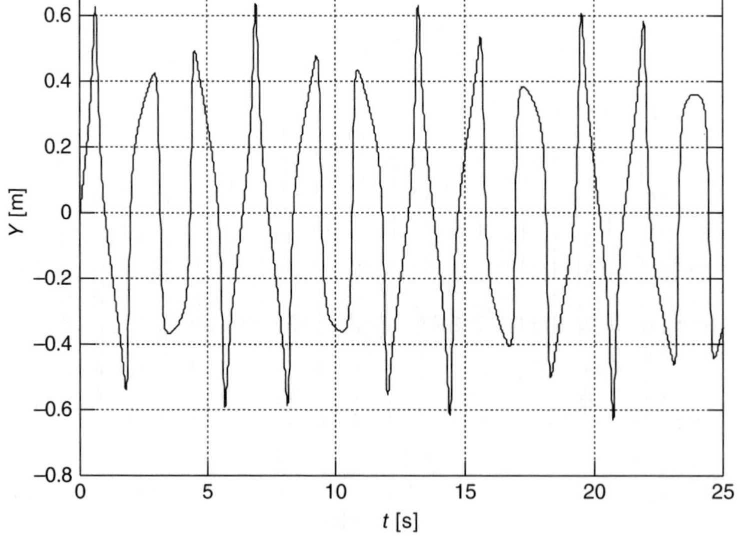

Figure 8.49 Time history $Y = Y(t)$ for $0 \le t \le 25$ s.

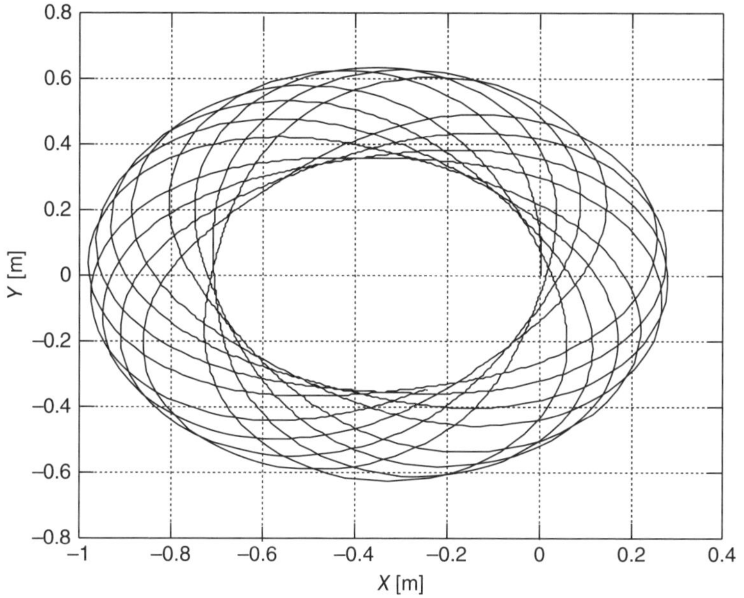

Figure 8.50 Trajectory of the contact point for the first 25 seconds.

From the diagrams presented above it is obvious the quasi-periodic character of the motion, the periods of the different parameters of the motion not being equal; these quasi-periods have the value of approximate 7 s for the all represented parameters, excepting the angle θ for which the quasi-period is equal to approximate 1.25 s.

The trajectory of the contact point is a bounded curve, but not a closed one. This fact justify once more the statement about the quasi-periodic character of the motion of wheel.

The same statements concerning the boundedness, but not closeness, may be made for the trajectory of the center of wheel (the point O).

The angle θ has values between approximate $0.5\,\text{rad} = 28.65^0$ and approximate $1.3\,\text{rad} = 74.48^0$, describing an oscillation (swinging) of the wheel relative to the vertical that passes through the contact point.

8.7 The Rolling without Sliding of a Rigid Solid Supported on Two Fixed Surfaces

8.7.1 General Aspects

We consider that in the reference system O_0XYZ the equations of the fixed surfaces are given in implicit form

$$F_i(X,Y,Z) = 0, i = 1, 2, \tag{8.151}$$

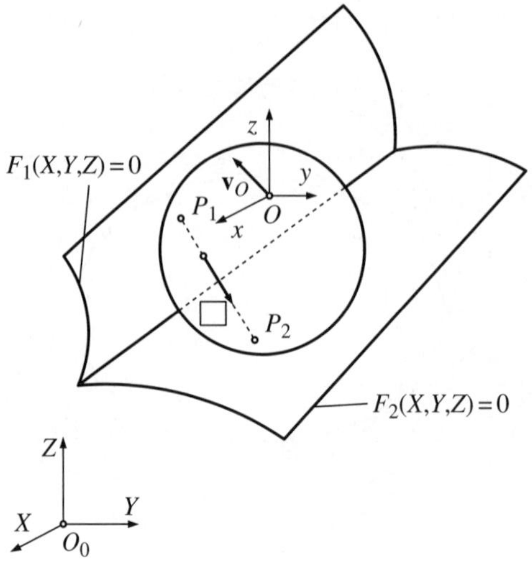

Figure 8.51 The rigid solid rolling without friction on two fixed surfaces.

while the equations of the bounding surface of the rigid solid are given in parametric form in the local reference system

$$x = x(\xi,\eta),\ y = y(\xi,\eta),\ z = z(\xi,\eta). \tag{8.152}$$

The contact points P_1, P_2 (Fig. 8.51) have, in the reference system $Oxyz$, the coordinates x_i, y_i, z_i, $i = 1, 2$, given by the relations

$$x_i = x(\xi_i,\eta_i),\ y_i = y(\xi_i,\eta_i),\ z_i = z(\xi_i,\eta_i), i = 1,\ 2, \tag{8.153}$$

while in the reference system O_0XYZ they have the coordinates X_i, Y_i, Z_i, $i = 1, 2$.

Using the notations

$$\{\mathbf{R}_i\} = [X_i \quad Y_i \quad Z_i]^{\mathrm{T}}, \{\mathbf{r}_i\} = [x_i \quad y_i \quad z_i]^{\mathrm{T}}, i = 1,\ 2, \tag{8.154}$$

one obtains the contact conditions

$$\{\mathbf{R}_i\} = \{\mathbf{R}_O\} + [\mathbf{A}]\{\mathbf{r}_i\}, i = 1,\ 2, \tag{8.155}$$

and the tangency conditions

$$\{\mathbf{F}_{ip}\}^{\mathrm{T}}[\mathbf{A}]\{\mathbf{r}_{i\xi}\} = 0, \{\mathbf{F}_{ip}\}^{\mathrm{T}}[\mathbf{A}]\{\mathbf{r}_{i\eta}\} = 0, \tag{8.156}$$

where $\{\mathbf{F}_{ip}\}$ are the column matrices of the partial derivatives of the functions F_i, $i = 1, 2$, calculated at the points $P_i(X_i, Y_i, Z_i)$, $i = 1, 2$, while $\{\mathbf{r}_{i\xi}\}$, and $\{\mathbf{r}_{i\eta}\}$ are the partial derivatives of the matrix $\{\mathbf{r}_i\}$ calculated with respect to ξ_i, and η_i, respectively, $i = 1, 2$.

By derivation the relations (8.155) with respect to time and taking into account the conditions of rolling without sliding

$$\{\dot{\mathbf{R}}_O\} + [\mathbf{A}][\mathbf{r}_i]^{\mathrm{T}}[\mathbf{Q}]\{\dot{\boldsymbol{\beta}}\} = \{\mathbf{0}\}, i = 1, 2, \tag{8.157}$$

where

$$[\mathbf{r}_i] = \begin{bmatrix} 0 & -z_i & y_i \\ z_i & 0 & -x_i \\ -y_i & x_i & 0 \end{bmatrix}, i = 1, 2, \tag{8.158}$$

one obtains the equalities

$$\{\dot{\mathbf{R}}_i\} = \dot{\xi}_i[\mathbf{A}]\{\mathbf{r}_{i\xi}\} + \dot{\eta}_i[\mathbf{A}]\{\mathbf{r}_{i\eta}\}, i = 1, 2. \tag{8.159}$$

The relations (8.157) are equivalent to the matrix equalities

$$\{\boldsymbol{\omega}\} = [\mathbf{Q}]\{\dot{\boldsymbol{\beta}}\} = \zeta\{\{\mathbf{r}_1\} - \{\mathbf{r}_2\}\}, \tag{8.160}$$

$$\{\dot{\mathbf{R}}_O\} = \zeta[\mathbf{A}][\mathbf{r}_2]\{\mathbf{r}_1\}, \tag{8.161}$$

and to the vector equalities (Fig. 8.51)

$$\boldsymbol{\omega} = \zeta(\mathbf{r}_1 - \mathbf{r}_2), \ \mathbf{v}_O = \zeta(\mathbf{r}_2 \times \mathbf{r}_1), \tag{8.162}$$

where ζ is a factor of proportionality.

In paragraph 8.4 we proved that if one denotes by λ_i, $i = \overline{1, 6}$, the components of the torsor at O of the constraint forces, in the local reference system $Oxyz$, then the relation (8.86) holds true

$$\left[\{\mathbf{r}_1\}^{\mathrm{T}}[\mathbf{r}_2]^{\mathrm{T}} \quad \{\{\mathbf{r}_1\} - \{\mathbf{r}_2\}\}^{\mathrm{T}}\right]\{\boldsymbol{\lambda}\} = 0, \tag{8.163}$$

where

$$\{\boldsymbol{\lambda}\} = [\lambda_1 \ \ \lambda_2 \ \ \lambda_3 \ \ \lambda_4 \ \ \lambda_5 \ \ \lambda_6]^{\mathrm{T}}. \tag{8.164}$$

8.7.2 The Differential Equations of Motion

If we denote by $\{\mathbf{F_s}\}$ the column matrix of the components of the given forces in the reference system O_0XYZ, by $\{\mathbf{F}_{\boldsymbol{\beta}}\}$ the column matrix of the components of the resultant moment of the given forces, calculated at the point O, in the local reference system $Oxyz$, and if we make the hypothesis that the point O coincides to the center of weight of the rigid solid, then, from the theorems of momentum and moment of momentum, one obtains the equations

$$[\mathbf{m}]\{\ddot{\mathbf{R}}_O\}=\{\mathbf{F}_\mathrm{s}\}+[\mathbf{A}][\lambda_1\ \ \lambda_2\ \ \lambda_3]^\mathrm{T},[\mathbf{J}]\{\dot{\boldsymbol{\omega}}\}=\{\mathbf{F}_\beta\}+[\lambda_4\ \ \lambda_5\ \ \lambda_6]-[\boldsymbol{\omega}][\mathbf{J}]\{\boldsymbol{\omega}\}. \quad (8.165)$$

Keeping into account the relations (8.160) and (8.161), and the notations

$$\left\{\widetilde{\mathbf{F}}_\mathrm{s}\right\}=-\zeta[\mathbf{m}]\{[\dot{\mathbf{A}}][\mathbf{r}_2]\{\mathbf{r}_1\}+[\mathbf{A}][\dot{\mathbf{r}}_2]\{\mathbf{r}_1\}+[\mathbf{A}][\mathbf{r}_2]\{\dot{\mathbf{r}}_1\}\}, \quad (8.166)$$

$$\left\{\widetilde{\mathbf{F}}_\beta\right\}=-\zeta^2[[\mathbf{r}_1]-[\mathbf{r}_2]][\mathbf{J}]\{\{\mathbf{r}_1\}-\{\mathbf{r}_2\}\}-\zeta[\mathbf{J}]\{\{\dot{\mathbf{r}}_1\}-\{\dot{\mathbf{r}}_2\}\}, \quad (8.167)$$

where

$$\{\dot{\mathbf{r}}_i\}=\dot{\xi}_i\{\mathbf{r}_{i\xi}\}+\dot{\eta}_i\{\mathbf{r}_{i\eta}\},[\dot{\mathbf{r}}_i]=\dot{\xi}_i[\mathbf{r}_{i\xi}]+\dot{\eta}_i[\mathbf{r}_{i\eta}],i=1,2, \quad (8.168)$$

the equations (8.165) write in the form

$$m\dot{\zeta}[\mathbf{A}][\mathbf{r}_2]\{\mathbf{r}_1\}-[\mathbf{A}][\lambda_1\ \ \lambda_2\ \ \lambda_3]^\mathrm{T}=\{\mathbf{F}_\mathrm{s}\}+\left\{\widetilde{\mathbf{F}}_\mathrm{s}\right\},\dot{\zeta}\{\{\mathbf{r}_1\}-\{\mathbf{r}_2\}\}-[\lambda_4\ \ \lambda_5\ \ \lambda_6]^\mathrm{T}=\{\mathbf{F}_\beta\}+\left\{\widetilde{\mathbf{F}}_\beta\right\}. \quad (8.169)$$

In these conditions from the relations (8.169) and (8.168) one obtains the equalities

$$\dot{\zeta}=\frac{\{\mathbf{r}_1\}^\mathrm{T}[\mathbf{r}_2]^\mathrm{T}[\mathbf{A}]^\mathrm{T}\left\{\{\mathbf{F}_\mathrm{s}\}+\left\{\widetilde{\mathbf{F}}_\mathrm{s}\right\}\right\}+\{\{\mathbf{r}_1\}-\{\mathbf{r}_2\}\}^\mathrm{T}\left\{\{\mathbf{F}_\beta\}+\left\{\widetilde{\mathbf{F}}_\beta\right\}\right\}}{m\{\mathbf{r}_1\}^\mathrm{T}[\mathbf{r}_2]^\mathrm{T}[\mathbf{r}_2]\{\mathbf{r}_1\}+\{\{\mathbf{r}_1\}-\{\mathbf{r}_2\}\}^\mathrm{T}[\mathbf{J}]\{\{\mathbf{r}_1\}-\{\mathbf{r}_2\}\}}, \quad (8.170)$$

$$\{\boldsymbol{\lambda}\}=\dot{\zeta}\begin{bmatrix} m[\mathbf{r}_2]\{\mathbf{r}_1\} \\ [\mathbf{J}]\{\{\mathbf{r}_1\}-\{\mathbf{r}_2\}\}\end{bmatrix}-\begin{bmatrix}[\mathbf{A}]^\mathrm{T}\left\{\{\mathbf{F}_\mathrm{s}\}+\left\{\widetilde{\mathbf{F}}_\mathrm{s}\right\}\right\} \\ \{\mathbf{F}_\beta\}+\left\{\widetilde{\mathbf{F}}_\beta\right\}\end{bmatrix}. \quad (8.171)$$

Further on, one determines, by numerical integration, the parameter ζ which permits the determination, also by numerical integration, of the new values for the parameters X_O, Y_O, Z_O, ψ, θ, φ in the relations (8.160) and (8.161).

To be out to determine the new values for the parameters ξ_i, η_i, $i=1, 2$, one solves, using the new values of the parameters X_O, Y_O, Z_O, ψ, θ, φ, the system

$$C_{i\xi\xi}\dot{\xi}_i+C_{i\xi\eta}\dot{\eta}_i+C_{i\xi}=0,C_{i\eta\xi}\dot{\xi}_i+C_{i\eta\eta}\dot{\eta}_i+C_{i\eta}=0,i=1,2. \quad (8.172)$$

The last system was obtained by deriving the relations (8.156), with respect to time. The notations are as follows

$$\begin{aligned}
&C_{i\xi\xi}=\{\mathbf{r}_{i\xi}\}^{\mathrm{T}}[\mathbf{A}]^{\mathrm{T}}[\mathbf{F}_{is}][\mathbf{A}]\{\mathbf{r}_{i\xi}\}+\{\mathbf{F}_{ip}\}^{\mathrm{T}}[\mathbf{A}]\{\mathbf{r}_{i\xi\xi}\},\\
&C_{i\xi\mu}=C_{i\mu\xi}=\{\mathbf{r}_{i\xi}\}^{\mathrm{T}}[\mathbf{A}]^{\mathrm{T}}[\mathbf{F}_{is}][\mathbf{A}]\{\mathbf{r}_{i\eta}\}+\{\mathbf{F}_{ip}\}^{\mathrm{T}}[\mathbf{A}]\{\mathbf{r}_{i\xi\eta}\},\\
&C_{i\eta\eta}=\{\mathbf{r}_{i\eta}\}^{\mathrm{T}}[\mathbf{A}]^{\mathrm{T}}[\mathbf{F}_{is}][\mathbf{A}]\{\mathbf{r}_{i\eta}\}+\{\mathbf{F}_{ip}\}^{\mathrm{T}}[\mathbf{A}]\{\mathbf{r}_{i\eta\eta}\},C_{i\xi}=\{\mathbf{F}_{ip}\}^{\mathrm{T}}[\dot{\mathbf{A}}]\{\mathbf{r}_{i\xi}\},C_{i\eta}=\{\mathbf{F}_{ip}\}^{\mathrm{T}}[\dot{\mathbf{A}}]\{\mathbf{r}_{i\eta}\}.
\end{aligned} \tag{8.173}$$

8.7.3 *The Determination of the Forces of Constraints*

The constraint forces at the points P_1, P_2 (Fig. 8.51) decompose in components situated onto the directions of the normals at the fixed surfaces, onto the directions of the tangents of the curves described by the points P_1, and P_2 on the fixed surfaces, and onto the direction of the straight line P_1P_2, respectively.

Denoting by Q_{in}, Q_{it}, Q_{ip}, $i=1, 2$, the magnitudes of these components, it results that they can be determined by solving the first five equations of the system

$$\begin{aligned}
\begin{bmatrix}[\mathbf{A}] & [\mathbf{0}]\\ [\mathbf{R}_O][\mathbf{A}] & [\mathbf{A}]\end{bmatrix}
&=\sum_{i=1}^{2}\frac{Q_{in}}{\sqrt{F_{iX}^2+F_{iY}^2+F_{iZ}^2}}\begin{bmatrix}F_{iX}\\ F_{iY}\\ F_{iZ}\\ Y_iF_{iZ}-Z_iF_{iY}\\ Z_iF_{iX}-X_iF_{iZ}\\ X_iF_{iY}-Y_iF_{iX}\end{bmatrix}\\
&+\sum_{i=1}^{2}\frac{Q_{it}}{\sqrt{\dot{X}_i^2+\dot{Y}_i^2+\dot{Z}_i^2}}\begin{bmatrix}\dot{X}_i\\ \dot{Y}_i\\ \dot{Z}_i\\ Y_i\dot{Z}_i-Z_i\dot{Y}_i\\ Z_i\dot{X}_i-X_i\dot{Z}_i\\ X_i\dot{Y}_i-Y_i\dot{X}_i\end{bmatrix}\\
&+\frac{Q_{1p}+Q_{2p}}{\sqrt{(X_1-X_2)^2+(Y_1-Y_2)^2+(Z_1-Z_2)^2}}\begin{bmatrix}X_1-X_2\\ Y_1-Y_2\\ Z_1-Z_2\\ Z_1Y_2-Y_1Z_2\\ X_1Z_2-Z_1X_2\\ Y_1X_2-X_1Y_2\end{bmatrix},
\end{aligned} \tag{8.174}$$

where

$$[\mathbf{R}_O] = \begin{bmatrix} 0 & -Z_O & Y_O \\ Z_O & 0 & -X_O \\ -Y_O & X_O & 0 \end{bmatrix}. \tag{8.175}$$

The relation (8.174) shows that the model of rigid solid can assure only the determination of the sum of components of the constraint forces onto the direction P_1P_2.

8.7.4 The Rolling without Sliding of an Ellipsoid Acted only by its Own Weight on Two Plans

We choose the reference system O_0XYZ so that the straight line of intersection of the two plans (Fig. 8.52) coincides to the O_0Y axis; in these conditions, the equations of the two plans read

$$F_i = X\cos\alpha_i - Z\sin\alpha_i = 0,\ i = 1, 2, \tag{8.176}$$

α_i, $i = 1, 2$, being the angles between the O_0X axis and the two plans.

We also consider the equations of the ellipsoid

$$x = a\sin\xi\cos\eta,\ y = b\sin\xi\sin\eta,\ z = -c\cos\xi. \tag{8.177}$$

and we denote by α_0 the angle made by the axis O_0Z with the vertical direction.

Example 8.7.1 One consider the values $a = 0.2\,\mathrm{m}$, $b = 0.22\,\mathrm{m}$, $c = 0.18\,\mathrm{m}$, $m = 200\,\mathrm{kg}$, $\alpha_1 = 60^0$, $\alpha_2 = 120^0$, $\alpha_0 = 30^0$, and $\alpha_0 = 45^0$, while the initial conditions at $t = 0\,\mathrm{s}$ are $X_O = Y_O = 0\,\mathrm{m}$, $Z_O = \dfrac{\sqrt{a^2\sin^2\alpha_1 + c^2\cos^2\alpha_1}}{\cos\alpha_1}$, $\psi = \theta = \varphi = 0\,\mathrm{rad}$, $\xi_i = \arctan\left(\dfrac{a}{c}\tan\alpha_i\right)$, $i = 1, 2$, $\eta_i = 0^0$, $i = 1, 2$, $\dot{X}_O = \dot{Y}_O = \dot{Z}_O = 0\,\mathrm{m/s}$, $\dot{\psi} = \dot{\theta} = \dot{\varphi} = 0\,\mathrm{rad/s}$, $\dot{\xi}_i = \dot{\eta}_i = 0\,\mathrm{rad/s}$, $\zeta = 0\,\mathrm{m^{-1}s^{-1}}$.

One obtains the next diagrams (Fig. 8.53 to Fig. 8.92). In the case $\alpha_0 = 30^0$ the ellipsoid oscillates and does not descent along the ditch defined by the two plans,

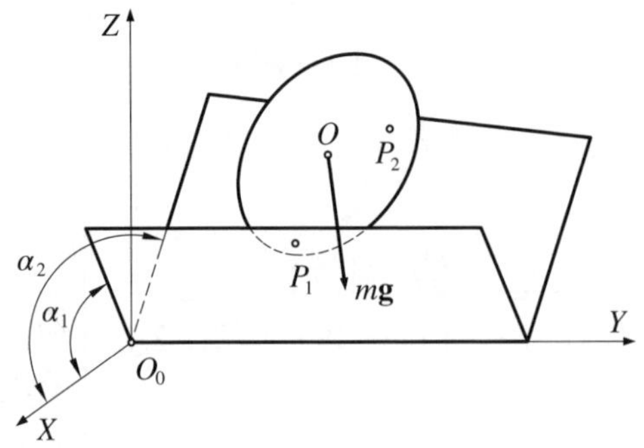

Figure 8.52 The ellipsoid rolling without sliding on two fixed plans.

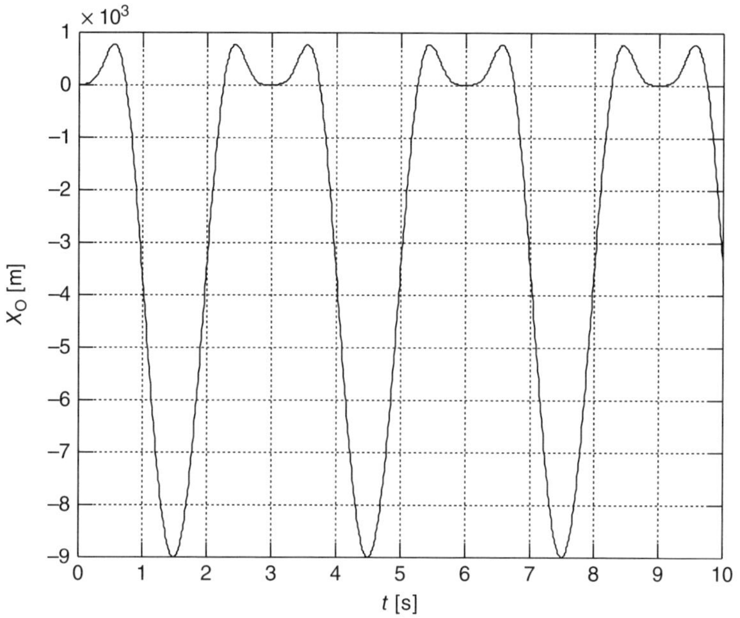

Figure 8.53 Time history $X_O = X_O(t)$ for $0 \leq t \leq 10\,\mathrm{s}$ and $\alpha_0 = 30^0$.

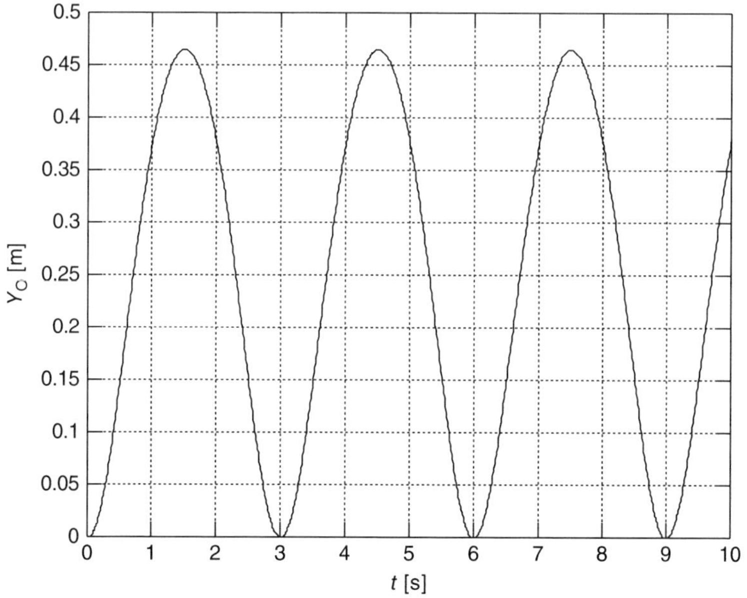

Figure 8.54 Time history $Y_O = Y_O(t)$ for $0 \leq t \leq 10\,\mathrm{s}$ and $\alpha_0 = 30^0$.

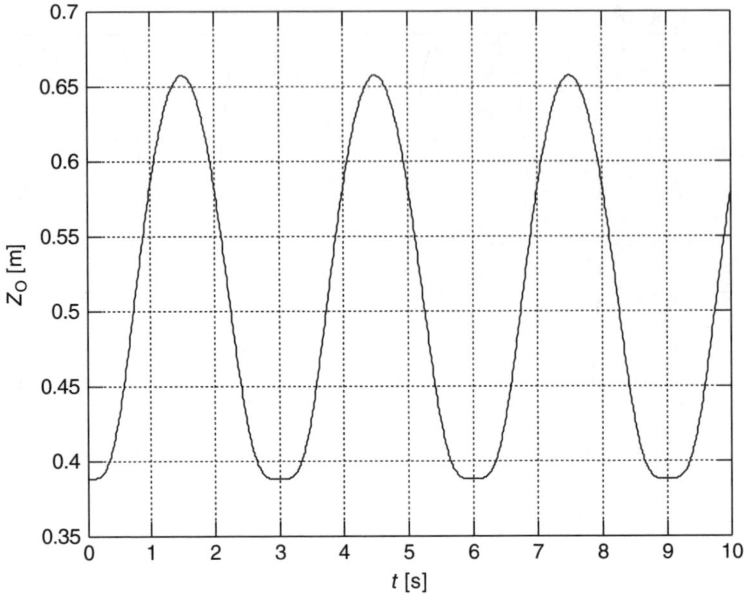

Figure 8.55 Time history $Z_O = Z_O(t)$ for $0 \leq t \leq 10\,\text{s}$ and $\alpha_0 = 30^0$.

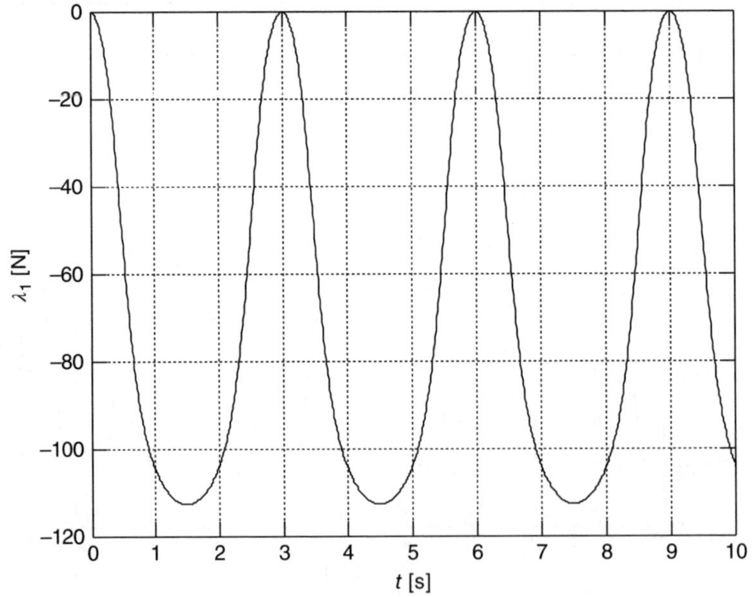

Figure 8.56 Time history $\lambda_1 = \lambda_1(t)$ for $0 \leq t \leq 10\,\text{s}$ and $\alpha_0 = 30^0$.

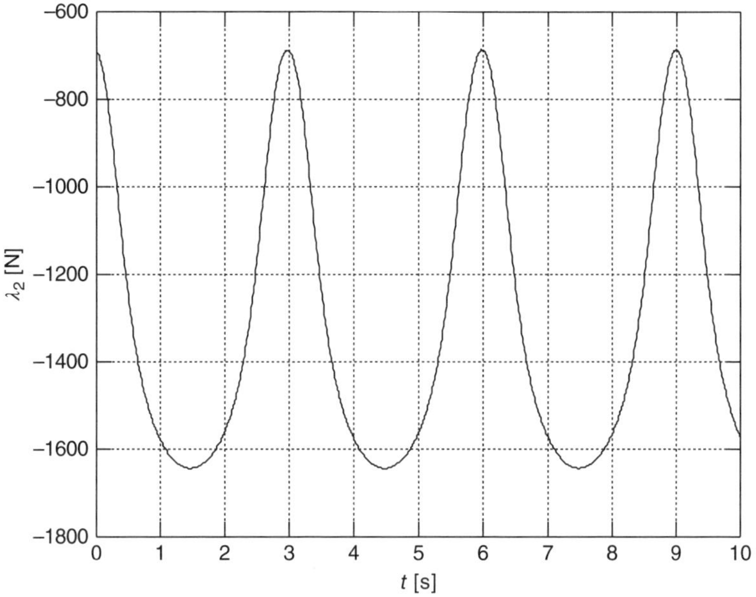

Figure 8.57 Time history $\lambda_2 = \lambda_2(t)$ for $0 \le t \le 10\,\text{s}$ and $\alpha_0 = 30^0$.

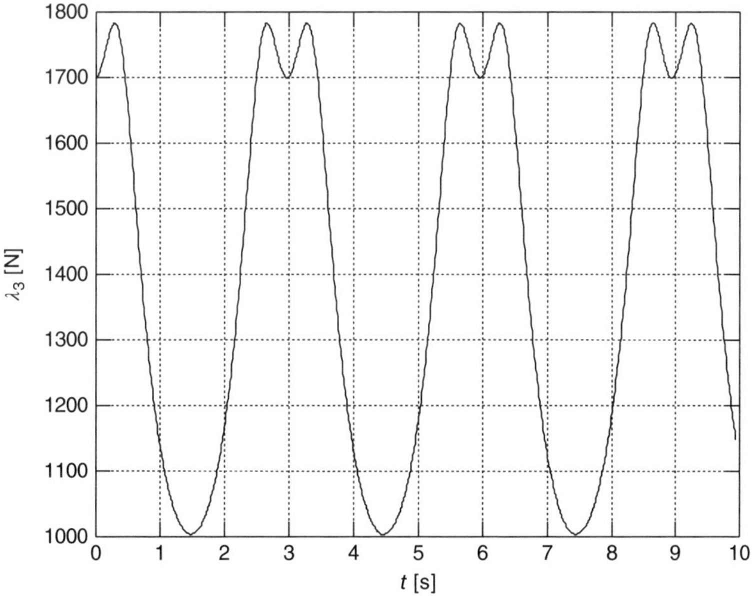

Figure 8.58 Time history $\lambda_3 = \lambda_3(t)$ for $0 \le t \le 10\,\text{s}$ and $\alpha_0 = 30^0$.

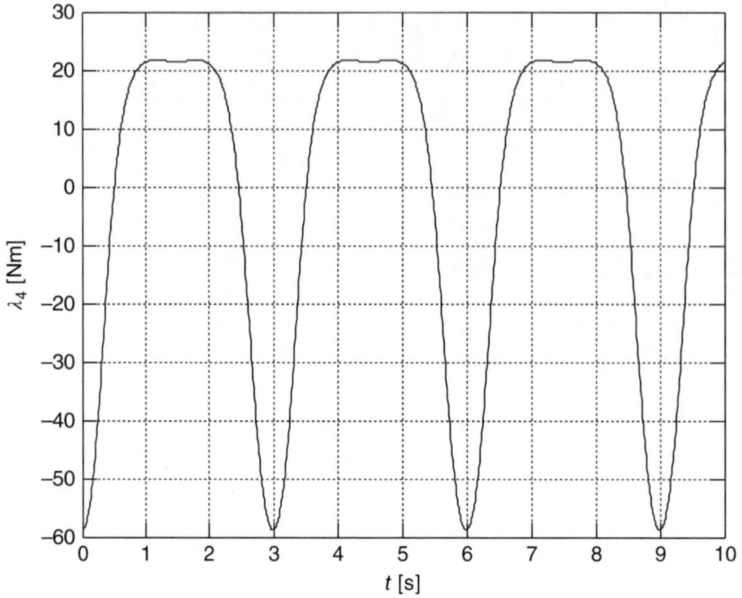

Figure 8.59 Time history $\lambda_4 = \lambda_4(t)$ for $0 \le t \le 10$ s and $\alpha_0 = 30^0$.

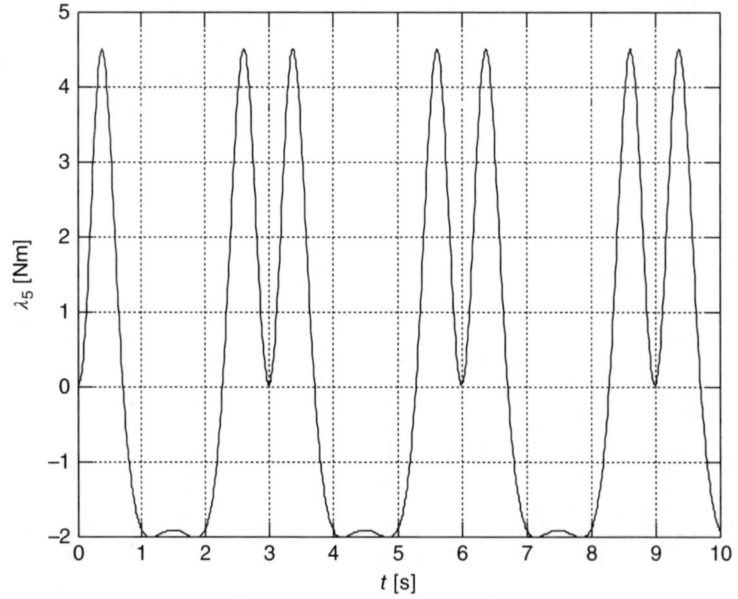

Figure 8.60 Time history $\lambda_5 = \lambda_5(t)$ for $0 \le t \le 10$ s and $\alpha_0 = 30^0$.

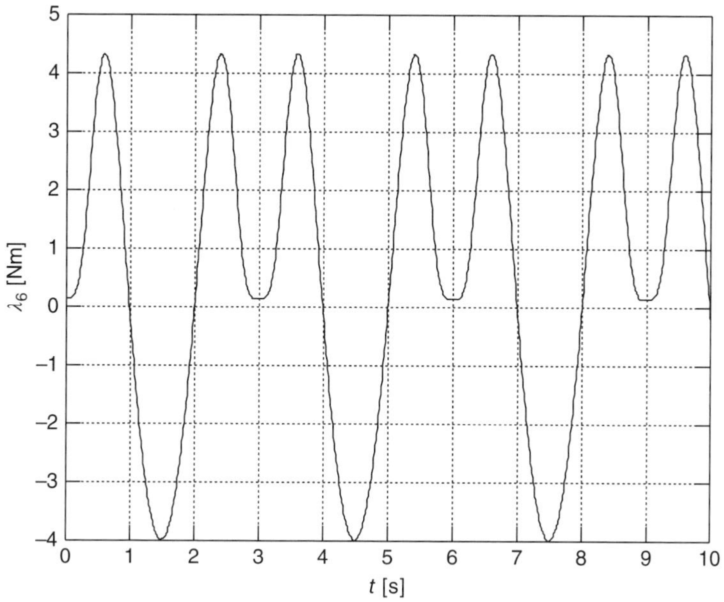

Figure 8.61 Time history $\lambda_6 = \lambda_6(t)$ for $0 \le t \le 10$ s and $\alpha_0 = 30^0$.

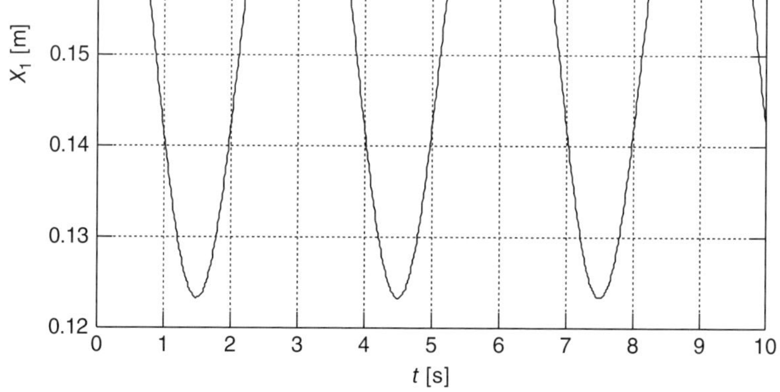

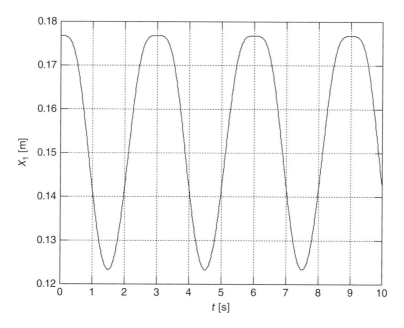

Figure 8.62 Time history $X_1 = X_1(t)$ for $0 \le t \le 10$ s and $\alpha_0 = 30^0$.

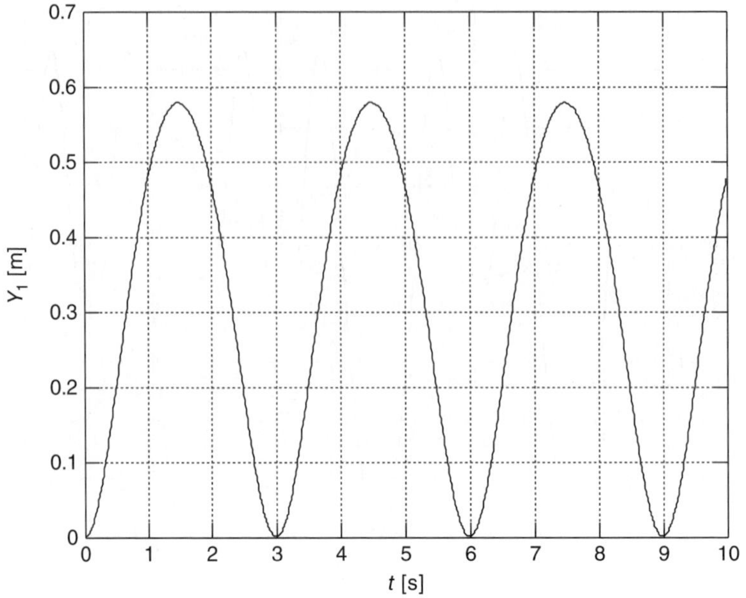

Figure 8.63 Time history $Y_1 = Y_1(t)$ for $0 \le t \le 10\,\text{s}$ and $\alpha_0 = 30^0$.

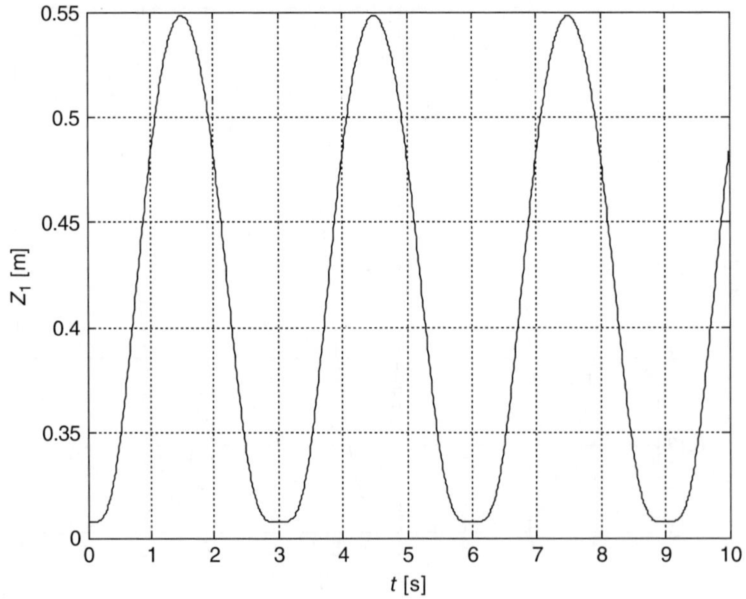

Figure 8.64 Time history $Z_1 = Z_1(t)$ for $0 \le t \le 10\,\text{s}$ and $\alpha_0 = 30^0$.

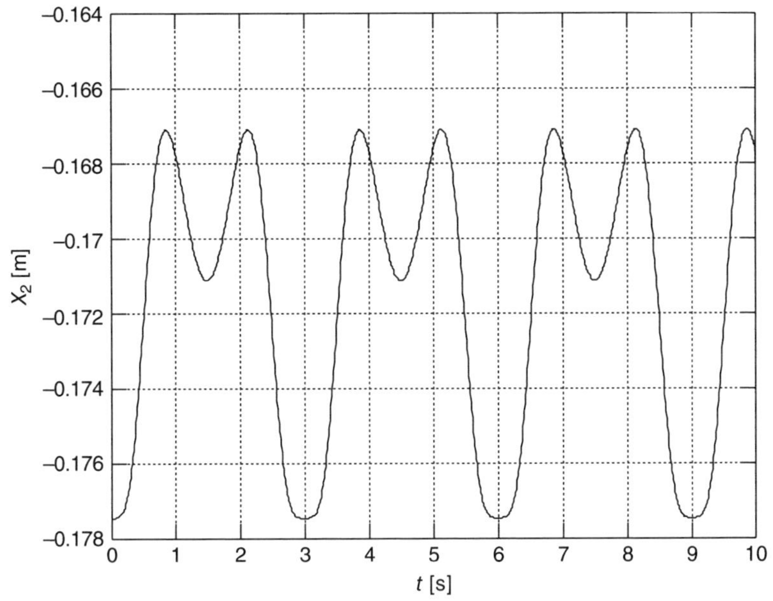

Figure 8.65 Time history $X_2 = X_2(t)$ for $0 \le t \le 10\,\text{s}$ and $\alpha_0 = 30^0$.

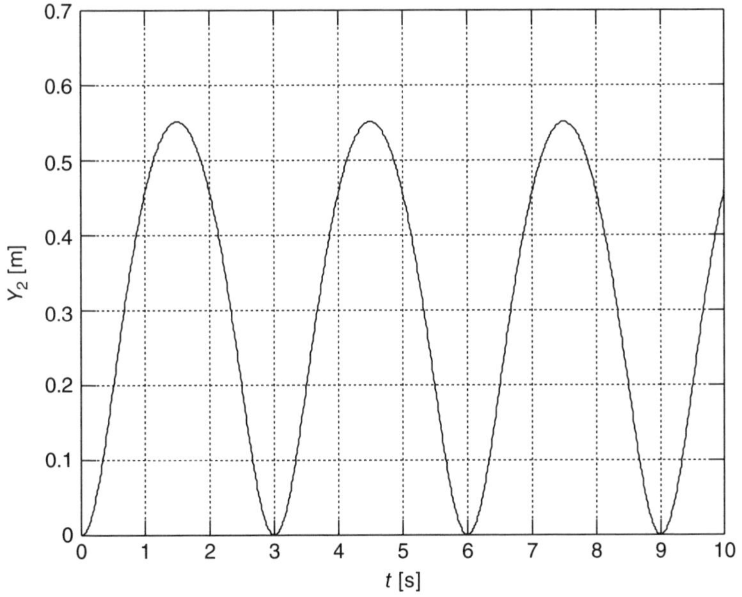

Figure 8.66 Time history $Y_2 = Y_2(t)$ for $0 \le t \le 10\,\text{s}$ and $\alpha_0 = 30^0$.

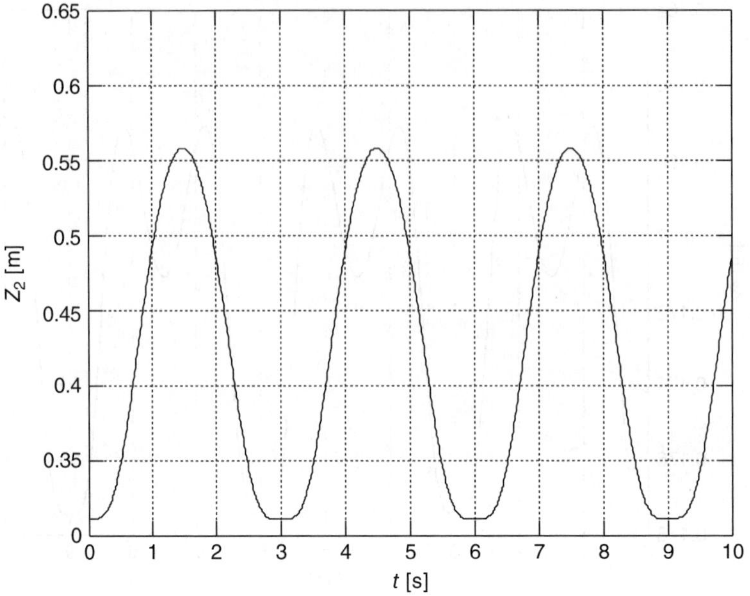

Figure 8.67 Time history $Z_2 = Z_2(t)$ for $0 \le t \le 10\,\mathrm{s}$ and $\alpha_0 = 30^0$.

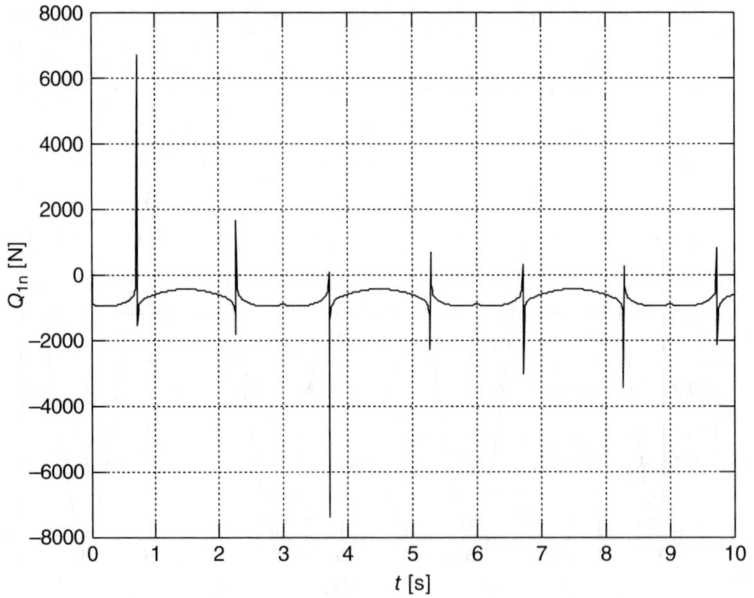

Figure 8.68 Time history $Q_{1n} = Q_{1n}(t)$ for $0 \le t \le 10\,\mathrm{s}$ and $\alpha_0 = 30^0$.

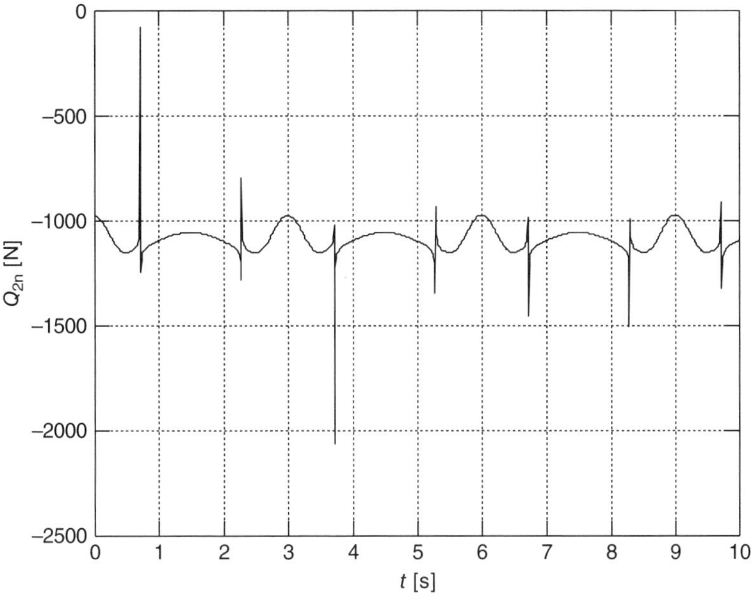

Figure 8.69 Time history $Q_{2n} = Q_{2n}(t)$ for $0 \le t \le 10\,\mathrm{s}$ and $\alpha_0 = 30^0$.

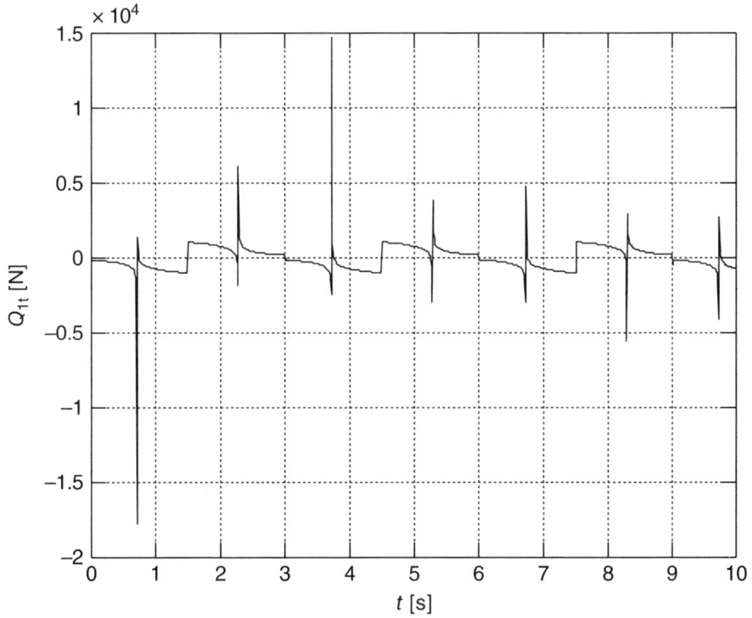

Figure 8.70 Time history $Q_{1t} = Q_{1t}(t)$ for $0 \le t \le 10\,\mathrm{s}$ and $\alpha_0 = 30^0$.

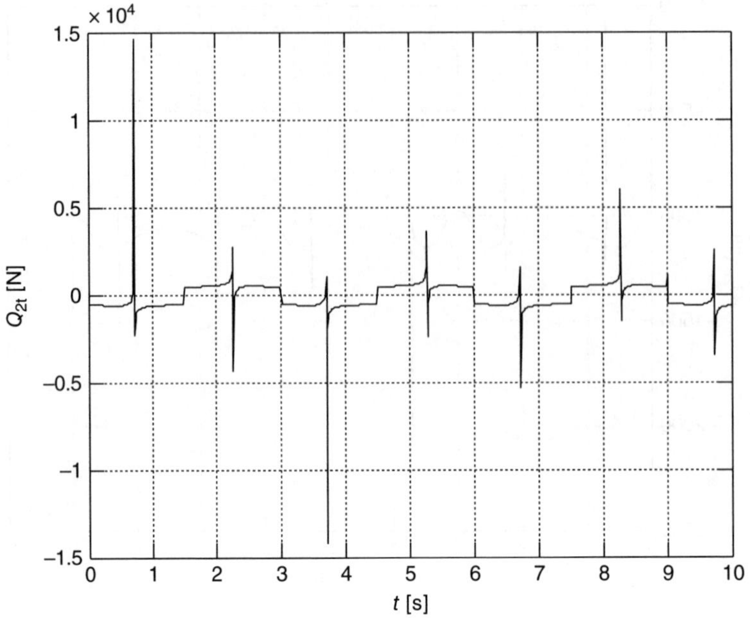

Figure 8.71 Time history $Q_{2t} = Q_{2t}(t)$ for $0 \le t \le 10\,\mathrm{s}$ and $\alpha_0 = 30^0$.

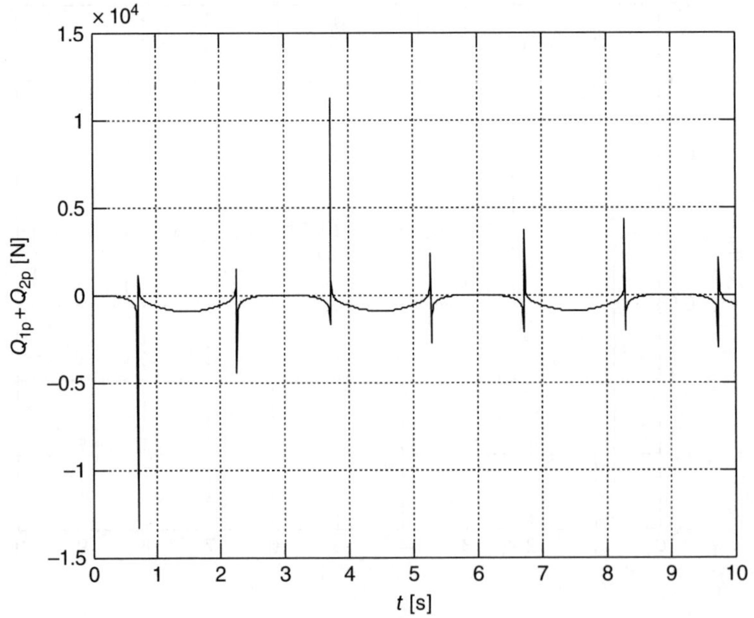

Figure 8.72 Time history $Q_{1p} + Q_{2p} = \left(Q_{1p} + Q_{2p}\right)(t)$ for $0 \le t \le 10\,\mathrm{s}$ and $\alpha_0 = 30^0$.

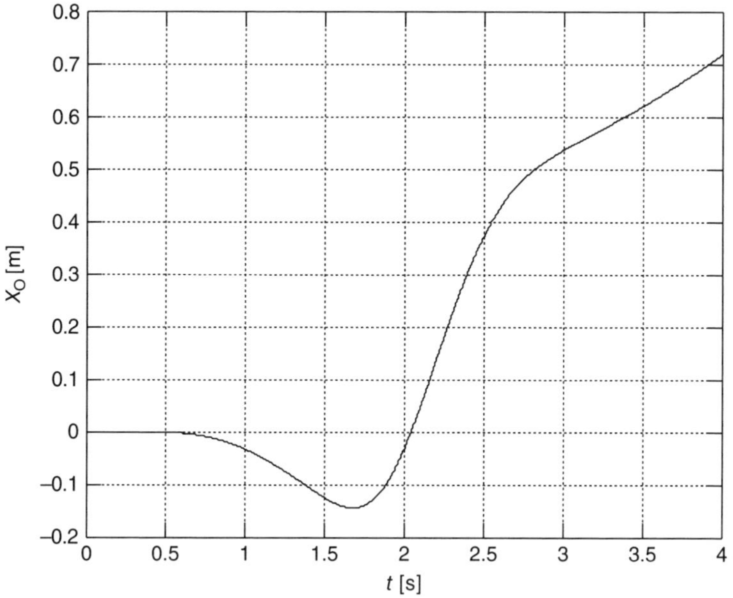

Figure 8.73 Time history $X_O = X_O(t)$ for $0 \le t \le 4$ s and $\alpha_0 = 45^0$.

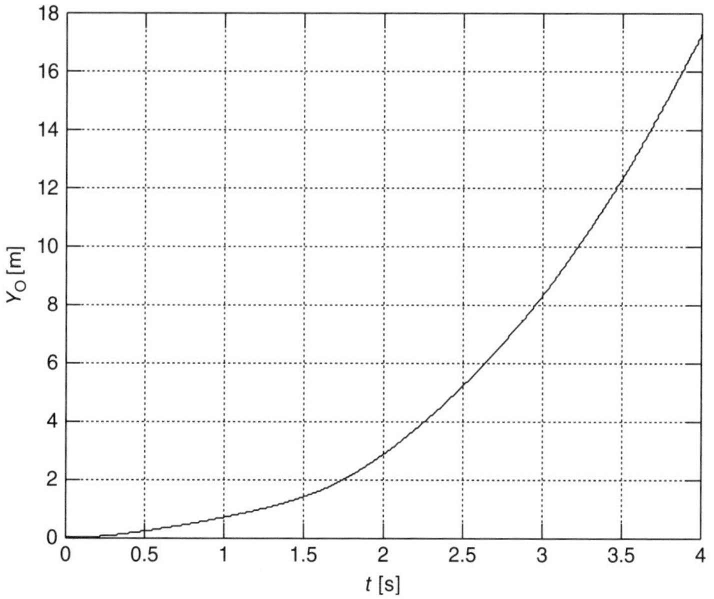

Figure 8.74 Time history $Y_O = Y_O(t)$ for $0 \le t \le 4$ s and $\alpha_0 = 45^0$.

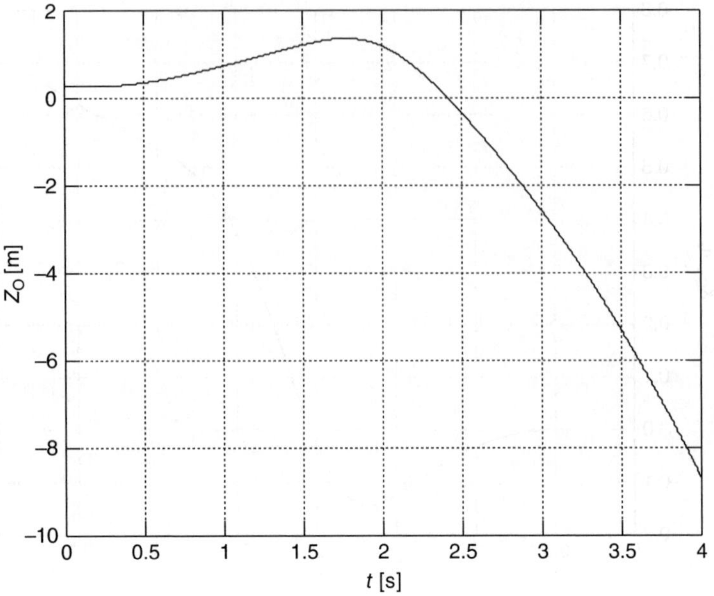

Figure 8.75 Time history $Z_O = Z_O(t)$ for $0 \le t \le 4$ s and $\alpha_0 = 45^0$.

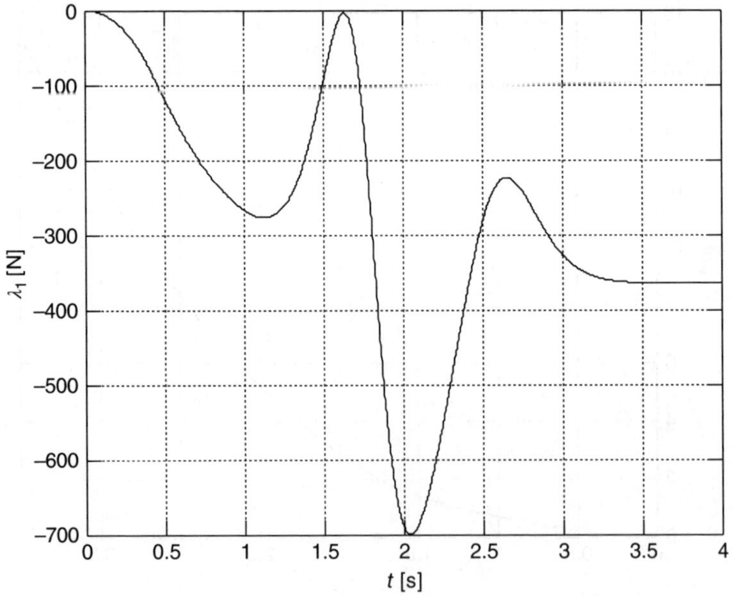

Figure 8.76 Time history $\lambda_1 = \lambda_1(t)$ for $0 \le t \le 4$ s and $\alpha_0 = 45^0$.

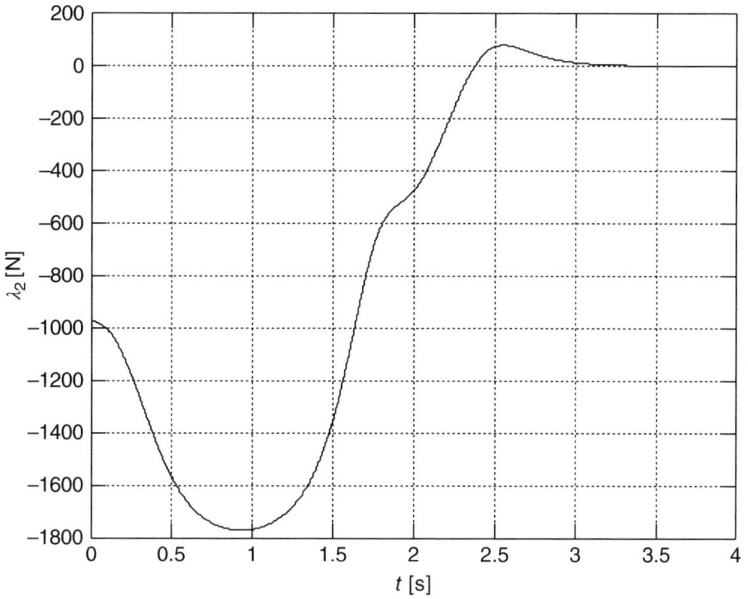

Figure 8.77 Time history $\lambda_2 = \lambda_2(t)$ for $0 \le t \le 4$ s and $\alpha_0 = 45^0$.

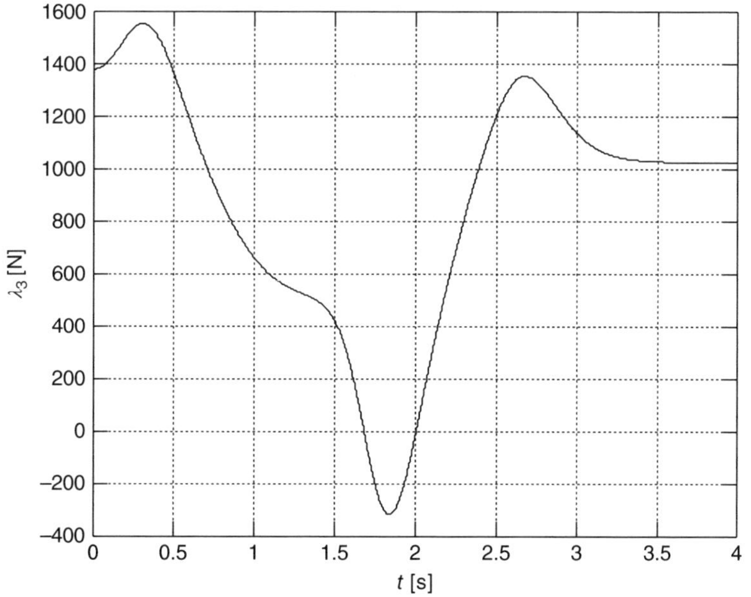

Figure 8.78 Time history $\lambda_3 = \lambda_3(t)$ for $0 \le t \le 4$ s and $\alpha_0 = 45^0$.

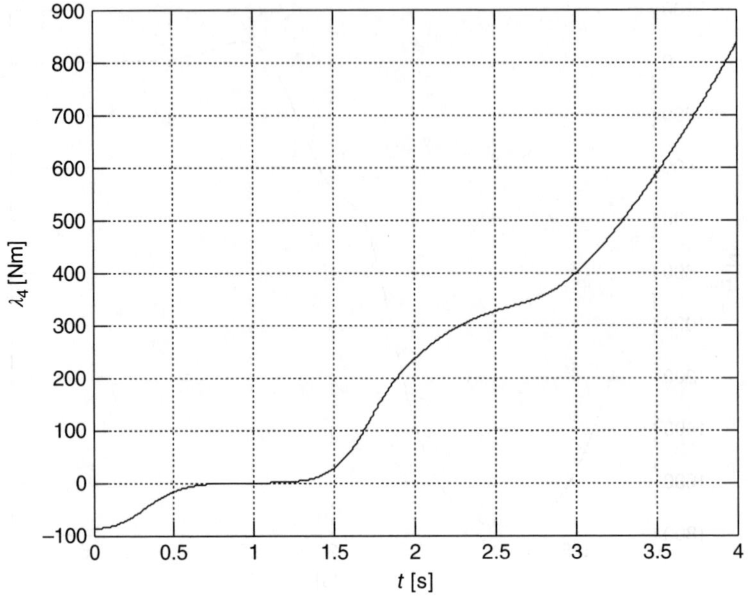

Figure 8.79 Time history $\lambda_4 = \lambda_4(t)$ for $0 \leq t \leq 4$ s and $\alpha_0 = 45^0$.

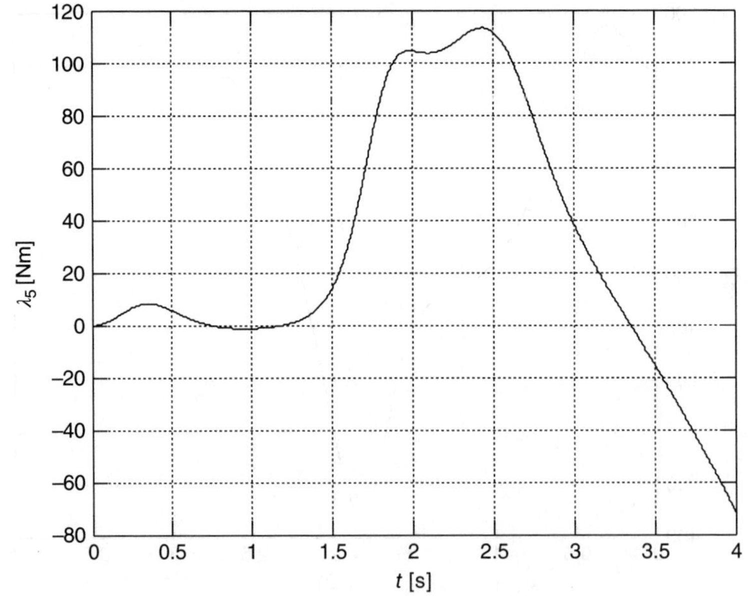

Figure 8.80 Time history $\lambda_5 = \lambda_5(t)$ for $0 \leq t \leq 4$ s and $\alpha_0 = 45^0$.

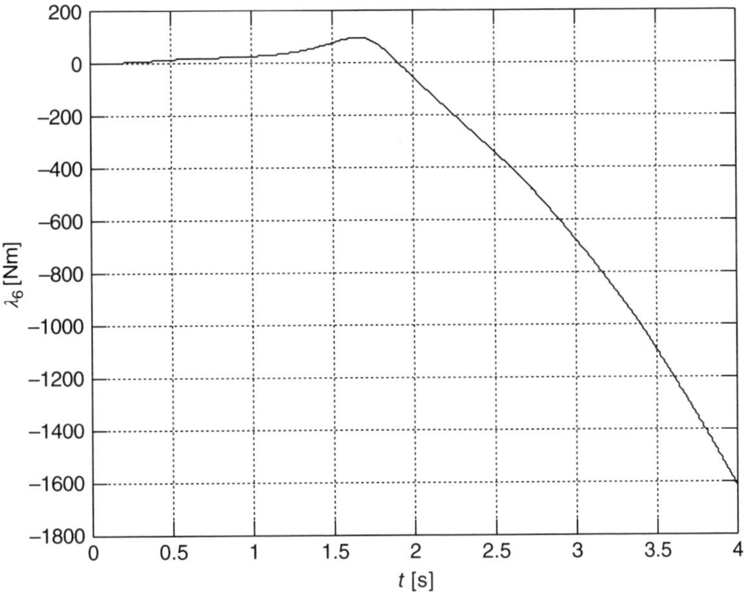

Figure 8.81 Time history $\lambda_6 = \lambda_6(t)$ for $0 \le t \le 4\,\mathrm{s}$ and $\alpha_0 = 45^0$.

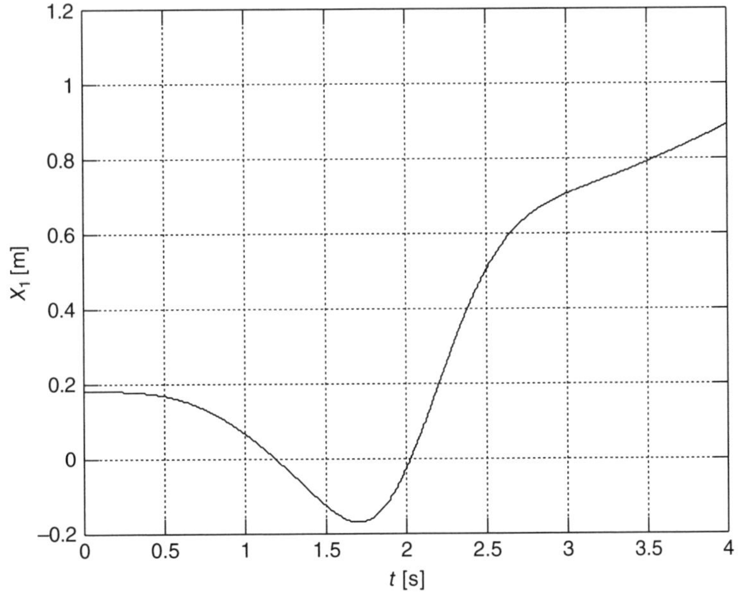

Figure 8.82 Time history $X_1 = X_1(t)$ for $0 \le t \le 4\,\mathrm{s}$ and $\alpha_0 = 45^0$.

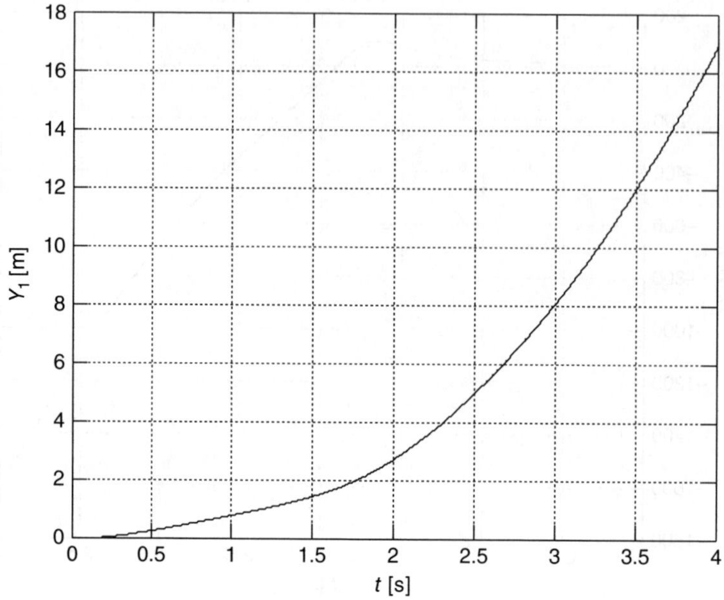

Figure 8.83 Time history $Y_1 = Y_1(t)$ for $0 \leq t \leq 4$ s and $\alpha_0 = 45^0$.

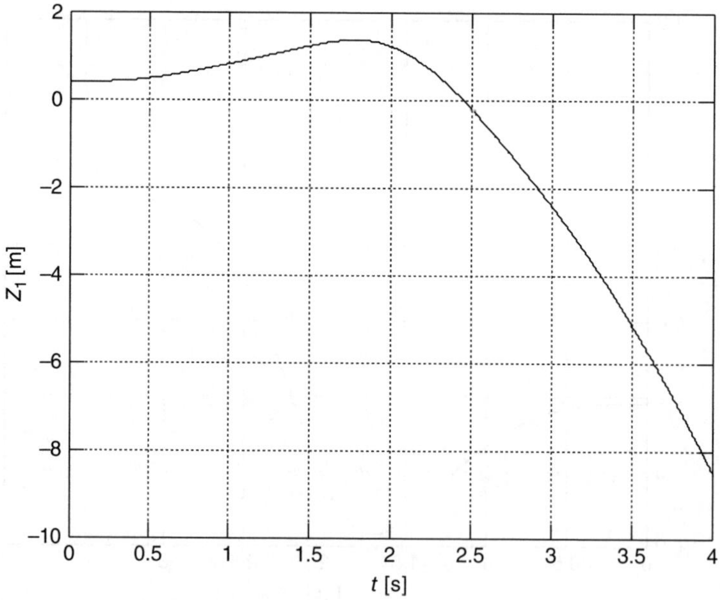

Figure 8.84 Time history $Z_1 = Z_1(t)$ for $0 \leq t \leq 4$ s and $\alpha_0 = 45^0$.

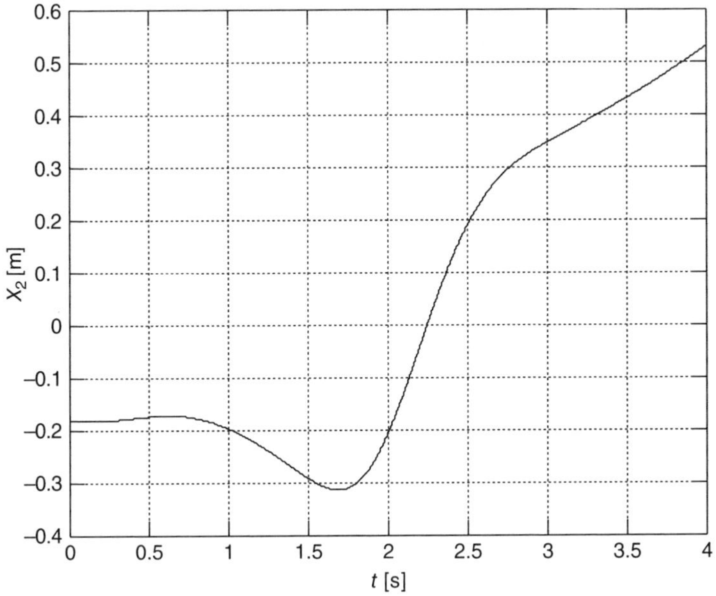

Figure 8.85 Time history $X_2 = X_2(t)$ for $0 \leq t \leq 4$ s and $\alpha_0 = 45^0$.

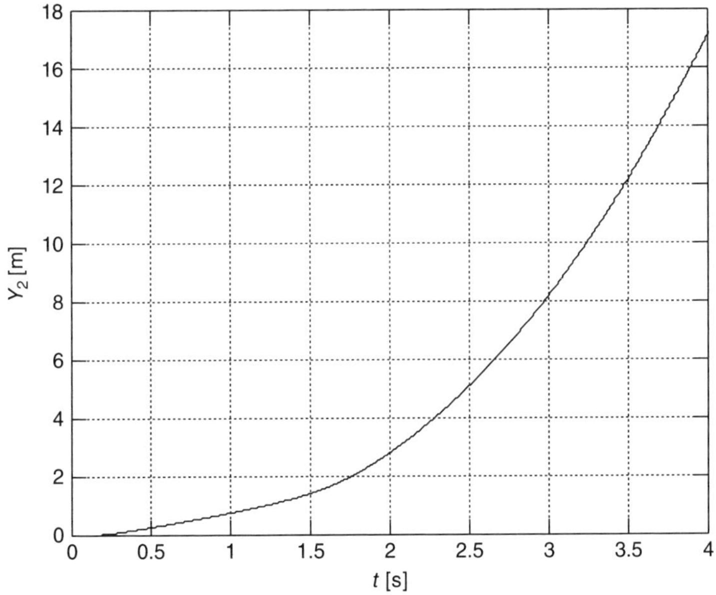

Figure 8.86 Time history $Y_2 = Y_2(t)$ for $0 \leq t \leq 4$ s and $\alpha_0 = 45^0$.

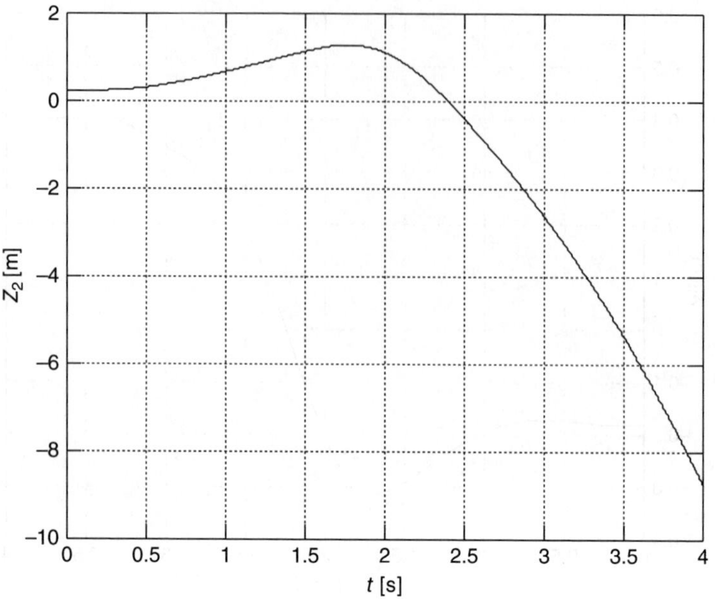

Figure 8.87 Time history $Z_2 = Z_2(t)$ for $0 \le t \le 4\,\mathrm{s}$ and $\alpha_0 = 45^0$.

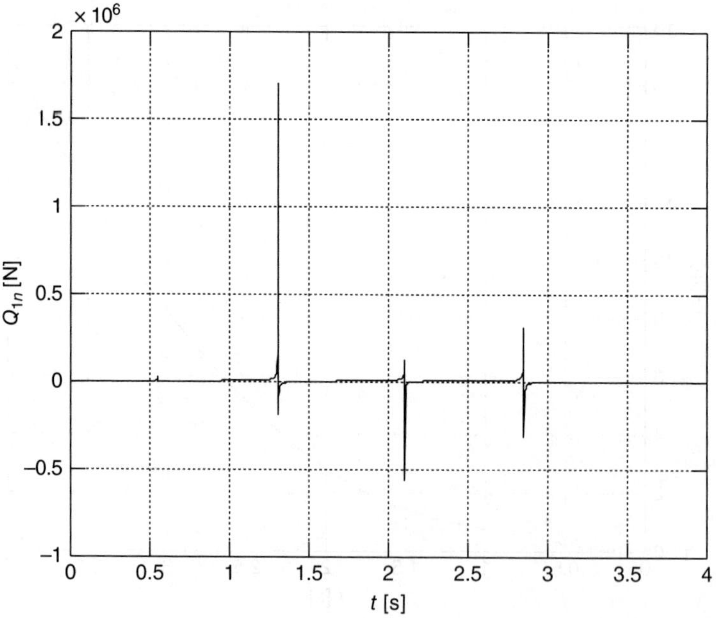

Figure 8.88 Time history $Q_{1n} = Q_{1n}(t)$ for $0 \le t \le 4\,\mathrm{s}$ and $\alpha_0 = 45^0$.

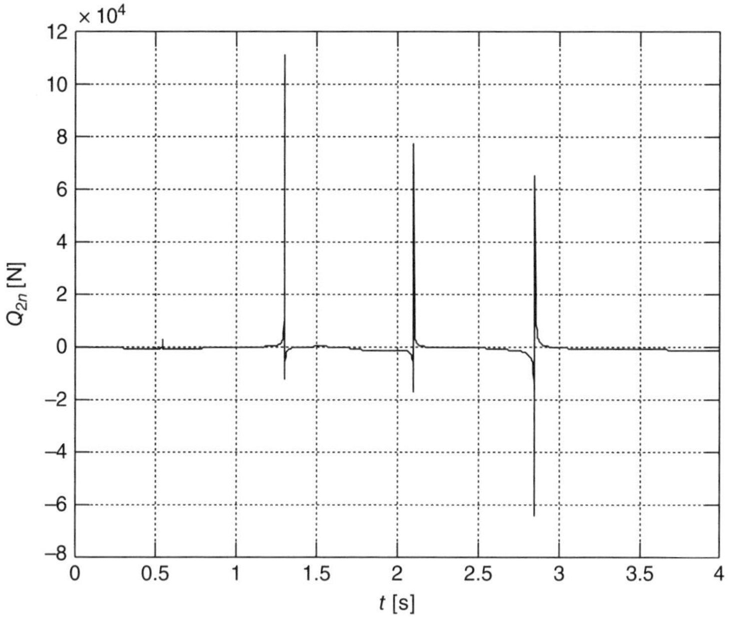

Figure 8.89 Time history $Q_{2n} = Q_{2n}(t)$ for $0 \le t \le 4$ s and $\alpha_0 = 45^0$.

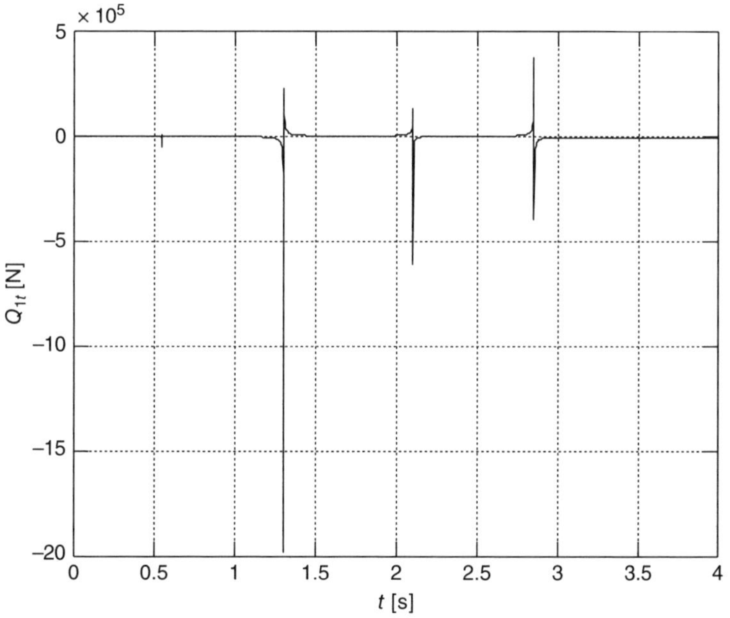

Figure 8.90 Time history $Q_{1t} = Q_{1t}(t)$ for $0 \le t \le 4$ s and $\alpha_0 = 45^0$.

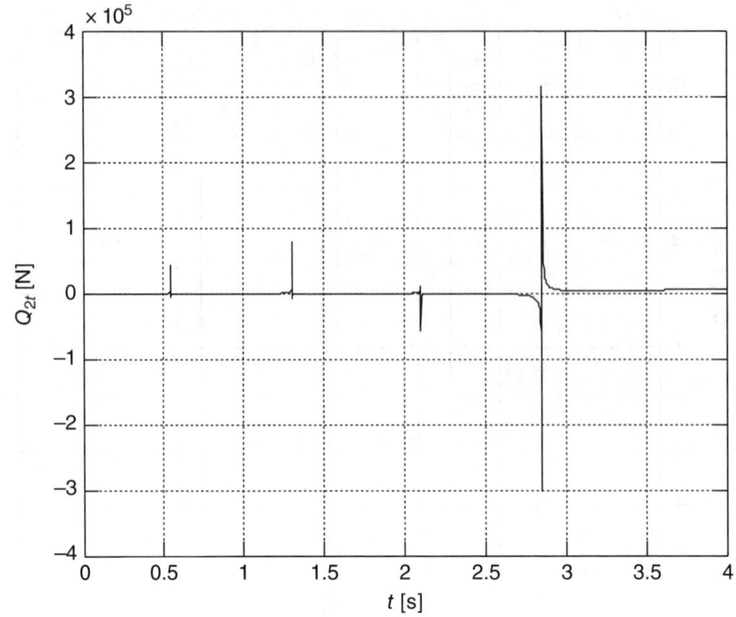

Figure 8.91 Time history $Q_{2t} = Q_{2t}(t)$ for $0 \le t \le 4$ s and $\alpha_0 = 45^0$.

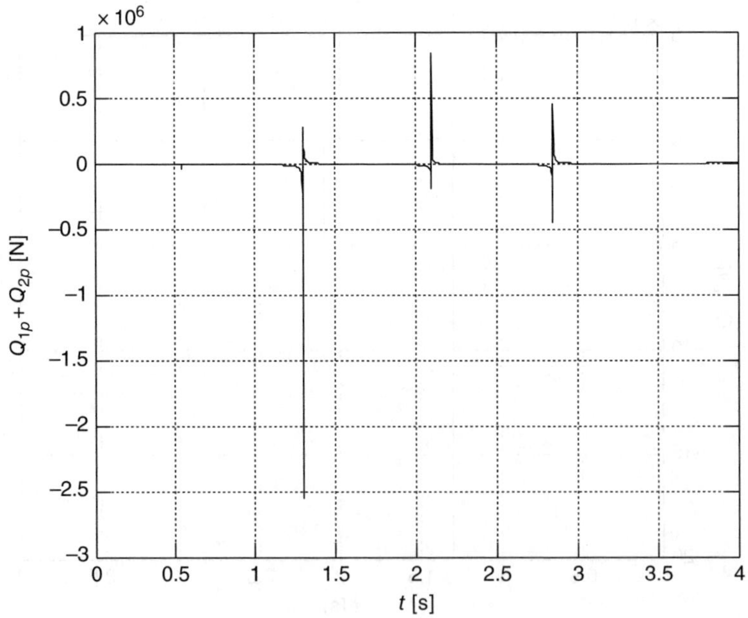

Figure 8.92 Time history $Q_{1p} + Q_{2p} = (Q_{1p} + Q_{2p})(t)$ for $0 \le t \le 4$ s and $\alpha_0 = 45^0$.

while in the case $\alpha_0 = 45^0$, the ellipsoid descents along the ditch. In the first situation the periodic character of the motion is obvious, the period being of approximate 3 s; the forces have a maximum value of approximate $1.8 \cdot 10^4$ N (about 9.2 times greater than the weight of the ellipsoid). In the second situation the motion looses its periodic character; the forces have a maximum value of approximate $2.5 \cdot 10^6$ N (about 1275 times greater than the weight of the ellipsoid). The changes of sign for the forces are possible only for bilateral constraints.

8.8 The Rolling without Sliding of a Rigid Solid Supported at Two Points on a Fixed Surface

8.8.1 General Aspects

We consider that the equation of the fixed surface (Fig. 8.93) is

$$F(X,Y,Z) = 0, \tag{8.178}$$

while the parametric equations of the portions of surfaces that bound the rigid solid, in the reference system $Oxyz$, are

$$x_i = x(\xi_i, \eta_i), y_i = y(\xi_i, \eta_i), z_i = z(\xi_i, \eta_i), i = 1, 2. \tag{8.179}$$

The contact points P_1, P_2 (Fig. 8.93) have, in the local reference system $Oxyz$, the coordinates x_i, y_i, z_i, $i = 1, 2$, while in the reference system O_0XYZ they have the coordinates X_i, Y_i, Z_i, $i = 1, 2$; the following column matrices are defined

$$\{\mathbf{R}_i\} = [X_i \;\; Y_i \;\; Z_i]^{\mathrm{T}}, \{\mathbf{r}_i\} = [x_i \;\; y_i \;\; z_i]^{\mathrm{T}}, i = 1, 2. \tag{8.180}$$

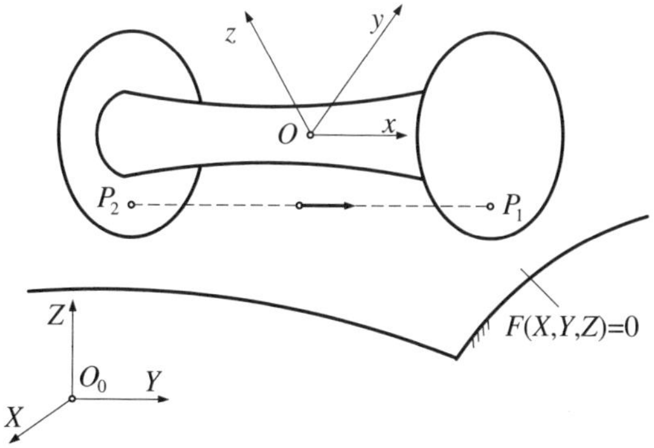

Figure 8.93 The rigid solid rolling without sliding and supporting at two points on a fixed surface.

The contact conditions at the points P_1, P_2 write in the form

$$\{\mathbf{R}_i\} = \{\mathbf{R}_O\} + [\mathbf{A}]\{\mathbf{r}_i\}, i = 1, 2, \tag{8.181}$$

while the tangency conditions read

$$\{\mathbf{F}_{ip}\}^{\mathrm{T}}[\mathbf{A}]\{\mathbf{r}_{i\xi}\} = 0, \{\mathbf{F}_{ip}\}^{\mathrm{T}}[\mathbf{A}]\{\mathbf{r}_{i\eta}\} = 0, \tag{8.182}$$

where $\{\mathbf{F}_{ip}\}$ is the column matrix of the partial derivatives of the function F, calculated at the points P_i, $i = 1, 2$, while $\{\mathbf{r}_{i\xi}\}$, and $\{\mathbf{r}_{i\eta}\}$ are the partial derivatives of the matrices $\{\mathbf{r}_i\}$ with respect to ξ_i, and η_i, respectively, $i = 1, 2$.

Deriving the relations (8.181) with respect to time, and taking into account the conditions of rolling without sliding

$$\{\dot{\mathbf{R}}_O\} + [\mathbf{A}][\mathbf{r}_i]^{\mathrm{T}}[\mathbf{Q}]\{\dot{\boldsymbol{\beta}}\} = \{\mathbf{0}\}, i = 1, 2, \tag{8.183}$$

where

$$[\mathbf{r}_i] = \begin{bmatrix} 0 & -z_i & y_i \\ z_i & 0 & -x_i \\ -y_i & x_i & 0 \end{bmatrix}, i = 1, 2, \tag{8.184}$$

one obtains the equalities

$$\{\dot{\mathbf{R}}_i\} = \dot{\xi}_i[\mathbf{A}]\{\mathbf{r}_{i\xi}\} + \dot{\eta}_i[\mathbf{A}]\{\mathbf{r}_{i\eta}\}, i = 1, 2. \tag{8.185}$$

The relations (8.183) are equivalent to the matrix equalities

$$\{\boldsymbol{\omega}\} = [\mathbf{Q}]\{\dot{\boldsymbol{\beta}}\} = \zeta\{\{\mathbf{r}_1\} - \{\mathbf{r}_2\}\}, \tag{8.186}$$

$$\{\dot{\mathbf{R}}_O\} = \zeta[\mathbf{A}][\mathbf{r}_2]\{\mathbf{r}_1\}, \tag{8.187}$$

and the vector equalities

$$\boldsymbol{\omega} = \zeta(\mathbf{r}_1 - \mathbf{r}_2), \mathbf{v}_O = \zeta(\mathbf{r}_2 \times \mathbf{r}_1), \tag{8.188}$$

where ζ is a factor of proportionality, while $\mathbf{v}_O$ is the velocity of the point O.

We proved in the paragraph 8.4 that if one denotes by λ_i, $i = \overline{1, 6}$, the components, in the local reference system $Oxyz$, of the torsor of the constraint forces at the point O, then the following relation is verified

$$\left[\{\mathbf{r}_1\}^{\mathrm{T}}[\mathbf{r}_2]^{\mathrm{T}} \quad \{\{\mathbf{r}_1\} - \{\mathbf{r}_2\}\}^{\mathrm{T}}\right]\{\boldsymbol{\lambda}\} = 0, \tag{8.189}$$

where

$$\{\boldsymbol{\lambda}\} = [\lambda_1\ \ \lambda_2\ \ \lambda_3\ \ \lambda_4\ \ \lambda_5\ \ \lambda_6]^{\mathrm{T}}. \tag{8.190}$$

8.8.2 *The Differential Equations of Motion: The Calculation of the Forces of Constraints*

Following the algorithm given in paragraph 8.7, using the notations

$$\{\dot{\mathbf{r}}_i\} = \dot{\xi}_i\{\mathbf{r}_{i\xi}\} + \dot{\eta}_i\{\mathbf{r}_{i\eta}\}, [\dot{\mathbf{r}}_i] = \dot{\xi}_i[\mathbf{r}_{i\xi}] + \dot{\eta}_i[\mathbf{r}_{i\eta}], i = 1, 2, \tag{8.191}$$

$$\{\widetilde{\mathbf{F}}_{\mathrm{s}}\} = -\zeta[\mathbf{m}]\{[\dot{\mathbf{A}}][\mathbf{r}_2]\{\mathbf{r}_1\} + [\mathbf{A}][\dot{\mathbf{r}}_2]\{\mathbf{r}_1\} + [\mathbf{A}][\mathbf{r}_2]\{\dot{\mathbf{r}}_1\}\}, \tag{8.192}$$

$$\{\widetilde{\mathbf{F}}_{\boldsymbol{\beta}}\} = -\zeta^2[[\mathbf{r}_1] - [\mathbf{r}_2]][\mathbf{J}]\{\{\mathbf{r}_1\} - \{\mathbf{r}_2\}\} - \zeta[\mathbf{J}]\{\{\dot{\mathbf{r}}_1\} - \{\dot{\mathbf{r}}_2\}\}, \tag{8.193}$$

and making the hypothesis that the center of weight of the rigid solid coincides to the point O, one obtains the equalities

$$\dot{\zeta} = \frac{\{\mathbf{r}_1\}^T[\mathbf{r}_2]^T[\mathbf{A}]^T\left\{\{\mathbf{F}_{\mathrm{s}}\} + \{\widetilde{\mathbf{F}}_{\mathrm{s}}\}\right\} + \{\{\mathbf{r}_1\} - \{\mathbf{r}_2\}\}^T\left\{\{\mathbf{F}_{\boldsymbol{\beta}}\} + \{\widetilde{\mathbf{F}}_{\boldsymbol{\beta}}\}\right\}}{m\{\mathbf{r}_1\}^T[\mathbf{r}_2]^T[\mathbf{r}_2]\{\mathbf{r}_1\} + \{\{\mathbf{r}_1\} - \{\mathbf{r}_2\}\}^T[\mathbf{J}]\{\{\mathbf{r}_1\} - \{\mathbf{r}_2\}\}}, \tag{8.194}$$

$$\{\boldsymbol{\lambda}\} = \dot{\zeta}\begin{bmatrix} m[\mathbf{r}_2]\{\mathbf{r}_1\} \\ [\mathbf{J}]\{\{\mathbf{r}_1\} - \{\mathbf{r}_2\}\} \end{bmatrix} - \begin{bmatrix} [\mathbf{A}]^{\mathrm{T}}\left\{\{\mathbf{F}_{\mathrm{s}}\} + \{\widetilde{\mathbf{F}}_{\mathrm{s}}\}\right\} \\ \{\mathbf{F}_{\boldsymbol{\beta}}\} + \{\widetilde{\mathbf{F}}_{\boldsymbol{\beta}}\} \end{bmatrix}, \tag{8.195}$$

where $\{\mathbf{F}_{\mathrm{s}}\}$ is the column matrix of the components of the resultant of the given forces in the reference system O_0XYZ, while $\{\mathbf{F}_{\boldsymbol{\beta}}\}$ is the column matrix of the components of the resultant moment of the given forces, calculated at the point O, in the reference system $Oxyz$.

By numerical integration, from the relation (8.194) one determines the parameter ζ, while from the relations (8.186) and (8.187) one determines the new values of the parameters X_O, Y_O, Z_O, ψ, θ, and φ.

Further on, using the notations (8.173), from the system (8.172) one determines, by numerical integration, the new values of the parameters ξ_i, η_i, $i = 1, 2$, while from the system (8.174) one determines the components Q_{1n}, Q_{2n}, Q_{1t}, Q_{2t}, and $Q_{1p} + Q_{2p}$ of the constraint forces.

Further Reading

Amirouche FML (1992). Computational Methods in Multibody Dynamics. Englewood Cliffs: Prentice Hall.
Arnold VI (1997). Mathematical Methods of Classical Mechanics (Graduate Texts in Mathematics, Vol. 60). 2nd ed. New York: Springer.
Bloch AM, Baillieul J, Crouch P, Marsden J (2007). Nonholonomic Mechanics and Control (Interdisciplinary Applied Mathematics). Berlin: Springer.
Cortes J (2002). Geometric, Control and Numerical Aspects of Nonholonomic Systems. Berlin: Springer.
Coutinho MG (2001). Dynamic Simulations of Multibody Systems. New York: Springer.
de Jalón JG, Bayo E (2011). Kinematic and Dynamic Simulation of Multibody Systems: The Real-Time Challenge. Berlin: Springer Verlag.
Douglas Gregory R (2006). Classical Mechanics. Cambridge: Cambridge University Press.
Edelen DG (2010). Lagrangian Mechanics of Nonconservative Nonholonomic Systems. Leyden: Kluwer.
Eich-Soellner E, Führer C (2013). Numerical Methods in Multibody Dynamics. Wiesbaden: Springer.
Fasano A, Marmi S, Pelloni B (2006). Analytical Mechanics: An Introduction. Oxford: Oxford University Press.
Fearstone R (2007). Rigid body Dynamics Algorithms. New York: Springer.
Gattringer H, Gerstmayr J (2013). Multibody Systems Dynamics, Robotics and Control. Wien: Springer.
Greenwood DT (2006). Advanced Dynamics. Cambridge: Cambridge University Press.
Hahn H (2002). Rigid Body Dynamics of Mechanisms 1: Theoretical Basis. Berlin: Springer.
Hahn H (2010). Rigid Body Dynamics of Mechanisms 2: Applications. Berlin: Springer.
Hand LN, Finch JD (1998). Analytical Mechanics. Cambridge: Cambridge University Press.
Lagrange JL (1996). Analytical Mechanics. Dordrecht: Kluwer Academic Publishers,
Lurie AI (2002). Analytical Mechanics. Berlin: Springer.
Neimark IuI, Fufaev NA (1972). Dynamics of Nonholonomic Systems (Translations of Mathematical Monographs, V. 33). Providence, Rhode Island: American Mathematical Society.
Nikravesh PE (2007). Planar Multibody Dynamics: Formulation, Programming and Applications. Boca Raton: CRC Press.
Pandrea N, Stănescu ND (2002). Mecanica. Bucureşti: Editura Didactică şi Pedagogică (in Romanian).
Papastavridis JG (2002). Analytical Mechanics: A Comprehensive Treatise on the Dynamics of Constrained Systems; For Engineers, Physicists, and Mathematicians. Oxford: Oxford University Press.
Pfeiffer F, Glocker C (1996). Multibody Dynamics with Unilateral Contacts. New York: John Wiley & Sons, Inc.
Roberson RE, Schwertassek R (1988). Dynamics of Multibody Systems. Berlin: Springer Verlag.
Shabana AA (2008). Computational Dynamics. New York: John Wiley & Sons.
Shabana AA (2013). Dynamics of Multibody Systems. Cambridge: Cambridge University Press.
Soltakhanov ShKh, Yushkov MPP Zegzhda SA (2009). Mechanics of non-nolonomic systems: A New Class of control systems (Foundations of Engineering Mechanics). Berlin: Springer.
Stănescu ND (2013). Mecanica sistemelor. Bucureşti: Editura Didactică şi Pedagogică (in Romanian)
Stănescu ND, Munteanu L, Chiroiu V, Pandrea N (2007). Sisteme dinamice. Teorie şi aplicaţii vol. I. Bucureşti: Editura Academiei Române (in Romanian)
Stănescu ND, Munteanu L, Chiroiu V, Pandrea N (2011). Sisteme dinamice. Teorie şi aplicaţii vol. II. Bucureşti: Editura Academiei Române (in Romanian).

Teodorescu PP (2009). Mechanical Systems, Classical Models: Volume 3: Analytical Mechanics (Mathematical and Analytical Techniques with Applications to Engineering). Dordrecht: Springer.

Teodorescu PP, Stănescu ND, Pandrea N (2013). Numerical Analysis with Applications in Mechanics and Engineering. Hoboken: Wiley.

Udwadia FE, Kalaba RE (2007). Analytical Dynamics: A New Approach. Cambridge: Cambridge University Press.

Wittenburg J (2007). Dynamics of Multibody Systems. Berlin: Springer.

Appendix

Solution of the Algebraic-Differential Systems

We consider the sistem

$$\begin{bmatrix} [\mathbf{A}_1] & [\mathbf{A}_2] \\ [\mathbf{A}_3] & [\mathbf{A}_4] \end{bmatrix} \begin{bmatrix} \{\ddot{\mathbf{x}}\} \\ \{\mathbf{y}\} \end{bmatrix} = \begin{bmatrix} \{\mathbf{E}_1\} \\ \{\mathbf{E}_2\} \end{bmatrix}, \tag{A.1}$$

in which $[\mathbf{A}_1]$ is a square matrix of order n, $[\mathbf{A}_2]$ is a rectangular matrix with n rows and p columns, $p \leq n$, $[\mathbf{A}_3]$ is a rectangular matrix with p rows and n columns, $[\mathbf{A}_4]$ is a square matrix of order p, $\{\ddot{\mathbf{x}}\}$ and $\{\mathbf{E}_1\}$ are column matrices with n rows, while $\{\mathbf{y}\}$ and $\{\mathbf{E}_2\}$ are column matrices with p rows. In addition, $[\mathbf{A}_i]$, $i = \overline{1, 4}$, and $\{\mathbf{E}_j\}$, $j = \overline{1, 2}$, may depend on $\{\mathbf{x}\}$, $\{\dot{\mathbf{x}}\}$, but they do not depend on $\{\mathbf{y}\}$, or its derivatives.

We denote by $[\mathbf{A}]$ the matrix

$$[\mathbf{A}] = \begin{bmatrix} [\mathbf{A}_1] & [\mathbf{A}_2] \\ [\mathbf{A}_3] & [\mathbf{A}_4] \end{bmatrix}. \tag{A.2}$$

For the beginning, let us assume that the matrix $[\mathbf{A}]$ is an invertible one, and let $[\mathbf{B}] = [\mathbf{A}]^{-1}$ be its inverse. We search the matrix $[\mathbf{B}]$ in the form

$$[\mathbf{B}] = \begin{bmatrix} [\mathbf{B}_1] & [\mathbf{B}_2] \\ [\mathbf{B}_3] & [\mathbf{B}_4] \end{bmatrix}, \tag{A.3}$$

Dynamics of the Rigid Solid with General Constraints by a Multibody Approach, First Edition.
Nicolae Pandrea and Nicolae-Doru Stănescu.

Companion website: www.wiley.com/go/pandrea

in which the matrices $[\mathbf{B}_i]$ are of the same orders as the matrices $[\mathbf{A}_i]$, $i=\overline{1,4}$.

From the relation

$$[\mathbf{A}][\mathbf{B}] = [\mathbf{B}][\mathbf{A}] = \left[\mathbf{I}_{n+p,\ n+p}\right], \tag{A.4}$$

where $\left[\mathbf{I}_{n+p,\ n+p}\right]$ is the unity matrix of order $n+p$, one deduces the equations

$$\begin{aligned}
&[\mathbf{A}_1][\mathbf{B}_1] + [\mathbf{A}_2][\mathbf{B}_3] = [\mathbf{I}_{n,\ n}],\ [\mathbf{A}_1][\mathbf{B}_2] + [\mathbf{A}_2][\mathbf{B}_4] = \left[\mathbf{0}_{n,\ p}\right],\\
&[\mathbf{A}_3][\mathbf{B}_1] + [\mathbf{A}_4][\mathbf{B}_3] = \left[\mathbf{0}_{p,\ n}\right],\ [\mathbf{A}_3][\mathbf{B}_2] + [\mathbf{A}_4][\mathbf{B}_4] = \left[\mathbf{I}_{p,\ p}\right],
\end{aligned} \tag{A.5}$$

$$\begin{aligned}
&[\mathbf{B}_1][\mathbf{A}_1] + [\mathbf{B}_2][\mathbf{A}_3] = [\mathbf{I}_{n,\ n}],\ [\mathbf{B}_1][\mathbf{A}_2] + [\mathbf{B}_2][\mathbf{A}_4] = \left[\mathbf{0}_{n,\ p}\right],\\
&[\mathbf{B}_3][\mathbf{A}_1] + [\mathbf{B}_4][\mathbf{A}_3] = \left[\mathbf{0}_{p,\ n}\right],\ [\mathbf{B}_3][\mathbf{A}_2] + [\mathbf{B}_4][\mathbf{A}_4] = \left[\mathbf{I}_{p,\ p}\right].
\end{aligned} \tag{A.6}$$

The two inferior indices have the signification of the numbers of rows, and columns, respectively, for that matrix. For instance, $[\mathbf{0}_{p,\ n}]$ signifies a matrix with p rows, n columns and all entries equal to zero.

If the matrix $[\mathbf{A}_1]$ is invertible, then from the second relation (A.5) one finds

$$[\mathbf{B}_2] = -[\mathbf{A}_1]^{-1}[\mathbf{A}_2][\mathbf{B}_4]; \tag{A.7}$$

replacing in the fourth relation (A.5), one obtains

$$-[\mathbf{A}_3][\mathbf{A}_1]^{-1}[\mathbf{A}_2][\mathbf{B}_4] + [\mathbf{A}_4][\mathbf{B}_4] = \left[\mathbf{I}_{p,\ p}\right]. \tag{A.8}$$

If the matrix $[\mathbf{A}_4]-[\mathbf{A}_3][\mathbf{A}_1]^{-1}[\mathbf{A}_2]$ is invertible, then from the relation (A.8) we get

$$[\mathbf{B}_4] = \left[[\mathbf{A}_4]-[\mathbf{A}_3][\mathbf{A}_1]^{-1}[\mathbf{A}_2]\right]^{-1}. \tag{A.9}$$

The first relation (A.6) leads to

$$[\mathbf{B}_1] = [[\mathbf{I}_{n,\ n}]-[\mathbf{B}_2][\mathbf{A}_3]][\mathbf{A}_1]^{-1}, \tag{A.10}$$

while the third relation (A.6) offers

$$[\mathbf{B}_3] = -[\mathbf{B}_4]^{-1}[\mathbf{A}_3][\mathbf{A}_1]^{-1}. \tag{A.11}$$

In addition, it results that the matrix $[\mathbf{A}]$ is invertible if the matrices $[\mathbf{A}_1]$ and $[\mathbf{A}_4]-[\mathbf{A}_3][\mathbf{A}_1]^{-1}[\mathbf{A}_2]$ are also invertible.

Example A.1 Determine the inverse of the matrix

$$\mathbf{A}=\begin{bmatrix}5 & 1 & 0\\ 2 & 9 & -1\\ 0 & 1 & 2\end{bmatrix}.$$

Solution: We consider

$$[\mathbf{A}_1]=\begin{bmatrix}5 & 1\\ 2 & 9\end{bmatrix},\ [\mathbf{A}_2]=\begin{bmatrix}0\\ -1\end{bmatrix},\ [\mathbf{A}_3]=[0\ \ 1],\ [\mathbf{A}_4]=[2].$$

We find

$$[\mathbf{A}_1]^{-1}=\frac{1}{43}\begin{bmatrix}9 & -1\\ -2 & 5\end{bmatrix},$$

$$[\mathbf{A}_4]-[\mathbf{A}_3][\mathbf{A}_1]^{-1}[\mathbf{A}_2]=\begin{bmatrix}\frac{91}{43}\end{bmatrix},$$

$$[\mathbf{B}_4]=\begin{bmatrix}\frac{43}{91}\end{bmatrix},$$

$$[\mathbf{B}_2]=-\frac{1}{91}\begin{bmatrix}1\\ -5\end{bmatrix},$$

$$[\mathbf{B}_1]=\frac{1}{91}\begin{bmatrix}19 & -2\\ -4 & 10\end{bmatrix},$$

$$[\mathbf{B}_3]=-\frac{1}{91}[-2\ \ 5].$$

One obtains

$$[\mathbf{A}]^{-1}=\frac{1}{91}\begin{bmatrix}19 & -2 & -1\\ -4 & 10 & 5\\ 2 & -5 & 43\end{bmatrix}.$$

A very often particular case is that in which $[\mathbf{A}_4]=[\mathbf{0}]$. From the formulae (A.7)–(A.11) result

$$\begin{aligned}
&[\mathbf{B}_4]=-\Big[[\mathbf{A}_3][\mathbf{A}_1]^{-1}[\mathbf{A}_2]\Big]^{-1},\ [\mathbf{B}_2]=[\mathbf{A}_1]^{-1}[\mathbf{A}_2]\Big[[\mathbf{A}_3][\mathbf{A}_1]^{-1}[\mathbf{A}_2]\Big]^{-1},\\
&[\mathbf{B}_1]=\Big[[\mathbf{I}_{n,\,n}]-[\mathbf{A}_1]^{-1}[\mathbf{A}_2]\Big[[\mathbf{A}_3][\mathbf{A}_1]^{-1}[\mathbf{A}_2]\Big]^{-1}[\mathbf{A}_3]\Big][\mathbf{A}_1]^{-1},\\
&[\mathbf{B}_3]=\Big[[\mathbf{A}_3][\mathbf{A}_1]^{-1}[\mathbf{A}_2]\Big]^{-1}[\mathbf{A}_3][\mathbf{A}_1]^{-1}.
\end{aligned}\tag{A.12}$$

Example A.2 We ask the inverse of the matrix

$$\mathbf{A}=\begin{bmatrix}1&3&2\\0&1&5\\2&1&0\end{bmatrix}.$$

Solution: Denoting

$$[\mathbf{A}_1]=\begin{bmatrix}1&3\\0&1\end{bmatrix},[\mathbf{A}_2]=\begin{bmatrix}2\\5\end{bmatrix},\ [\mathbf{A}_3]=[2\ \ 1],\ [\mathbf{A}_4]=[\mathbf{0}],$$

one obtains the values

$$[\mathbf{A}_1]^{-1}=\begin{bmatrix}1&-3\\0&1\end{bmatrix},$$

$$[\mathbf{A}_3][\mathbf{A}_1]^{-1}[\mathbf{A}_2]=[-21],$$

$$\left[[\mathbf{A}_3][\mathbf{A}_1]^{-1}[\mathbf{A}_2]\right]^{-1}=-\begin{bmatrix}\frac{1}{21}\end{bmatrix},$$

$$[\mathbf{B}_4]=\frac{1}{21}[1],$$

$$[\mathbf{B}_2]=-\begin{bmatrix}1&-3\\0&1\end{bmatrix}\begin{bmatrix}2\\5\end{bmatrix}=\frac{1}{21}\begin{bmatrix}13\\-5\end{bmatrix},$$

$$[\mathbf{B}_1]=\frac{1}{21}\begin{bmatrix}-5&2\\10&-4\end{bmatrix},$$

$$[\mathbf{B}_3]=\frac{1}{21}[-2\ \ 5],$$

the inverse being

$$[\mathbf{A}_1]^{-1}=-\frac{1}{21}\begin{bmatrix}-5&2&13\\10&-4&-5\\-2&5&1\end{bmatrix}.$$

It the relations $[\mathbf{A}_4]=[\mathbf{0}]$ and $[\mathbf{A}_2]=-[\mathbf{A}_3]^{\mathrm{T}}$ hold simultaneously true, then the condition of the inversability of matrix $[\mathbf{A}_4]-[\mathbf{A}_3][\mathbf{A}_1]^{-1}[\mathbf{A}_2]$ becomes the condition of the inversability of matrix

$$[\mathbf{H}]=[\mathbf{A}_3][\mathbf{A}_1]^{-1}[\mathbf{A}_3]^{\mathrm{T}}. \tag{A.13}$$

Assuming that the matrix $[\mathbf{A}_3]$ with p rows and n columns, $p \le n$, is of complete rank, it results from a well known theorem in linear algebra that the matrix $[\mathbf{H}]$ is an invertible square matrix of order p.

In this case the invertibility conditions reduce to condition of inversability for the matrix $[\mathbf{A}_1]$.

Example A.3 Determine the inverse of the matrix

$$\mathbf{A} = \begin{bmatrix} 2 & 1 & -3 \\ 0 & 4 & 1 \\ 3 & -1 & 0 \end{bmatrix}.$$

Solution: With the notations

$$[\mathbf{A}_1] = \begin{bmatrix} 2 & 1 \\ 0 & 4 \end{bmatrix},\ [\mathbf{A}_2] = \begin{bmatrix} -3 \\ 1 \end{bmatrix},\ [\mathbf{A}_3] = [3 \quad -1],\ [\mathbf{A}_4] = [0]$$

one successively obtains

$$[\mathbf{A}_1]^{-1} = \frac{1}{8}\begin{bmatrix} 4 & -1 \\ 0 & 2 \end{bmatrix},$$

$$[\mathbf{A}_3][\mathbf{A}_1]^{-1}[\mathbf{A}_2] = \frac{1}{8}[-41],$$

$$\left[[\mathbf{A}_3][\mathbf{A}_1]^{-1}[\mathbf{A}_2]\right]^{-1} = -\left[\frac{8}{41}\right],$$

$$[\mathbf{B}_4] = \frac{8}{41}[1],$$

$$[\mathbf{B}_2] = \frac{1}{41}\begin{bmatrix} 13 \\ -2 \end{bmatrix},$$

$$[\mathbf{B}_1] = \frac{1}{41}\begin{bmatrix} 1 & 3 \\ 3 & 9 \end{bmatrix},$$

$$[\mathbf{B}_3] = -\frac{1}{41}[12 \quad -5].$$

It results

$$[\mathbf{A}_1]^{-1} = \frac{1}{41}\begin{bmatrix} 1 & 3 & 13 \\ 3 & 9 & -2 \\ -12 & 5 & 8 \end{bmatrix}.$$

The system (A.1) transforms is

$$\begin{bmatrix}\{\ddot{\mathbf{x}}\}\\ \{\mathbf{y}\}\end{bmatrix}=\begin{bmatrix}[\mathbf{B}_1] & [\mathbf{B}_2]\\ [\mathbf{B}_3] & [\mathbf{B}_4]\end{bmatrix}\begin{bmatrix}\{\mathbf{E}_1\}\\ \{\mathbf{E}_2\}\end{bmatrix}, \tag{A.14}$$

wherefrom

$$\{\ddot{\mathbf{x}}\}=[\mathbf{B}_1]\{\mathbf{E}_1\}+[\mathbf{B}_2]\{\mathbf{E}_2\},\ \{\mathbf{y}\}=[\mathbf{B}_3]\{\mathbf{E}_1\}+[\mathbf{B}_4]\{\mathbf{E}_2\}. \tag{A.15}$$

Example A.4 Consider the system

$$\begin{bmatrix}2 & 1 & -1\\ 1 & 2 & -1\\ 1 & 1 & 0\end{bmatrix}\begin{bmatrix}\ddot{\mathbf{x}}_1\\ \ddot{\mathbf{x}}_2\\ y\end{bmatrix}=\begin{bmatrix}x_1\\ x_2\\ 0\end{bmatrix} \tag{A.16}$$

Determine x_1, x_2 and y.

Solution: Denoting

$$[\mathbf{A}_1]=\begin{bmatrix}2 & 1\\ 1 & 2\end{bmatrix},\ [\mathbf{A}_2]=\begin{bmatrix}-1\\ -1\end{bmatrix},\ [\mathbf{A}_3]=[1\ \ 1],\ [\mathbf{A}_4]=[0],$$

it results

$$[\mathbf{A}_1]^{-1}=\frac{1}{3}\begin{bmatrix}2 & -1\\ -1 & 2\end{bmatrix},$$

$$[\mathbf{B}_4]=\frac{3}{2},$$

$$[\mathbf{B}_2]=\frac{1}{2}\begin{bmatrix}1\\ 1\end{bmatrix},$$

$$[\mathbf{B}_1]=\frac{1}{2}\begin{bmatrix}1 & -1\\ -1 & 1\end{bmatrix},$$

$$[\mathbf{B}_3]=-\frac{1}{2}[1\ \ 1],$$

so that

$$[\mathbf{B}]=[\mathbf{A}_1]^{-1}=\frac{1}{2}\begin{bmatrix}1 & -1 & 1\\ -1 & 1 & 1\\ -1 & -1 & 3\end{bmatrix}.$$

The system (A.16) becomes

$$\begin{bmatrix} \ddot{x}_1 \\ \ddot{x}_2 \\ y \end{bmatrix} = \frac{1}{2} \begin{bmatrix} 1 & -1 & 1 \\ -1 & 1 & 1 \\ -1 & -1 & 3 \end{bmatrix} \begin{bmatrix} x_1 \\ x_2 \\ 0 \end{bmatrix} = \frac{1}{2} \begin{bmatrix} x_1 - x_2 \\ -x_1 + x_2 \\ -x_1 - x_2 \end{bmatrix}$$

wherefrom

$$\ddot{x}_1 = \frac{1}{2}(x_1 - x_2),\ \ddot{x}_2 = \frac{1}{2}(-x_1 + x_2),\ y = -\frac{1}{2}(x_1 + x_2). \tag{A.17}$$

From the first relation (A.17) one obtains

$$x_2 = x_1 - 2\ddot{x}_1,\ \ddot{x}_2 = \ddot{x}_1 - x_1^{(iv)},$$

so that the second relation (A.17) leads to

$$x_1^{(iv)} - \ddot{x}_1 = 0. \tag{A.18}$$

The characteristic equation of the differential equation (A.18) reads $r^4 - r^2 = 0$ and has the solutions $r_1 = 0$, $r_2 = 0$, $r_3 = 1$, $r_4 = -1$. It results that the solution of the differential equation (A.18) is

$$x_1 = C_1 + C_2 t + C_3 e^t + C_4 e^{-t},$$

where C_1, C_2, C_3, and C_4 are constants of integration determined from the initial conditions.

We also obtain

$$x_2 = C_1 + C_2 t - C_3 e^t - C_4 e^{-t},$$

$$y = -C_1 - C_2 t.$$

For the case $[\mathbf{A}_4] = [\mathbf{0}]$, $[\mathbf{A}_3] = -[\mathbf{A}_3]^{\mathrm{T}}$ we may give a direct solution of the system (A.1).

Indeed, in this situation the system (A.1) reads

$$[\mathbf{A}_1]\{\ddot{\mathbf{x}}\} - [\mathbf{A}_3]^{\mathrm{T}}\{\mathbf{y}\} = \{\mathbf{E}_1\},\ [\mathbf{A}_3]\{\ddot{\mathbf{x}}\} = \{\mathbf{E}_2\}; \tag{A.19}$$

assuming that the matrix $[\mathbf{A}_1]$ is invertible, from the first relation (A.19) one obtains

$$\{\ddot{\mathbf{x}}\} = [\mathbf{A}_1]^{-1}\left\{\{\mathbf{E}_1\} + [\mathbf{A}_3]^{\mathrm{T}}\{\mathbf{y}\}\right\} = \{\mathbf{E}_2\}. \tag{A.20}$$

Replacing the expression (A.20) in the second relation (A.19), it results

$$[\mathbf{A}_3][\mathbf{A}_1]^{-1}\left\{\{\mathbf{E}_1\}+[\mathbf{A}_3]^{\mathrm{T}}\{\mathbf{y}\}\right\}=\{\mathbf{E}_2\}, \tag{A.21}$$

wherefrom

$$\left[[\mathbf{A}_3][\mathbf{A}_1]^{-1}[\mathbf{A}_3]^{\mathrm{T}}\right]\{\mathbf{y}\}=\{\mathbf{E}_2\}-[\mathbf{A}_3][\mathbf{A}_1]^{-1}\{\mathbf{E}_1\}. \tag{A.22}$$

Assuming that the matrix $[\mathbf{A}_3]$ is of complete rank, we deduce that the matrix $[\mathbf{A}_3][\mathbf{A}_1]^{-1}[\mathbf{A}_3]^{\mathrm{T}}$ is invertible and from the relation (A.22) we find

$$\{\mathbf{y}\}=\left[[\mathbf{A}_3][\mathbf{A}_1]^{-1}[\mathbf{A}_3]^{\mathrm{T}}\right]^{-1}\left\{\{\mathbf{E}_2\}-[\mathbf{A}_3][\mathbf{A}_1]^{-1}\{\mathbf{E}_1\}\right\}; \tag{A.23}$$

hence

$$\{\ddot{\mathbf{x}}\}=[\mathbf{A}_1]^{-1}\left\{\{\mathbf{E}_1\}+[\mathbf{A}_3]^{\mathrm{T}}\left[[\mathbf{A}_3][\mathbf{A}_1]^{-1}[\mathbf{A}_3]^{\mathrm{T}}\right]^{-1}\left\{\{\mathbf{E}_2\}-[\mathbf{A}_3][\mathbf{A}_1]^{-1}\{\mathbf{E}_1\}\right\}\right\}. \tag{A.24}$$

Further on, we analyze the case when the matrix $[\mathbf{A}_1]$ is singular, and the matrices $[\mathbf{A}_2]$ and $[\mathbf{A}_3]$ are of complete rank, $[\mathbf{A}_2]=-[\mathbf{A}_3]^{\mathrm{T}}$.

Let $[\mathbf{H}_{i,j}]$ be a square matrix of order m obtained from the unity matrix of order m, $[\mathbf{I}_m]$, by interchanging the rows i and j,

$$[\mathbf{H}_{i,j}]=\begin{bmatrix} 1 & \cdots & 0 & 0 & \cdots & 0 & 0 & \cdots & 0 \\ \cdots & \cdots & \cdots & \cdots & \cdots & \cdots & \cdots & \cdots & \cdots \\ 0 & \cdots & 1 & 0 & \cdots & 0 & 0 & \cdots & 0 \\ 0 & \cdots & 0 & 0 & \cdots & 1 & 0 & \cdots & 0 \\ \cdots & \cdots & \cdots & \cdots & \cdots & \cdots & \cdots & \cdots & \cdots \\ 0 & \cdots & 0 & 1 & \cdots & 0 & 0 & \cdots & 0 \\ 0 & \cdots & 0 & 0 & \cdots & 0 & 1 & \cdots & 0 \\ \cdots & \cdots & \cdots & \cdots & \cdots & \cdots & \cdots & \cdots & \cdots \\ 0 & \cdots & 0 & 0 & \cdots & 0 & 0 & \cdots & 1 \end{bmatrix}. \tag{A.25}$$

If $[\mathbf{G}]$ is a matrix with m rows and an arbitrary number of columns, then the product $[\mathbf{H}_{i,j}][\mathbf{G}]$ has as result the interchanging of the rows i and j in the matrix $[\mathbf{G}]$. Analogically, if $[\mathbf{G}]$ is a matrix with an arbitrary number of rows and m columns, then the matrix $[\mathbf{G}][\mathbf{H}_{i,j}]$ has as result the interchanging of the columns i and j in the matrix $[\mathbf{G}]$.

If the matrix $[\mathbf{A}_2]$ is of complete rank, then it results that one can find a non-zero determinant formed by p of the n rows in the matrix $[\mathbf{A}_2]$. We may consider that this determinant is formed by the first p rows of the matrix $[\mathbf{A}_2]$ (eventually, multiplying the matrix $[\mathbf{A}_2]$ at the left by matrices of $[\mathbf{H}_{i,\ j}]$ type). Let us denote by $[\mathbf{D}^*]$ the product of these $[\mathbf{H}_{i,j}]$ matrices. It results that the matrix $[\mathbf{A}^*] = [\mathbf{D}^*][\mathbf{A}_2]$ has the first p rowslinear independent. Analogically, one may obtain the first p columns linear independent in the matrix $[\mathbf{A}_3]$ multiplying this matrix at the right by some matrices of $[\mathbf{H}_{i,j}]$ type.

One may realize a partition of the matrices $[\mathbf{A}_1]$, $[\mathbf{A}_2]$, and $[\mathbf{A}_3]$ in the form

$$[\overline{\mathbf{A}}_1] = \begin{bmatrix} [\mathbf{A}_1^*] \\ [\mathbf{A}_1^{**}] \end{bmatrix},\ [\overline{\mathbf{A}}_3] = \left[[\mathbf{A}_3^*] \quad [\mathbf{A}_3^{**}] \right],\ [\overline{\mathbf{A}}_2] = \begin{bmatrix} [\mathbf{A}_2^*] \\ [\mathbf{A}_2^{**}] \end{bmatrix} = -\begin{bmatrix} [\mathbf{A}_3^*]^{\mathrm{T}} \\ [\mathbf{A}_3^{**}]^{\mathrm{T}} \end{bmatrix}. \quad \text{(A.26)}$$

It also results

$$\{\mathbf{x}\} = [\mathbf{D}]\{\mathbf{x}^*\}, \quad \text{(A.27)}$$

where

$$[\overline{\mathbf{A}}_1] = [\mathbf{A}_1][\mathbf{D}],\ [\overline{\mathbf{A}}_2] = [\mathbf{A}_2][\mathbf{D}],\ [\overline{\mathbf{A}}_3] = -[\overline{\mathbf{A}}_2]^{\mathrm{T}}. \quad \text{(A.28)}$$

Analogically, we have

$$\{\overline{\mathbf{E}}_1\} = \begin{bmatrix} \{\mathbf{E}_{11}^*\} \\ \{\mathbf{E}_{12}^*\} \end{bmatrix},\ \{\mathbf{E}_1\} = [\mathbf{D}]\begin{bmatrix} \{\mathbf{E}_{11}^*\} \\ \{\mathbf{E}_{12}^*\} \end{bmatrix}. \quad \text{(A.29)}$$

The system (A.1) leads to

$$[\mathbf{A}_1^*]\{\ddot{\mathbf{x}}^*\} - [\mathbf{A}_3^*]^{\mathrm{T}}\{\mathbf{y}\} = \{\mathbf{E}_{11}^*\},\ [\mathbf{A}_1^{**}]\{\ddot{\mathbf{x}}\} - [\mathbf{A}_3^{**}]^{\mathrm{T}}\{\mathbf{y}\} = \{\mathbf{E}_{12}^*\}. \quad \text{(A.30)}$$

Since the matrix $[\mathbf{A}_3]$ is of complete rank and taking into account the transformations made, it results that the matrix $[\mathbf{A}_3^*]$ is invertible and the first relation (A.30) reads

$$\{\mathbf{y}\} = \left[[\mathbf{A}_3^*]^{\mathrm{T}}\right]^{-1}\left\{[\mathbf{A}_1^*]\{\ddot{\mathbf{x}}^*\} - \{\mathbf{E}_{11}^*\}\right\}. \quad \text{(A.31)}$$

The second relation (A.30) leads to

$$[\mathbf{A}_1^{**}]\{\ddot{\mathbf{x}}\} - [\mathbf{A}_3^{**}]^{\mathrm{T}}\left[[\mathbf{A}_3^*]^{\mathrm{T}}\right]^{-1}[\mathbf{A}_1^*]\{\ddot{\mathbf{x}}\} + [\mathbf{A}_3^{**}]\left[[\mathbf{A}_3^*]^{\mathrm{T}}\right]^{-1}\{\mathbf{E}_{11}^*\} = \{\mathbf{E}_{12}^*\}, \quad \text{(A.32)}$$

wherefrom

$$\left[\left[\mathbf{A}_1^{**}\right]-\left[\mathbf{A}_3^{**}\right]^{\mathrm{T}}\left[\left[\mathbf{A}_3^{*}\right]^{\mathrm{T}}\right]^{-1}\left[\mathbf{A}_1^{*}\right]\right]\{\ddot{\mathbf{x}}\}=\{\mathbf{E}_{12}^{*}\}-\left[\mathbf{A}_3^{**}\right]\left[\left[\mathbf{A}_3^{*}\right]^{\mathrm{T}}\right]^{-1}\{\mathbf{E}_{11}^{*}\}. \tag{A.33}$$

The system consisting in equations (A.30) and (A.31) may be written as

$$[\mathbf{C}]\{\ddot{\mathbf{x}}^{*}\}=\begin{bmatrix}\{\mathbf{E}_{12}^{*}\}-\left[\mathbf{A}_3^{**}\right]\left[\left[\mathbf{A}_3^{*}\right]^{\mathrm{T}}\right]^{-1}\{\mathbf{E}_{11}^{*}\}\\ \{\mathbf{E}_2\}\end{bmatrix}, \tag{A.34}$$

in which

$$[\mathbf{C}]=\begin{bmatrix}\left[\left[\mathbf{A}_1^{**}\right]-\left[\mathbf{A}_3^{**}\right]^{\mathrm{T}}\left[\left[\mathbf{A}_3^{*}\right]^{\mathrm{T}}\right]^{-1}\left[\mathbf{A}_1^{*}\right]\right]\\ \left[\bar{\mathbf{A}}_3\right]\end{bmatrix}. \tag{A.35}$$

Example A.5 Consider the system

$$\begin{bmatrix}1 & 2 & -1\\ 2 & 2 & -1\\ 1 & 1 & 0\end{bmatrix}\begin{bmatrix}\ddot{x}_1\\ \ddot{x}_2\\ y\end{bmatrix}=\begin{bmatrix}x_1\\ x_2\\ 0\end{bmatrix}. \tag{A.36}$$

Determine x_1, x_2, x_3.

Solution: Taking into account that $[\mathbf{A}_3]=[1 \ \ 1]$ and it has the rank equal to one, it results

$$\left[\mathbf{A}_1^{**}\right]=[2 \ \ 2],\ \left[\mathbf{A}_3^{**}\right]=[1],\ \left[\mathbf{A}_3^{*}\right]=[1],\ \left[\mathbf{A}_1^{*}\right]=[1 \ \ 1],$$
$$\left\{\bar{\mathbf{E}}_1\right\}=[x_1 \ \ x_2]^{\mathrm{T}},\ [\mathbf{E}_2]=0,\ \left[\bar{\mathbf{A}}_3\right]=[1 \ \ 1],$$

so that

$$\left[\bar{\mathbf{C}}\right]=\begin{bmatrix}1 & 1\\ 1 & 1\end{bmatrix}.$$

One obtains the system

$$\ddot{x}_1+\ddot{x}_2=x_2-x_1,\ \ddot{x}_1+\ddot{x}_2=0,$$

wherefrom $\ddot{x}_1=-\ddot{x}_2$, $x_1=-x_2+at+b$, a and b being two constants of integration, while t is the time. We also deduce the relation $x_1=x_2$ and therefore $x_1=x_2=\frac{a}{2}t+\frac{b}{2}$. Moreover, $y=-x_1=-x_2$.

Example A.6 We consider the system

$$\begin{bmatrix} 1 & 1 & -1 \\ 2 & 2 & -3 \\ 1 & 3 & 0 \end{bmatrix} \begin{bmatrix} \ddot{x}_1 \\ \ddot{x}_2 \\ y \end{bmatrix} = \begin{bmatrix} x_1 \\ x_2 \\ 0 \end{bmatrix} \tag{A.37}$$

and we ask to determine x_1, x_2, y.

Solution: Proceeding in an analogous way, one obtains the system

$$\begin{bmatrix} -1 & -1 \\ 1 & 3 \end{bmatrix} \begin{bmatrix} \ddot{x}_1 \\ \ddot{x}_2 \end{bmatrix} = \begin{bmatrix} x_2 - 3x_1 \\ 0 \end{bmatrix}.$$

Since

$$\begin{bmatrix} -1 & -1 \\ 1 & 3 \end{bmatrix}^{-1} = \frac{1}{2} \begin{bmatrix} -3 & -1 \\ 1 & 1 \end{bmatrix}$$

it results

$$\begin{bmatrix} \ddot{x}_1 \\ \ddot{x}_2 \end{bmatrix} = \frac{1}{2} \begin{bmatrix} 9x_1 - 3x_2 \\ x_2 - 3x_1 \end{bmatrix},$$

that is,

$$\ddot{x}_1 = \frac{1}{2}(9x_1 - 3x_2),\ \ddot{x}_2 = \frac{1}{2}(x_2 - 3x_1). \tag{A.38}$$

From the first equation (A.38) we find

$$x_2 = 3x_1 - \frac{2\ddot{x}_1}{3},\ \ddot{x}_2 = 3\ddot{x}_1 - \frac{2x_1^{(iv)}}{3},$$

which, replaced in the second relation (A.38), offer

$$x_1^{(iv)} - 5\ddot{x}_1 = 0.$$

The characteristic equation reads

$$r^4 - 5r^2 = 0$$

and has the solutions $r_1 = 0$, $r_2 = 0$, $r_3 = \sqrt{5}$, $r_4 = -\sqrt{5}$, wherefrom

$$x_1 = C_1 + C_2 t + C_3 e^{\sqrt{5}} + C_4 e^{-\sqrt{5}},\ \ddot{x}_1 = 5C_3 e^{\sqrt{5}t} + 5C_4 e^{-\sqrt{5}t},$$
$$x_2 = \frac{7}{3}(C_1 + C_2 t) - \frac{1}{3}\left(C_3 e^{\sqrt{5}t} + C_4 e^{-\sqrt{5}t}\right),\ \ddot{x}_2 = -\frac{5}{3}\left(C_3 e^{\sqrt{5}t} + C_4 e^{-\sqrt{5}t}\right).$$

Using the first (or the second) equation (A.37), it results

$$y = \ddot{x}_1 + \ddot{x}_2 - x_1 = \frac{7}{3}\left(C_3 e^{\sqrt{5}t} + C_4 e^{-\sqrt{5}t}\right) - C_1 - C_2 t.$$

Example A.7 Let the system be

$$\begin{bmatrix} 1 & 1 & -1 \\ 2 & 2 & -2 \\ 1 & 2 & 0 \end{bmatrix} \begin{bmatrix} \ddot{x}_1 \\ \ddot{x}_2 \\ y \end{bmatrix} = \begin{bmatrix} x_1 \\ x_2 \\ 0 \end{bmatrix}. \tag{A.39}$$

Determine x_1, x_2, y.

Solution: In a complete analogous way we find the system

$$\begin{bmatrix} 0 & 0 \\ 1 & 2 \end{bmatrix} \begin{bmatrix} \ddot{x}_1 \\ \ddot{x}_2 \end{bmatrix} = \begin{bmatrix} x_2 - 2x_1 \\ 0 \end{bmatrix},$$

from which we deduce the equations

$$x_2 - 2x_1 = 0,\ \ddot{x}_1 + 2\ddot{x}_2 = 0. \tag{A.40}$$

Deriving twice the first relation (A.40), we find

$$\ddot{x}_2 = 2\ddot{x}_1,$$

which, replaced in the second equation (A.40) offers

$$\ddot{x}_1 = 0;$$

hence, $x_1 = at + b$, $x_2 = 2at + 2b$, where a and b are two constants of integration which result from the initial conditions.

The last equation (A.39) leads to $y = 0$.

The previous examples highlight a few aspects:

- the matrix [**C**] given by the relation (A.35) may be or may be not an invertible one;
- if the matrix [**C**] is invertible, then one may determine the all $2n$ integration constants, where n is the order of the matrix [$\mathbf{A}_1$];

- if the matrix [**C**] is singular, then there exist certain relations between the integration constants. Moreover, the initial conditions can not be arbitrary; they have to take into account these relations;
- Example A.7 can not be considered a possible case of the equations of motion because the first and the second equations (A.39) are identical. If such a system would be the result of a real mechanic system, then it would result that the equations of motion obtained from the theorem of momentum and theorem of moment of momentum are not independent, which is absurd.

We saw that, in the case of the dynamics of rigid solid, the matrix [$\mathbf{A}_1$] is in fact the matrix [**M**] given by

$$[\mathbf{M}] = \begin{bmatrix} [\mathbf{m}] & [\mathbf{A}][\mathbf{S}]^{\mathrm{T}}[\mathbf{Q}] \\ [\mathbf{Q}]^{\mathrm{T}}[\mathbf{S}][\mathbf{A}]^{\mathrm{T}} & [\mathbf{Q}]^{\mathrm{T}}[\mathbf{J}_O][\mathbf{Q}] \end{bmatrix}. \tag{A.41}$$

For the clarity, we assume that the system of mobile axes has its origin in the center of weight of the rigid solid, and the mobile axes are central principal axes of inertia. It results

$$[\mathbf{S}] = \begin{bmatrix} 0 & 0 & 0 \\ 0 & 0 & 0 \\ 0 & 0 & 0 \end{bmatrix},\ [\mathbf{J}_O] = \begin{bmatrix} J_x & 0 & 0 \\ 0 & J_y & 0 \\ 0 & 0 & J_z \end{bmatrix}, \tag{A.42}$$

so that the matrix [**M**] takes the form

$$[\mathbf{M}] = \begin{bmatrix} [\mathbf{m}] & [\mathbf{0}_{3,3}] \\ [\mathbf{0}_{3,3}] & [\mathbf{Q}]^{\mathrm{T}}[\mathbf{J}_O][\mathbf{Q}] \end{bmatrix}. \tag{A.43}$$

Since

$$[\mathbf{m}] = \begin{bmatrix} m & 0 & 0 \\ 0 & m & 0 \\ 0 & 0 & m \end{bmatrix}, \tag{A.44}$$

one gets

$$\det[\mathbf{M}] = m^3(\det[\mathbf{Q}])^2 \det[\mathbf{J}_O]. \tag{A.45}$$

Assuming that the central principal moments of inertia are not zero, from the relation (A.45) it results that the equation $\det[\mathbf{M}]=0$ is equivalent to the equation $\det[\mathbf{Q}]=0$[1].

The most used rotational schemata are the Euler schema and the Bryan schema for which the matrices [**Q**] read

$$[\mathbf{Q}_E]=\begin{bmatrix}\sin\theta\sin\varphi & \cos\varphi & 0\\ \sin\theta\cos\varphi & -\sin\varphi & 0\\ \cos\theta & 0 & 1\end{bmatrix}, \tag{A.46}$$

$$[\mathbf{Q}_B]=\begin{bmatrix}\cos\theta\cos\varphi & \sin\varphi & 0\\ -\cos\theta\sin\varphi & \cos\varphi & 0\\ 0 & 0 & 1\end{bmatrix}, \tag{A.47}$$

with

$$\det[\mathbf{Q}_E]=-\sin\theta,\ \det[\mathbf{Q}_B]=\cos\theta. \tag{A.48}$$

In addition, the matrices of rotation for the two schemata have the expressions

$$[\mathbf{A}_E]=\begin{bmatrix}\cos\psi\cos\varphi-\sin\psi\cos\theta\sin\varphi & -\cos\psi\sin\varphi-\sin\psi\cos\theta\cos\varphi & \sin\psi\sin\theta\\ \sin\psi\cos\varphi+\cos\psi\cos\theta\sin\varphi & -\sin\psi\sin\varphi+\cos\psi\cos\theta\cos\varphi & -\cos\psi\sin\theta\\ \sin\theta\sin\varphi & \sin\theta\cos\varphi & \cos\theta\end{bmatrix}, \tag{A.49}$$

$$[\mathbf{A}_B]=\begin{bmatrix}\cos\theta\cos\varphi & -\cos\theta\sin\varphi & \sin\theta\\ \sin\psi\sin\theta\cos\varphi+\cos\psi\sin\varphi & -\sin\psi\sin\theta\sin\varphi+\cos\psi\cos\varphi & -\sin\psi\cos\theta\\ -\cos\psi\sin\theta\cos\varphi+\sin\psi\sin\varphi & \cos\psi\sin\theta\sin\varphi+\sin\psi\cos\varphi & \cos\psi\cos\theta\end{bmatrix}. \tag{A.50}$$

Let us assume that one used the Euler rotational schema. The non-invertibility of the matrix [**M**] leads to the equation $\sin\theta=0$, according to the first relation (A.48). For the clarity we will consider $\theta=0$. The case $\theta=\pi$ may be discussed in a complete analogous way.

---

[1] This statement holds also true if the mobile axes are not central principal axes of inertia, but they are only central axes of inertia. In this situation we have to assume that the matrix $[\mathbf{J}_O]$ has not an eigenvalue equal to zero.

From the relation (A.49) we get

$$[\mathbf{A}_E] = \begin{bmatrix} \cos\psi\cos\varphi - \sin\psi\sin\varphi & -\cos\psi\sin\varphi - \sin\psi\cos\varphi & 0 \\ \sin\psi\cos\varphi + \cos\psi\sin\varphi & -\sin\psi\sin\varphi + \cos\psi\cos\varphi & 0 \\ 0 & 0 & 1 \end{bmatrix}. \qquad (A.51)$$

We denote by $\bar{\psi}$, $\bar{\theta}$ and $\bar{\varphi}$ the rotational angles corresponding to the Bryan schema. Equating the matrix $[\mathbf{A}_E]$ in the relation (A.51) with the matrix $[\mathbf{A}_B]$ in the relation (A.50), one deduces the equations

$$\sin\bar{\theta} = 0,\ -\sin\bar{\psi}\cos\bar{\theta} = 0,\ \cos\bar{\psi}\cos\bar{\theta} = 1. \qquad (A.52)$$

For instance, choosing $\bar{\theta} = 0$, it results $\bar{\psi} = 0$ and the matrix $[\mathbf{A}_B]$ reads

$$[\mathbf{A}_B] = \begin{bmatrix} \cos\bar{\varphi} & -\sin\bar{\varphi} & 0 \\ \sin\bar{\varphi} & \cos\bar{\varphi} & 0 \\ 0 & 0 & 1 \end{bmatrix}. \qquad (A.53)$$

Equating now the relations (A.51) and (A.53), one gets $\bar{\varphi}$.

On the other hand, the angular velocities relative to the two mobile reference systems, obtained using the rotational Euler and Bryan schemata, must be equal (the two reference systems coincide). We obtain

$$[\mathbf{Q}_E]\begin{bmatrix} \dot{\psi} \\ \dot{\theta} \\ \dot{\varphi} \end{bmatrix} = [\mathbf{Q}_B]\begin{bmatrix} \dot{\bar{\psi}} \\ \dot{\bar{\theta}} \\ \dot{\bar{\varphi}} \end{bmatrix}, \qquad (A.54)$$

wherefrom

$$\begin{bmatrix} 0 & \cos\varphi & 0 \\ 0 & -\sin\varphi & 0 \\ 1 & 0 & 1 \end{bmatrix}\begin{bmatrix} \dot{\psi} \\ \dot{\theta} \\ \dot{\varphi} \end{bmatrix} = \begin{bmatrix} \cos\bar{\varphi} & \sin\bar{\varphi} & 0 \\ -\sin\bar{\varphi} & \cos\bar{\varphi} & 0 \\ 0 & 0 & 1 \end{bmatrix}\begin{bmatrix} \dot{\bar{\psi}} \\ \dot{\bar{\theta}} \\ \dot{\bar{\varphi}} \end{bmatrix}; \qquad (A.55)$$

it results the system

$$\cos\bar{\varphi}\dot{\bar{\psi}} + \sin\bar{\varphi}\dot{\bar{\theta}} = \cos\varphi\dot{\theta},\ -\sin\bar{\varphi}\dot{\bar{\psi}} + \cos\bar{\varphi}\dot{\bar{\theta}} = -\sin\varphi\dot{\theta},\ \dot{\bar{\varphi}} = \dot{\psi} + \dot{\varphi}, \qquad (A.56)$$

with the solution

$$\dot{\bar{\psi}} = -\cos(\bar{\varphi} - \varphi)\dot{\theta},\ \dot{\bar{\theta}} = \sin(\bar{\varphi} - \varphi)\dot{\theta},\ \dot{\bar{\varphi}} = \dot{\psi} + \dot{\varphi}. \qquad (A.57)$$

One observes that $\det[\mathbf{Q}_B]=1$ and the matrix $[\mathbf{M}]$ is invertible.

Example A.8 The Euler angles ψ, θ, and φ vary after the laws (in radians):

$$\psi=\frac{\pi}{4}+t^3-3t+2,\ \theta=t^2-t,\ \varphi=\frac{\pi}{4}+t^4-4t^2+3.$$

At the moment $t=1$ we ask for:

i. the corresponding Bryan angles;
ii. the matrix $\{\boldsymbol{\omega}\}$;
iii. the derivatives $\dot{\bar{\psi}}$, $\dot{\bar{\theta}}$ and $\dot{\bar{\varphi}}$ of the Bryan angles.

Solution:

1. At the moment $t=1$ we have

$$\psi=\frac{\pi}{4},\ \theta=0,\ \varphi=\frac{\pi}{4},\ \dot{\psi}=0,\ \dot{\theta}=1,\ \dot{\varphi}=-4,$$

$$[\mathbf{A}_E]=\begin{bmatrix}0&-1&0\\1&0&0\\0&0&1\end{bmatrix},$$

$$\sin\bar{\theta}=0.$$

We choose $\bar{\theta}=0$ and it results

$$\sin\bar{\psi}=0,\ \cos\bar{\psi}=1,\ \bar{\psi}=0,$$

$$\cos\bar{\varphi}=0,\ \sin\bar{\varphi}=1,\ \bar{\varphi}=\frac{\pi}{2}.$$

2. From the relation

$$\{\boldsymbol{\omega}\}=[\mathbf{Q}_E]\begin{bmatrix}\dot{\psi}\\\dot{\theta}\\\dot{\varphi}\end{bmatrix},$$

where

$$[\mathbf{Q}_E]=\begin{bmatrix}0&\frac{\sqrt{2}}{2}&0\\0&-\frac{\sqrt{2}}{2}&0\\1&0&1\end{bmatrix},$$

one gets

$$\{\boldsymbol{\omega}\} = \begin{bmatrix} 0 & \frac{\sqrt{2}}{2} & 0 \\ 0 & -\frac{\sqrt{2}}{2} & 0 \\ 1 & 0 & 1 \end{bmatrix} \begin{bmatrix} 0 \\ 1 \\ -4 \end{bmatrix} = \begin{bmatrix} \frac{\sqrt{2}}{2} \\ -\frac{\sqrt{2}}{2} \\ -4 \end{bmatrix}.$$

3. We obtain

$$\dot{\bar{\psi}} = -\frac{\sqrt{2}}{2}, \dot{\bar{\theta}} = \frac{\sqrt{2}}{2}, \dot{\bar{\varphi}} = -4.$$

Obviously, by passing from a rotational schema to another one, the matrix of constraints changes, but it remains of complete rank, too.

The modification of the matrix $[\mathbf{A}_1]$ in the system (A.1) leads to the modification of the rest of the matrices, but the form of the system remains unchanged. The transformation described above is neither linear, nor uniquely determined.

The determination of the new angles and their derivatives with respect to time is necessary for the integration process.

Index

Dynamics of the Rigid Solid with General Constraints by a Multibody Approach, First Edition.
Nicolae Pandrea and Nicolae-Doru Stănescu.

Companion website: www.wiley.com/go/pandrea